Molecular Spectroscopy and Chemical Analysis

Molecular Spectroscopy and Chemical Analysis

Dr Vijaykumar Deshmukh

ARCLER
PRESS

www.arclerpress.com

Molecular Spectroscopy and Chemical Analysis

Dr Vijaykumar Deshmukh

Arcler Press
4164 Lakeshore Road
Burlington, ON L7L 1A4
Canada
www.arclerpress.com
Email: orders@arclereducation.com

ISBN: 978-1-77956-961-5 (Hardcover)

ABOUT THE AUTHOR

Dr. Vijaykumar D Deshmukh is presently serving as Assistant Professor in Department of Plant Biotechnology, MGM College of Agricultural Biotechnology, Chh. Sambhajinagar (Maharashtra). He obtained his M.Sc. in (Agril. Biotech.) and Ph.D. in (Biotech.). He started his career as an Assistant Professor in 2005. His field of specialisation is Plant Biotechnology and Molecular Biology. He has published Six books and more than 25 research papers in National and international journals. He has participated in International, National and State Level Seminar and Symposiums.

TABLE OF CONTENTS

LIST OF FIGURES

PREFACE

The examination of molecules and their interactions with electromagnetic radiation constitutes one of the most potent methods for comprehending matter at a fundamental level. Spectroscopy, in its various manifestations, has become essential in contemporary science for investigating atomic and molecular structures, chemical bonding, and dynamic processes. Scientists can learn more about a molecule's composition, shape, electronic structure, and energy levels by watching how it absorbs, emits, or scatters radiation. This field is not limited to laboratories; its applications encompass medicine, environmental monitoring, materials science, astronomy, and numerous other domains of research and industry.

Chemical analysis, on the other hand, is a systematic way to find and measure the amounts of substances in a sample. Chemical analysis relies heavily on spectroscopic methods, whether the goal is to make sure a pharmaceutical compound is pure, find trace pollutants in water, or check the composition of advanced materials. Spectroscopy and chemical analysis work together to create a system that lets scientists see things that the naked eye can't see. This system connects the very small world of atoms with the very large properties of matter. Spectroscopic techniques are getting more advanced all the time, which means they can detect and measure things more accurately. This makes them important tools for both academic research and real-world uses.

The book *Molecular Spectroscopy and Chemical Analysis* bring these themes together in a systematic and comprehensive way. It is designed to provide readers with both foundational knowledge and detailed insights into the different spectroscopic techniques used in modern chemical analysis. The chapters are organized in a sequence that begins with general principles before moving into specialized methods, ensuring that readers first gain a solid grounding in the basics before exploring advanced applications. Each technique is explained not only in terms of theory but also in terms of practical applications, instrumentation, and relevance to chemical analysis.

Chapter 1 gives an overview of molecular spectroscopy. It explains what spectroscopy is, how it has changed over time, and why light-matter interactions are so important for understanding chemical systems. The chapter talks about how spectroscopy can help figure out the structure of molecules and introduces the ideas of energy levels, absorption, emission, and scattering, which are important for all spectroscopic methods.

Chapter 2 focuses on the chemical analysis process. It introduces the steps involved in analyzing a sample, from sampling and preparation to data interpretation. The chapter highlights how spectroscopic methods integrate into this process and compares qualitative versus quantitative analysis. This foundation prepares readers to appreciate how different spectroscopic tools serve specific roles in the broader analytical workflow.

In Chapter 3, we talk about microwave spectroscopy. It talks about how the rotational transitions of molecules in the microwave range can tell us about bond lengths, bond angles, and molecular symmetry. Microwave spectroscopy is used to figure out the structure of molecules and study how they move. The focus is on gas-phase analysis, which is where this method works best.

Chapter 4 explores atomic spectroscopy. This chapter introduces the principles behind atomic absorption, emission, and fluorescence spectroscopy. It describes how these methods are used to detect and quantify elemental composition in various samples. Real-world applications, including environmental testing, metallurgy, and clinical diagnostics, are used to illustrate the importance of atomic spectroscopy in routine chemical analysis.

Chapter 5 looks at infrared (IR) spectroscopy. The main point is vibrational transitions and how IR spectra can be used to identify different molecules. The chapter talks about instruments, how to prepare samples, and how to read spectra. It shows how good the method is at finding functional groups and describing organic and inorganic compounds.

Chapter 6 talks about nuclear magnetic resonance (NMR) spectroscopy. It explains the basics of nuclear spin and resonance and how NMR can give very detailed information about the structure, connectivity, and dynamics of molecules. The chapter talks about both one-dimensional and two-dimensional NMR methods and how they are used in materials science, biochemistry, and organic chemistry.

Chapter 7 talks about mass spectrometry, which is a method that is often used with spectroscopy. It talks about ionization methods, mass analyzers, and detectors, showing how mass spectrometry can tell you about molecular weight, fragmentation patterns, and structure. This analytical tool is useful in many fields, such as proteomics, pharmaceuticals, and environmental science.

Chapter 8 talks about Raman spectroscopy. It talks about the basic ideas behind inelastic light scattering and how Raman spectra work with infrared data. The chapter talks about the benefits of Raman spectroscopy, like how little sample preparation it needs and how it can look at water-based systems. There are detailed applications in chemistry, biology, and materials research.

Chapter 9 talks about molecular luminescence spectroscopy. It talks about fluorescence and phosphorescence and how excited-state processes can tell us about molecular environments, dynamics, and interactions. The chapter also talks about fluorescence quenching, lifetime measurements, and how luminescence can be used for chemical sensing and imaging.

Chapter 10 ends with X-ray diffraction and spectroscopy. It tells how X-rays can look at both electronic structures and crystal lattices. We talk about techniques like X-ray absorption spectroscopy (XAS), X-ray photoelectron spectroscopy (XPS), and X-ray diffraction (XRD), as well as how they can be used in crystallography, materials characterization, and surface analysis.

Students of chemistry, physics, and materials science will find it a structured introduction to spectroscopy and chemical analysis. Researchers and laboratory professionals will find it a useful reference for understanding instrumentation and methods in greater detail. For professionals in applied fields—such as pharmaceuticals, environmental monitoring, or forensic science—the text offers practical insights into how spectroscopy supports real-world analysis. Combining theory, technique, and application, this text aims to serve as both a teaching resource and a reliable reference, supporting the advancement of scientific knowledge and its practical application in modern society.

CHAPTER 1

INTRODUCTION TO MOLECULAR SPECTROSCOPY

Contents

INTRODUCTION

Molecular spectroscopy is a key area that connects physical chemistry and analytical sciences. It is mainly about how electromagnetic radiation interacts with matter. This field includes a number of methods that are each designed to study molecular properties by looking at how they absorb, emit, or scatter light at different wavelengths. The main idea behind molecular spectroscopy is that molecules have unique spectral fingerprints, which are patterns that show up when they vibrate and rotate as they interact with energy inputs. These spectral fingerprints help scientists learn about a sample's molecular composition, structure, and dynamics in both qualitative and quantitative ways. By comprehending these interactions, researchers can clarify the behavior of molecules in different states, yielding insights essential for progress in disciplines such as environmental science, materials science, and biochemistry. Molecular spectroscopy is a broad field that includes many different methods, such as infrared (IR), ultraviolet-visible (UV-Vis), and nuclear magnetic resonance (NMR) spectroscopy. Each method is based on different principles of electromagnetic interactions, which means that they all have their own pros and cons for different uses.

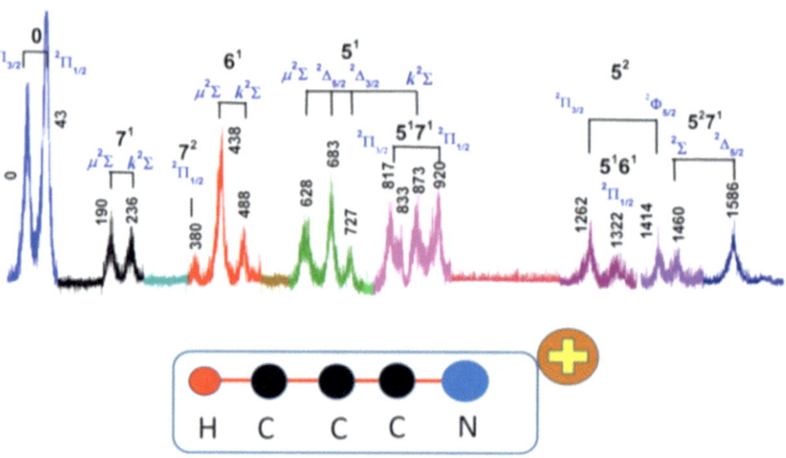

Figure 1.1. *Molecular spectroscopy.*

Source: https://publishing.aip.org/wp-content/uploads/2018/12/Mo.jpg.

For example, UV-Vis spectroscopy is mainly used to look at electronic transitions and figure out how much of a substance is in a solution. On the other hand, IR spectroscopy uses molecular vibrations to find functional groups. The choice of technique depends on the specific information needed for analysis, which shows how useful and flexible molecular spectroscopy is in research and industry. Utilizing these diverse methodologies not only augments our understanding of molecular interactions but also facilitates novel applications in pharmaceuticals, nanotechnology, and other fields, integrating theoretical insights with practical utility. Molecular spectroscopy also plays a role in interdisciplinary research, which encourages scientists from different fields to work together. The ModSysC2020 working group at University Montpellier 2 is an example of a group that is working to combine methods from molecular spectroscopy with other analytical fields. These kinds of cross-disciplinary projects are very important for solving difficult scientific problems, especially when it comes to understanding natural and biological systems. By utilizing molecular spectroscopy techniques and principles within expansive research contexts, scientists can derive novel data-driven insights that profoundly influence technological and societal progress. This amalgamation of knowledge signifies a pivotal transformation in scientific investigation, promoting a more collaborative methodology that synthesizes conventional spectroscopic analysis with nascent disciplines, as evidenced by recent reports that stress the imperative of this interdisciplinary emphasis (Arnaud *et al.,* 2012; Anonymous, 2003) (Figure 1.1).

Explanation of Molecular Spectroscopy

Molecular spectroscopy is a scientific method that looks at how electromagnetic radiation and matter interact, especially at the molecular level. This field is very important for figuring out how molecules absorb, emit, or scatter light, which can tell us a lot about their structure and electronic properties. Scientists can see the unique energy changes that happen in molecules by using different types of spectroscopies, like infrared (IR), ultraviolet-visible (UV-Vis), and nuclear magnetic resonance (NMR). These transitions can be seen in different situations, which helps scientists figure out the identity, concentration, and even the dynamics of molecules in complicated mixtures. Molecular spectroscopy provides knowledge that is applicable not only in chemistry but also in biochemistry and materials science, illustrating its interdisciplinary significance and utility in modern scientific research. In the larger picture of molecular spectroscopy, it is important to talk about how it is used in different scientific fields. Techniques developed in molecular spectroscopy have become very useful in areas like food chemistry, where they are very important for checking the safety and quality of food. For example, knowing how molecules vibrate can help you figure out what's in food and find dangerous substances. Baianu (2011) emphasizes that the amalgamation of physical chemistry with molecular spectroscopy facilitates meticulous observation of the chemical interactions within food systems. Molecular spectroscopy has the potential to improve research in food biotechnology and pharmaceuticals by making interactions clearer at a basic level. This will lead to better product development and quality assurance practices. Molecular spectroscopy is also important in microbiology, where it helps us understand genetic and biochemical processes. Bauer et al. (2019) asserts that progress in the quantification or characterization of biochemical substances, including proteins, nucleic acids, and carbohydrates, is heavily dependent on spectroscopic techniques. By modifying these methods for microbiological research, scientists can learn more about microbial genetics, metabolism, and how microorganisms interact with their surroundings. Molecular spectroscopy is also very useful because it can be used in many different areas of life sciences, such as creating new methods for accurate biochemical analysis. As a result, molecular spectroscopy not only helps us understand how molecules interact with each other better, but it also helps make progress in many different areas of science.

Overview of Different Types of Molecular Spectroscopy

Molecular spectroscopy includes a number of methods that let scientists' study how molecules behave and what they are made of by looking at how they interact with electromagnetic radiation. Infrared (IR), ultraviolet-visible (UV-Vis), nuclear magnetic resonance (NMR), and Raman spectroscopy are the four main types of molecular spectroscopy. Each method uses a different part of the electromagnetic spectrum to show different things about molecular structures, functional groups, and dynamics. For instance, IR spectroscopy is very good at finding functional groups in organic compounds by measuring molecular vibrations. On the other hand, UV-Vis spectroscopy looks at electronic transitions in molecules. Scientists can get different kinds of information from these techniques, which helps them understand complex molecular systems better. Molecular spectroscopy has made progress in making specialized methods for certain uses, in addition to traditional methods. Surface-enhanced Raman spectroscopy (SERS) is one of the techniques that makes Raman spectroscopy more powerful by boosting the signal from molecules on surfaces. This lets you analyze biological samples and very small amounts of substances with great sensitivity. In the same way, improvements in Fourier transform infrared (FTIR) spectroscopy have made it possible to map the spatial distributions of components in samples. This is very important in fields like materials science and conservation. The amalgamation of these advanced methodologies equips scientists with a more comprehensive toolkit for characterizing materials and phenomena, thereby expanding the scope of molecular analysis (Lennard *et al.,* 2017). The usefulness of these spectroscopic methods is shown even more by the fact that they are used in many different fields of science, such as environmental analysis, pharmaceuticals, and nanotechnology. For example, in the field of engineered nanoparticles (ENPs), various molecular spectroscopy techniques are utilized to delineate their physical and chemical characteristics, including size, morphology, and elemental composition. Electron and X-ray microscopy, along with separation techniques, can give us a lot of information about how nanoparticles act and interact on a very small scale (Anderson *et al.,* 2014). This thorough understanding is necessary for creating new materials and making current technologies better. So, the different kinds of molecular spectroscopy not only help us understand how molecules are put together, but they also help applied research and new ideas move forward in many fields.

1.1. HISTORICAL BACKGROUND

The history of molecular spectroscopy goes back to the early 1800s, when scientists who were the first to study how light and matter interact laid the groundwork for the field. Michael Faraday's experiments in the 1850s were important because they helped us understand how gases absorb light at certain wavelengths. Then, in the early 1800s, Joseph von Fraunhofer invented the spectroscope, which made it possible to systematically study spectral lines. This technological advancement enabled the transition from qualitative observations to quantitative measurements, signifying a substantial transformation in the examination of molecular interactions. The identification of different molecular species was made possible by the discovery of distinct absorption lines. This led to improvements in both laboratory methods and theoretical interpretations that eventually led to the modern framework of molecular spectroscopy. As the 20th century went on, new ideas in spectroscopy kept coming up. This was mostly because of better technology and a better understanding of how molecules act. Chemists got a great tool to study molecular vibrations when infrared spectroscopy was first used in the early 1930s. This tool taught them a lot about molecular structure and interactions. The Fourier transform infrared (FTIR) spectroscopy method changed this field forever by making it possible to quickly and accurately analyze many different compounds. These techniques have not only made it easier to do in-depth research in organic chemistry, but they have also been used in fields like art conservation and archaeology that are not usually associated with science. The continuing integration of vibrational and X-ray spectroscopic techniques, as highlighted in recent literature, demonstrates a commitment to enhancing analytical capabilities in material characterization and conservation, especially in contexts such as painted textiles, where comprehending material composition is essential for preservation (Lennard *et al.,* 2017). Additionally, the study of molecular spectroscopy has grown to include figuring out how complicated biological processes work, like photosynthesis. Investigating the function of carotenoids as vital pigments in this context demonstrates how spectroscopic methods elucidate energy transfer mechanisms at the molecular scale. The intricacies of molecular interactions that molecular spectroscopy aims to elucidate are exemplified by singlet–singlet excitation energy transfer, notwithstanding the complexities introduced by functional groups in carotenoid structures (Alastair *et al.,* 2018). This interdisciplinary approach has not only enhanced our understanding of molecular dynamics but has also underscored the significance of collaboration among chemists, biologists, and conservationists. Focusing on the history and current state of molecular spectroscopy makes it clear that this field is still changing and is important to many areas of science (Figure 1.2).

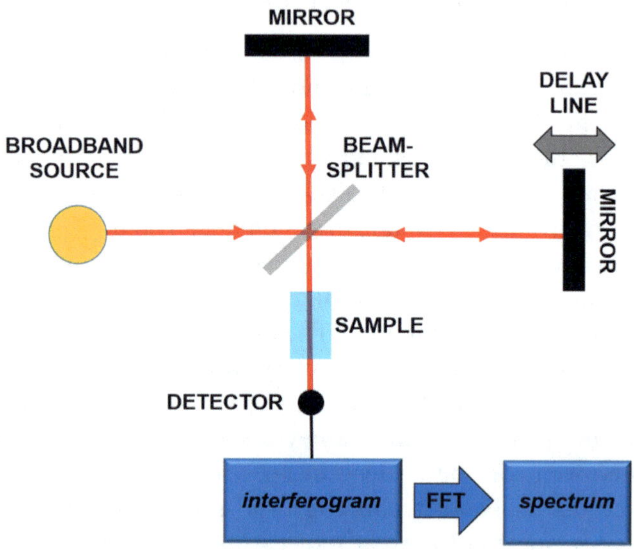

Figure 1.2. *Fourier transform infrared spectroscopy.*

Source: https://www.creative-biostructure.com/upload/image/5-4-2-9-guide-to-fourier-transform-infrared-spectroscopy-ftir-2.jpg.

1.1.1. Early Discoveries in Spectroscopy

The early discoveries in spectroscopy helped us learn about how molecules are built and how they interact with each other. In the early 1800s, scientists like Joseph von Fraunhofer started to look into the idea of light spreading out through prisms. This eventually led to the discovery of absorption lines in the solar spectrum. This event showed how light and matter are related by showing how different materials absorb certain wavelengths. Spectroscopy, which is the study of these interactions, quickly became an important way to analyze things. By mapping absorption lines, scientists could figure out what kinds of molecules and elements were in the atmospheres of stars. This changed the way we think about atomic and molecular composition in astrophysics. This change was a big step forward for both experimental methods and theoretical explanations, and it was the start of modern molecular spectroscopy. As the field progressed, the comprehension of spectroscopy evolved from basic absorption to include intricate molecular interactions. Scientists like Gustav Kirchhoff and Robert Bunsen were the first to talk about emission spectra, which is when heated elements give off light at certain wavelengths. This discovery not only supported current ideas about how atoms are put together, but it also made it easier to find new elements. The introduction of more advanced tools, like the spectroscope, made it possible to study these emissions and learn more about molecular signatures. Because of this, early spectroscopy had a big impact on chemistry and our understanding of how molecules interact with each other. It gave a systematic way to measure materials down to their atomic parts, which made it more useful in many scientific fields, such as environmental studies and drug research. The knowledge gained from early spectroscopy has had a big impact on modern research in many areas, especially when it comes to learning about the atmospheres of planets outside our solar system. Scientists can now use modern spectroscopy to look more closely at the atmospheres of celestial bodies and tell the difference between their gaseous and molecular layers. Recent literature has elucidated that the capacity to analyze molecular and atomic abundances in extraterrestrial atmospheres is essential for comprehending energetic processes and chemical interactions beyond Earth. The ability to study the atmospheres of planets outside of our solar system opens up new possibilities for figuring out how they formed and whether they could support life. Instruments created for this purpose, like the upcoming launches of the Herschel and ALMA observatories, are expected to change the way we understand molecular interactions in the far-infrared range. This shows how early discoveries in spectroscopy have led to ongoing progress in astrophysics and molecular spectroscopy (Blake *et al.,* 2009; Crossfield *et al.,* 2015).

1.1.2. Key Figures in the Development of Molecular Spectroscopy

Many pioneering figures have had a big impact on the development of molecular spectroscopy. Their work laid the groundwork for modern methods. Robert Hooke is one of these people. In the 17th century, he was one of the first to study light and how it interacts with matter. Hooke's early work with lenses and light helped scientists figure out how molecular vibrations could change how light is absorbed. Joseph Fraunhofer built on these ideas in the early 1800s by coming up with the idea of spectral lines. His careful work with prism spectrometry not only found certain wavelengths of light but also showed how important molecular absorption spectra are. Hooke and Fraunhofer's ideas made it clear that spectral analysis needs to be very accurate. They also showed how important it will be for light to interact with molecular structures in future spectroscopic methods. In the 20th century, techniques became more advanced, thanks in large part to the work of people like Max Planck and Albert Einstein. Planck's introduction of quantized energy levels transformed the comprehension of molecular transitions, establishing a theoretical framework for the emission and absorption of light by molecules at specific frequencies. Einstein's research on the photoelectric effect clarified the connection between light and electrons, which is important for vibrational and rotational spectroscopy.

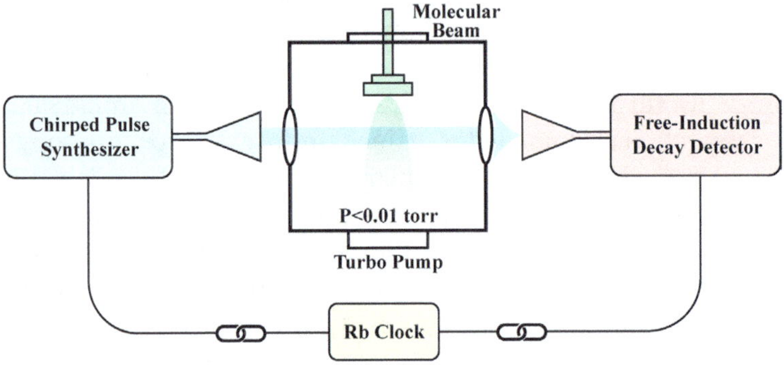

Figure 1.3. *Advancing molecular spectroscopy efficiency with extensive parallelism.*

Source: https://www.mdpi.com/metrology/metrology-04-00043/article_deploy/html/images/metrology-04-00043-g001-550.jpg.

These basic ideas made it possible to create methods like infrared spectroscopy, which uses molecular vibrations and rotations to learn about molecular structure. Planck and Einstein's work together not only improved quantum theory, but also made big improvements in molecular spectroscopy, which lets scientists look closely at how molecules interact and what they are made of. Richard Feynman and Gérard Mourou, along with Planck and Einstein, have also had a big effect on the technological progress in molecular spectroscopy. Feynman's work on quantum mechanics and how it affects how electrons behave laid the groundwork for many modern spectroscopic methods that depend on electronic transitions. Mourou's work on the chirped pulse amplification technique has changed laser spectroscopy forever by making it possible to produce very short laser pulses. This major breakthrough makes it possible to look closely at molecular structures with never-before-seen levels of time and space resolution. Furthermore, the amalgamation of diverse spectroscopic techniques, including Raman and infrared spectroscopy, alongside progress in computational algorithms and machine learning, perpetually augments the functionalities of molecular spectroscopy. Acknowledging the contributions of these pivotal individuals facilitates a thorough comprehension of the evolution of molecular spectroscopy into an essential instrument in chemical analysis and materials science, as demonstrated by significant studies cited in academic literature, including those addressing visible resonance Raman spectroscopy (Alfano *et al.,* 2019) and innovations in analytical FT-Raman spectroscopy (Nickless *et al.,* 2015) (Figure 1.3).

1.1.3. Evolution of Techniques and Technologies

The progress of molecular spectroscopy has been greatly influenced by technological advancements and novel methodologies that improve the accuracy and utility of spectroscopic techniques. Ultraviolet-visible (UV-Vis) spectroscopy and other early spectroscopic methods used simple tools to measure how much light was absorbed. But the need for higher sensitivity and resolution led to the creation of more advanced methods. New technologies have made it possible to use techniques like fluorescence resonance energy transfer (FRET) and nuclear magnetic resonance (NMR) spectroscopy, which give us a lot of information about how molecules interact and how their structures change. These improvements have not only made it easier to find things, but they have also made it easier to measure things over time, which is very important for capturing fast molecular events in biological systems. As the field keeps changing, it becomes more and more important to combine high-throughput and real-time data analysis methods. This puts molecular spectroscopy at the top of molecular science. The use of molecular spectroscopy in a variety of fields shows how it could change the game, especially in oncology and environmental science. For instance, in cancer research, modern techniques have progressed to the extent that gene expression data can be effectively employed to differentiate between various tumor types and forecast patient outcomes, thus guiding personalized treatment strategies. The swift growth of translational oncogenomics demonstrates the applicability of these molecular techniques to pharmaceutical and clinical challenges, yielding substantial implications for personalized medicine. Recent studies have shown that improvements in detection sensitivity and time resolution are pushing the limits of what can be done to find cellular events. It has become clear that we need methods that can fully map human cancer interactomes. This has led to the creation of high-efficiency and ultra-sensitive technologies that help us learn more about molecular biology (Baianu, 2011). Molecular spectroscopy is also very important in new fields like renewable energy, in addition to biomedical applications. For example, hybrid technologies that combine inorganic semiconductor light harvesters with biological catalysts show how modern scientific research crosses many fields. The goal of these methods is to turn solar energy into useful chemicals with better efficiency than natural photosynthesis. These kinds of new ideas require a deep understanding of the interface between materials and microorganisms to make sure that energy and charge transfer works well. The investigation of bio- and chemo-compatibility among various materials underscores the significance of collaborative research among spectroscopists, electrochemists, and biologists. As molecular spectroscopy continues to change, these insights will be very important for answering basic scientific questions and making technologies that can last and help solve global problems (Cestellos-Blanco *et al.,* 2018).

1.2. FUNDAMENTAL PRINCIPLES

The basic ideas behind molecular spectroscopy come from how electromagnetic radiation and matter interact at the molecular level. To understand these interactions, we need to look closely at how molecules take in, give off, or scatter light. When a molecule interacts with electromagnetic radiation, it can move between different energy states. This can create either absorption or emission spectra that give us important information about the structure and behavior of the molecule. These basic processes have energy levels that are quantized, which means that the difference in energy between levels is equal to certain wavelengths of light. This means that molecular spectroscopy can be used to find out what types of molecules are present or to learn more about their structure. This is mostly based on the physics of energy transitions. As a result, it is necessary to understand these principles in order to use spectroscopic methods in many scientific fields, such as chemistry, biology, and material science. Quantum mechanics is very important to molecular spectroscopy, especially the ideas of vibrational and rotational transitions. Every molecule has its own set of vibrational modes, each with its own energy levels that correspond to different ways the molecule moves. During these transitions, a photon is either absorbed or emitted, and its energy exactly matches the energy gap between the molecule's vibrational states. Raman spectroscopy can help us learn more about this phenomenon by looking at how light interacts with molecular vibrations to give us more information. The principles of Raman scattering, as detailed in contemporary scientific literature, suggest that quantum electrodynamics can enhance the understanding of energy exchange between photons during these transitions, offering a more profound insight into molecular interactions at the nanoscale (Andrews *et al.,* 2016). Consequently, the comprehension of these fundamental principles augments our capacity to elucidate complex molecular behaviors. Furthermore, the importance of physical chemistry principles in molecular spectroscopy cannot be overstated. This field studies how light interacts with molecules and the physical states and conditions in which these interactions happen. The research includes a range of methods and tools used to look into the physiochemical properties of materials. For example, the thermal and kinetic properties of a substance may affect its spectral properties, showing how physical and chemical principles work together (Baianu, 2011). By combining thermodynamic and kinetic theories with spectroscopic techniques, scientists can learn a lot about how molecules change during reactions and phase changes. Ultimately, comprehending these fundamental principles provides scientists with the essential tools to create novel applications in molecular spectroscopy, thereby promoting progress across various fields (Figure 1.4).

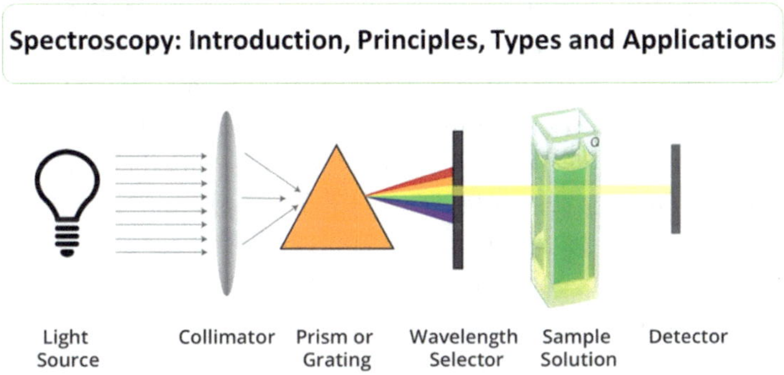

Figure 1.4. *Spectroscopy: Introduction, principles, types, and applications.*

Source: https://i.pinimg.com/originals/3c/96/69/3c96698e8a8a4b438c6344888c81e107.png.

1.2.1. Interaction of Light with Matter

The interaction of light with matter is a basic idea in molecular spectroscopy because it explains how energy moves between photons and molecules. Light can be absorbed, transmitted, or scattered when it hits a material. This depends on the electronic structure of the molecules in the material. When a photon has enough energy to move an electron from a lower energy state to a higher one, this is called absorption. This causes energy transitions to happen in discrete amounts. This process is the basis for spectroscopic methods that look at the composition and movement of molecules. You can learn a lot about molecular structures, bonding characteristics, and electronic configurations by looking at absorption spectra. The intricate interaction between light and matter elucidates essential molecular behaviors and facilitates the advancement of sophisticated applications in disciplines such as materials science and biochemistry. Coherent exciton-vibrational coupling is another complex way that light and matter interact that is important for molecular spectroscopy. This mechanism happens when the energy states of excitons, which are bound states of electrons and holes in a material, interact with the vibrational modes of the lattice or the molecules themselves. This kind of coupling makes excited-state coherence last longer. This has been shown in both biological systems and artificial light-harvesting structures, such as molecular J-aggregates. Researchers can see and study these coherent interactions using experimental methods like two-dimensional (2D) spectroscopy. This gives them a better understanding of how energy moves through complex systems. Controlling polarization in 2D spectroscopy makes it even better at figuring out how light interacts with matter, which is the basis for more advanced theoretical interpretations of molecular dynamics (Caycedo-Soler *et al.,* 2015). Investigating the interaction of light with molecular ensembles extends beyond conventional spectroscopy, including the examination of the influence of hypothetical particles on atomic and molecular transitions. Recent studies have examined the impact of low-mass spin-0 dark matter, capable of oscillation and inducing fluctuations in fundamental constants, on light-matter interactions. These interactions could have an effect on spectroscopic measurements, which would let researchers set new limits on how these particles affect atomic clocks and molecular systems. Scientists can learn a lot about basic physics by using spectroscopic methods to look at these interactions. For example, they can find out how strong the coupling between these particles and standard model components like the Higgs boson is. Consequently, the convergence of classical molecular spectroscopy with modern theoretical inquiries presents exhilarating opportunities for comprehending both molecular dynamics and the fundamental physics that dictate light-matter interactions (Flambaum *et al.,* 2016).

1.2.2. Energy Levels and Transitions in Molecules

Molecular spectroscopy is based on how molecules behave in terms of energy levels and transitions. This gives us information about their structure and how they change over time. Molecules have energy levels because of quantized states caused by electronic, vibrational, and rotational motions. These levels correspond to certain energies, and photons can help move between them by being absorbed or emitted. When a molecule absorbs light, it usually goes from a lower energy state to a higher one. This is usually accompanied by changes in how it vibrates and rotates. To understand spectra, you need to know how these transitions work. They show up as peaks or bands at certain wavelengths, which give you useful information about the composition and interactions of molecules. Complicated interactions show up most clearly in systems with excited states, as recent studies on high-energy transitions have shown. The dynamics of these processes can be further clarified by examining the effects of external interactions, such as those occurring in high-precision atomic beam spectroscopy. In this context, van der Waals forces are very important, especially for excited hydrogen atoms that are moving between different P states. The need to account for interactions between multiple states, like when going from 4P to 2S, shows how hard it is to model molecular behavior accurately. Observations indicate that these interactions can significantly increase transition probabilities, suggesting that even minor perturbations can result in considerable alterations in spectral outcomes. This underscores the necessity for rigorous theoretical frameworks in spectroscopy research (Adhikari *et al.,* 2019). Additionally, improvements in molecular spectroscopy methods, such as two-photon coupling in ultracold fermionic mixtures, have created new ways to study molecular transitions with great accuracy. Investigating vibrational levels in *Feshbach* molecules illustrates the substantial impact of interactions among electronic states on various spectral lines. For example, researchers have been able to determine binding energies with an accuracy never seen before by using dark state spectroscopy to study these transitions. These methods not only allow us to see chemically stable states, but they also show how important dipole moments are in how molecules behave, which helps us understand intermolecular forces better (Park *et al.,* 2015). In this context, the interaction of energy levels and transitions shows the complex web of molecular interactions that molecular spectroscopy tries to figure out (Figure 1.5).

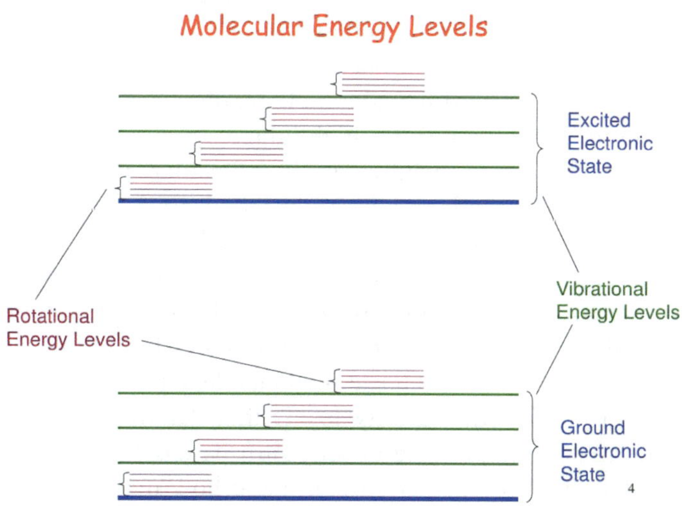

Figure 1.5. *Molecular energy levels and IR.*

Source: https://image.slideserve.com/705205/molecular-energy-levels1-l.jpg.

1.2.3. Quantization of Energy and Its Implications

Energy quantization signifies a pivotal transition from classical to quantum mechanics, especially within the realm of molecular spectroscopy. In classical physics, energy is regarded as a continuous variable; however, quantum mechanics posits that energy levels are discrete, fundamentally transforming our comprehension of atomic and molecular interactions. This discreteness is especially important in the vibrational and rotational spectra of molecules, where light absorption or emission causes transitions between quantized energy levels. These kinds of changes make unique spectral lines that can be used to identify molecules. The ionic Kratzer oscillator and other theoretical models have shown that it is possible to accurately predict molecular bond energies and vibrational states. This has given us new information about molecular structures and behaviors that classical models could not provide (Hooydonk *et al.,* 2008).

Energy quantization has consequences that go beyond just identifying molecules. It also has effects on the creation of new materials and technologies. For example, learning about quantum liquid-crystalline orders in resonant atomic gases could lead to new uses in quantum computing and superfluidity. These systems, which have a delicate balance of thermal and quantum fluctuations, show how quantized states can lead to complex behaviors like resonance pairing and smectic phase formation. This shows how quantum mechanics has a big effect on material science. This knowledge is not just theoretical; it has real-world uses, such as making advanced batteries better at storing energy and adjusting the properties of superconductors. The interplay between atom-molecule hybridization and quantum interactions is pivotal in designing future technologies, as highlighted by Agterberg et al. (2011). Moreover, the significance of quantization in molecular spectroscopy supports numerous experimental methodologies designed to investigate molecular dynamics and clarify essential behaviors. Infrared spectroscopy and nuclear magnetic resonance are two examples of spectroscopic methods that use the fact that energy states are quantized to study molecular vibrations and rotations. This gives us important information about how molecules interact with each other. The accuracy of these methods depends on the fact that each molecular system has its own energy levels that determine how it reacts to electromagnetic radiation. Because of this, shift patterns and spectral features can show us a lot about molecular structure and how the environment affects it. This has led to progress in many areas, including chemistry, biochemistry, and environmental science. The quantization of energy, which combines both theoretical frameworks and experimental techniques, is a key part of learning more about and understanding molecular systems.

1.3. TYPES OF MOLECULAR SPECTROSCOPY

Molecular spectroscopy is a field that includes many different methods, each with its own rules and uses that help us understand how molecules are structured and how they interact with each other. One of the most well-known methods is infrared (IR) spectroscopy, which measures how much infrared light a sample absorbs. This gives information about molecular vibrations and functional groups that are present. This method is very important for both qualitative and quantitative analysis in many fields, such as chemistry and materials science. Fourier Transform Infrared (FTIR) spectroscopy is another advancement that has greatly improved the ability to get high-resolution spectra. This has helped scientists learn more about how molecules are made up and how they interact with each other. The ability to map molecular distributions using methods like FTIR makes it more useful, especially for preserving and studying historical artifacts, as shown in the literature on vibrational spectroscopic techniques (Lennard *et al.,* 2017). Raman spectroscopy is another important type of molecular spectroscopy. It works with infrared methods to give information about molecular vibrations that IR can't pick up. This method uses inelastic scattering of monochromatic light, usually from lasers, to learn more about molecular vibrational modes. Raman spectroscopy is useful in many fields because it doesn't harm samples. For example, in pharmaceuticals, it can find active compounds without harming the samples. New technologies like surface-enhanced Raman spectroscopy (SERS) have also come out, making sensitivity much better, especially for trace analysis. SERS has been used in bioimaging and sensing, among other things, by using metallic surfaces to boost the Raman signal. This shows how useful this type of spectroscopy can be in both basic research and real-world situations (Lennard *et al.,* 2017). Nuclear magnetic resonance (NMR) spectroscopy is also a very useful tool for figuring out molecular structure. NMR uses the magnetic properties of nuclei in a magnetic field to give detailed information about how atoms are arranged in a molecule, such as how they are connected and where they are in space. NMR is different from IR and Raman in that it looks at the electronic environments around certain nuclei, like hydrogen or carbon. This makes it very useful in organic chemistry and biochemistry for figuring out the structure and dynamics of molecules. NMR technology has come a long way, with high-field magnets and multidimensional techniques making it a key part of molecular analysis, even though it takes longer than other methods. When used with other techniques that work well with NMR, like chromatography, its analytical power is even greater, which lets researchers answer difficult molecular questions in material and biomolecular sciences (Figure 1.6).

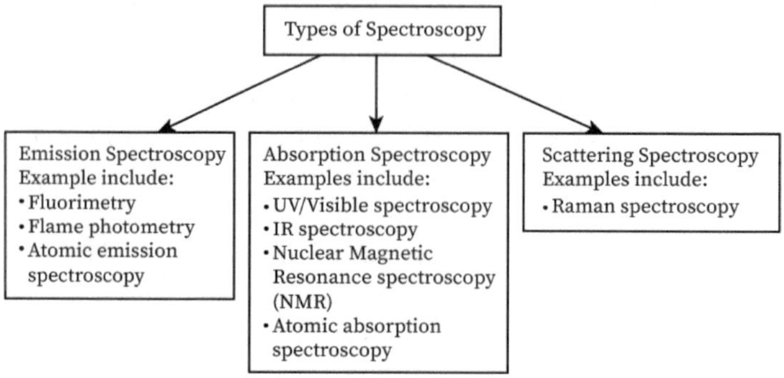

Figure 1.6. *Types of molecular spectroscopy.*

Source: https://cms-media.bartleby.com/wp-content/uploads/sites/ 2/2021 /09/09071 854/MM-1.jpg.

1.3.1. Infrared (IR) Spectroscopy

Infrared (IR) spectroscopy is an important part of molecular spectroscopy because it helps scientists figure out the structures and interactions of molecules. Researchers can learn about functional groups, bond strengths, and molecular shapes by measuring how molecular bonds absorb infrared radiation. This method is based on the idea that molecular vibrations happen at certain frequencies that match the energy of the IR radiation. When a molecule takes in this radiation, it moves between different vibrational energy levels, creating spectra that are like fingerprints for different compounds. Scientists can get both qualitative and quantitative information about a sample by looking at these spectra. This makes IR spectroscopy very useful in many fields, such as chemistry, biology, and materials science. The progress made in IR spectroscopy has been amazing, especially when it comes to synchrotron radiation (SR) applications and imaging methods. Because of its high intensity and tunability, SR has many advantages. It lets researchers' study very small samples and improves the spatial and temporal resolution of IR studies. This is especially important in mineralogy, where these advanced methods are used to study the hydrous parts of minerals in depth to learn more about geo-mineralogical processes. Recent literature has discussed how these new ideas make IR spectroscopy more useful than ever before, giving researchers better ways to study complicated materials. The use of SR technology is not yet common, but it shows how spectroscopic techniques are becoming more advanced and adaptable, which is part of the ongoing effort to improve our understanding of the molecular world (Bellatreccia *et al.,* 2013). IR spectroscopy has made a big impact on agriculture and food science, especially with the use of non-invasive Near Infrared (NIR) technology. It is also used in mineralogy. NIR spectroscopic methods have become cost-effective, quick, and non-destructive alternatives to traditional chemical analysis methods, which are often expensive and take a long time. NIR lets you do quantitative analysis on a wide range of food products, from grains to dairy, by following the rules set out by Lambert and Beer's law. Because of this ability, NIR spectroscopy has become an important tool for checking the quality and safety of food, which has had a big effect on the farming industry. NIR is useful for many things, including quickly checking the quality of food products and the composition of crop seeds. This shows how important it is to modern food science (Baianu *et al.,* 2011).

1.3.2. Nuclear Magnetic Resonance (NMR) Spectroscopy

The principles of Nuclear Magnetic Resonance (NMR) spectroscopy are based on how atomic nuclei act in a magnetic field. In such a field, certain nuclei vibrate at certain frequencies that depend on their magnetic environment. This property gives us very useful information about how molecules are built, how they move, and how they interact with each other. NMR spectroscopy uses radiofrequency pulses to make nuclei switch between energy levels. This lets researchers collect spectral data that shows how different atomic nuclei interact and change their chemical properties. Chemists can figure out bonding properties, molecular shapes, and even the presence of certain functional groups in a compound by noticing small changes in resonance frequencies. Recent improvements in NMR technology have made it much more useful in many areas of science. One of these advances is the hyperpolarization of nuclear spins, which makes NMR signals much more sensitive. To achieve hyperpolarization, it is usually necessary to use low temperatures or strong magnetic fields. But new methods have come up, like cross-relaxation induced polarization (CRIP) protocols that use single nitrogen-vacancy (NV) centers in diamond, that allow for very high levels of nuclear spin hyperpolarization at room temperature. These improvements not only push the limits of molecular imaging and spectroscopy, but they also promise to make NMR spectroscopy more useful in a wide range of fields, from clinical diagnostics to materials science (Broadway *et al.,* 2017). Theoretical proposals are also being looked into to use NV centers to make multidimensional NMR spectroscopy protocols work better. These methods can achieve Hartmann-Hahn resonances by using continuous microwave and radio-frequency driving fields along with magnetic field gradients. This makes it easier to control the coherence of nuclear spins. These kinds of new ideas make it less necessary to use a lot of nuclear spin polarization to get strong signals, which shortens the time it takes to make measurements in NMR experiments. Moreover, methods such as singular value thresholding matrix completion algorithms are being incorporated to enhance data acquisition, guaranteeing the identification of essential spectral features even with constrained sampling. The combination of cutting-edge technology and theoretical groundwork shows how powerful NMR spectroscopy can be in explaining how molecules behave (Cai *et al.*, 2015) (Figure 1.7).

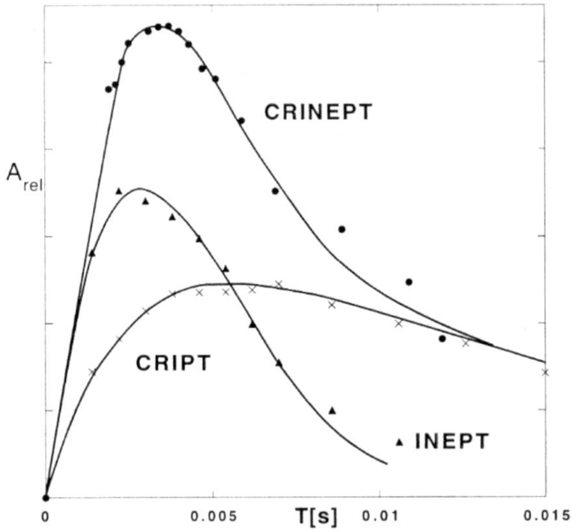

Figure 1.7. *Polarization transfer by cross-correlated relaxation in solution NMR with very large molecules.*

Source: https://www.pnas.org/cms/10.1073/pnas.96.9.4918/asset/ce932a38-cde6-4076-9ab6-55c48b59eba9/assets/graphic/pq0990753004.jpeg.

1.3.3. Ultraviolet-Visible (UV-Vis) Spectroscopy

Ultraviolet-visible (UV-Vis) spectroscopy is an important analytical method in molecular spectroscopy. It lets scientists figure out what substances are by looking at how they absorb ultraviolet and visible light. This method works on the idea that molecules absorb light at certain wavelengths, which causes their structures to change electronically. Researchers can make absorption spectra by measuring how much light is absorbed at different wavelengths. These spectra are like fingerprints for chemical compounds. The UV-Vis range, which usually goes from 200 to 800 nanometers, makes it possible to study a wide range of samples, such as organic compounds, inorganic materials, and biological macromolecules. Because it doesn't damage the sample, UV-Vis spectroscopy is very useful in situations where keeping the sample's integrity is very important.

This makes it important in both academic research and industry. One important use of UV-Vis spectroscopy is to look at cultural artifacts like manuscripts and old documents. In this context, the technique helps identify the compositions of inks and pigments, which helps protect and restore fragile materials. Studies have shown that UV-induced fluorescence imaging can make texts that have been damaged by ink fading or discoloration easier to read. This makes it easier to digitally restore important historical works. Being able to tell the different parts of inks, like iron gall ink or colored pigments like cinnabar and azurite, shows how useful the technique is for preserving art, since it protects both the material and cultural heritage of these objects. Combining UV-Vis spectroscopy with multispectral imaging techniques shows how flexible and useful it is for solving problems related to document preservation (Pronti *et al.,* 2018). Additionally, UV-Vis spectroscopy has become more popular in areas other than historical research, such as environmental monitoring and biomedical uses. Scientists use this method to check the quality of water by finding pollutants and to keep an eye on biochemical processes like drug interactions and enzyme activity. UV-Vis spectroscopy is an essential tool in labs all over the world because it can do quick and precise analyses. Notably, the technology has gone beyond its usual use by combining ideas from planetary science and the defense industry to improve tools and methods. As technology advances, the uses of UV-Vis spectroscopy keep growing. This shows how important it is for the field of molecular spectroscopy and for many other scientific fields (Appleby *et al.,* 2025).

1.4. INSTRUMENTATION AND TECHNIQUES

The investigation of instrumentation and methodologies in molecular spectroscopy provides a critical basis for comprehending molecular interactions and structure. Spectroscopic techniques, such as ultraviolet-visible (UV-Vis), infrared (IR), and nuclear magnetic resonance (NMR), function as non-invasive instruments for molecular characterization. Each method has its own pros and cons. For example, UV-Vis spectroscopy is good at finding electronic transitions and can give quick results for quantitative analyses. On the other hand, infrared spectroscopy is important for looking into vibrational energy levels, which helps scientists find molecular functional groups. New technologies, like two-dimensional IR spectroscopy, improve the resolution of spectral data, which helps us better understand complex molecular systems. The combination of these methods leads to a more complete way of studying molecules, where each method works well with the others, which increases the potential for analysis and helps with the design of future experiments (Lennard *et al.,* 2017). New ways to do molecular spectroscopy are always being developed, which shows how dynamic the field is. Recent progress includes the use of surface-enhanced Raman spectroscopy (SERS), which makes Raman spectroscopy much more sensitive. This method can find species that are not very common, which makes it especially useful for studying biological samples and environmental pollutants. Fourier-transform infrared (FTIR) mapping and other methods also let you look at samples in space, which can show you compositions that traditional spectroscopy might miss. These new ideas not only make the analysis process faster, but they also help us learn more about how molecules move and interact with each other. Furthermore, the integration of traditional spectroscopy with contemporary imaging techniques enhances the applicability of these methods across various fields, including materials science and conservation studies, thus expanding the scope of molecular spectroscopy (Lennard *et al.,* 2017). The interaction between instrumentation and analytical techniques is essential for the progress of molecular spectroscopy, especially in the investigation of intricate systems like planetary atmospheres. Research on extrasolar planets shows that spectroscopic tools are necessary for figuring out what the atmosphere is made of and how it behaves. These methods let scientists connect the initial conditions of the atmosphere to the chemical processes that are still going on by looking at the amounts of molecules and atoms. The difficulties of studying atmospheres on other planets show how important it is to have strong instruments that can work with different kinds of data. With the emergence of new observational facilities and technologies, the incorporation of advanced spectroscopic techniques is expected to provide unparalleled insights into the formation and evolution of planetary atmospheres, thereby enhancing the enthusiasm within the domain of molecular spectroscopy (Figure 1.8).

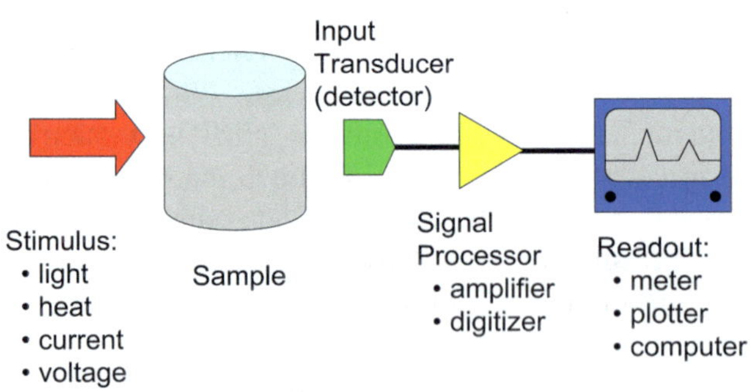

Figure 1.8. *Instrumentation and techniques.*

Source: https://upload.wikimedia.org/wikipedia/commons/b/b9/Analytical_instrument.png.

1.4.1. Components of a Spectroscopic Setup

To fully understand molecular spectroscopy, you need to look closely at its basic parts, which work together to help you study how molecules behave in different situations. A light source, a sample holder, a spectrometer, and a detector are the main parts of a spectroscopic setup. The light source, which is usually a laser or a lamp that gives off a wide range of light, gives off the electromagnetic radiation that the sample needs to interact with. This interaction makes the molecules absorb certain wavelengths, which raises their energy levels. The design and output of the light source are very important because they affect the sensitivity and resolution of the spectroscopic analysis, which affects how accurate the data is. The sample holder is just as important as the light source because it makes sure that the light and the molecules being studied interact in the best way possible. Different samples need different holder designs to stay stable and let light samples work well, whether they are in liquid, gas, or solid form. After the sample is lit up, the light that comes through or is emitted is sent to a spectrometer, where it is split into its different wavelengths. Spectrometers enhance the spectroscopy technique by enabling meticulous analysis of spectral lines that manifest as distinct peaks across the spectrum, utilizing prisms or diffraction gratings for dispersion. For instance, the application of multi-state analysis in the spectroscopy of cold cesium dimers illustrates the adaptability of spectrometers that function across a wider spectrum of energy levels, enabling profound molecular investigation (Allegrini *et al.,* 2010). The last part, the detector, is very important because it turns the optical signals into data that can be measured. Depending on the needs of the spectroscopy being done, detectors can be as simple as photodiodes or as complex as CCD cameras. The choice they make affects how sensitive and accurate the measurement is, which affects how well the spectroscopic setup can find and describe molecular features. Recent progress in optical detection methods, as shown in recent studies, shows that we are moving toward non-invasive ways to measure in tough environmental conditions. Such innovations are especially important for industries that need stricter rules about emissions because they make it possible to accurately measure gases without changing the samples' natural properties (Bauke *et al.,* 2018). So, putting these parts together—light source, sample holder, spectrometer, and detector—makes a strong framework for using spectroscopy to study molecular dynamics.

1.4.2. Sample Preparation Methods

The efficacy of molecular spectroscopy is predominantly dependent on the sample preparation techniques utilized prior to analysis. These methods improve the quality and reliability of the results because they directly affect how the sample interacts with the spectroscopic techniques. Distillation, filtration, and extraction are some of the most important ways to separate the parts of a sample matrix that you want. The better ways of preparing samples have made the analysis more sensitive and specific, which has improved the overall interpretive power of molecular spectroscopy. For instance, new methods like solid phase microextraction make it easier to get analytes from complex mixtures, which improves the quality of data in studies ranging from environmental monitoring to material science. These developments highlight the critical significance of customized sample preparation methodologies in obtaining precise molecular spectroscopic analyses (Lennard *et al.,* 2017).

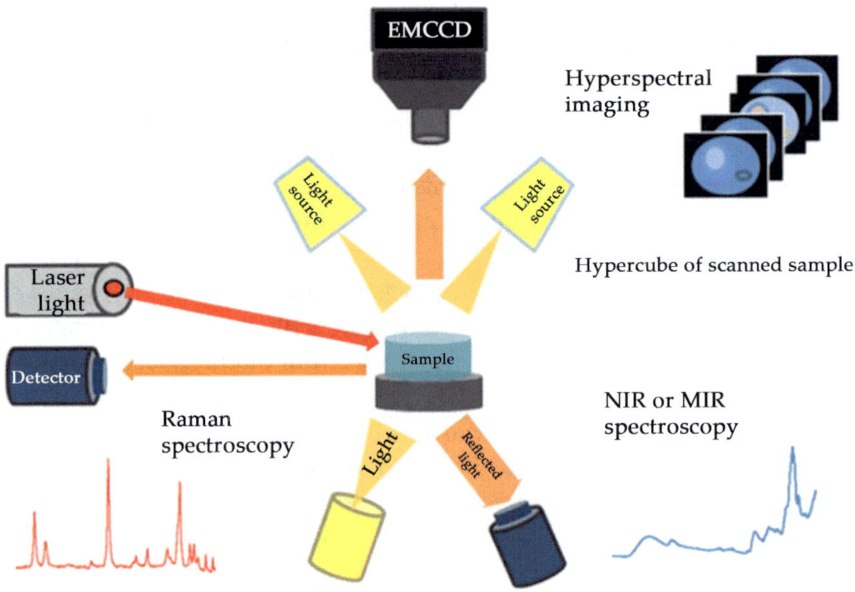

Figure 1.9. *Schematic diagram of several vibrational spectroscopic techniques.*

Source: https://www.researchgate.net/profile/Wen-Hao-Su/publication/332878514/figure/fig1/AS:962355043250177@1606454730376/Schematic-diagram-of-several-vibrational-spectroscopic-techniques-87.png.

The choice of a suitable sample preparation technique is frequently determined by the chemical composition and physical condition of the sample under examination. To make sure that the sample is safe and works with the spectroscopic equipment, different types of matrices, like solids, liquids, and gases, may need to be handled in different ways. For example, solid samples may need to be ground or milled to make their surface area bigger and make it easier for analytes to get to them. Liquid samples may need to be diluted or concentrated to get the best analyte concentration. Additionally, the impact of negative matrix effects must be acknowledged, as they can complicate spectroscopic analysis by obscuring or modifying signal responses. Researchers can effectively reduce these interferences by using advanced sample preparation methods, which improves the performance and accuracy of molecular spectroscopic methods (Cadorim *et al.,* 2017). As molecular spectroscopy keeps getting better, the importance of good sample preparation grows in interdisciplinary settings. The work that chemists, conservators, and materials scientists do together shows how sample preparation methods can change to meet different analytical needs. For example, optimized sample preparation methods make it much easier to use vibrational spectroscopy techniques like Fourier-transform infrared spectroscopy and Raman spectroscopy to get a detailed picture of complex materials. These changing relationships can be seen in areas like art conservation, where knowing the chemical makeup of pigments and binders is important for protecting cultural heritage. Recent literature has emphasized that advancements in analytical techniques, including secondary ion mass spectrometry and enhanced mapping capabilities, further necessitate the implementation of rigorous sample preparation processes to effectively support conservation initiatives and augment scholarly investigation into heterogeneous materials (Lennard *et al.,* 2017) (Figure 1.9).

1.4.3. Data Acquisition and Analysis

In molecular spectroscopy, data collection and analysis are two of the most important parts that determine how well an experiment works. The initial step involves gathering spectral data, utilizing diverse techniques such as infrared and Raman spectroscopy to extract molecular information from samples. The methodology selected influences both the nature of the data collected and its subsequent analysis. Researchers can effectively identify molecular bonds and concentrations with well-designed data acquisition systems that improve the resolution and accuracy of the spectra. These improvements in data collection methods are especially important in fields where it is important to understand how complex biological systems work and how they interact. This shows how important accuracy and precision are in molecular diagnostics and research. After data collection, the analytical phase uses advanced computer methods to make sense of the spectra that were taken. In this step, you usually need to use algorithms and software that make it easier to work with the complex datasets that come from spectroscopic analysis. Chemometrics and other techniques are very important at this stage because they help get the information you need while reducing noise and possible artifacts in spectral data. Good data analysis helps find hidden patterns and connections, which makes it easier to understand biochemical processes as a whole. Recent literature underscores that the incorporation of robust analytical tools guarantees the reliability and relevance of insights, thereby making substantial contributions to advancements in scientific research and clinical applications pertaining to molecular spectroscopy (Beechem *et al.,* 2012). Moreover, the development of multispectral technologies has improved the ability to collect and analyze data even more. Modern methods like Near Infrared (NIR) spectroscopy are faster and less invasive than traditional methods, which often use time-consuming and destructive techniques. This progress has given researchers in many fields, especially in the food and drug industries, the ability to get important information about molecular composition without harming samples. NIR spectroscopic measurements are fast and easy to use, which makes them perfect for high-throughput analysis, which is essential in today's fast-paced research environments (Baianu *et al.,* 2011). As a result, these technologies keep pushing the limits of molecular spectroscopy, showing how important they are for making research more efficient and accurate.

1.5. APPLICATIONS OF MOLECULAR SPECTROSCOPY

Molecular spectroscopy has a lot of uses in many different fields of science, and it makes both research and practical uses much better. One of the main uses of this method is to analyze biological samples, where it helps find and measure important biomolecules. This ability is especially important in clinical settings, like when diagnosing diseases or keeping an eye on treatment plans. For example, spectroscopic methods can find certain metabolites in urine or blood. This information is very useful for doctors because it helps them make better decisions about how to care for their patients. Molecular spectroscopy not only improves the accuracy of diagnoses, but it also helps us learn more about how biological processes work at the molecular level. This makes it an important tool in modern medicine and health research. Molecular spectroscopy has become a key technology in the fields of food and agricultural sciences for making sure that products are safe and of good quality. The use of Near Infrared (NIR) spectroscopy is a good example of this because it quickly and non-invasively checks the composition of food. This method makes it possible to look at a wide range of foods, such as meats, grains, and dairy, which makes it easier to keep an eye on food quality in real time. NIR spectroscopy is a quick way to do chemical analysis that follows Lambert and Beers law. This means that it makes it easy and accurate to take quantitative measurements (Baianu *et al.,* 2011). So, adding spectroscopy to food analysis not only makes it more efficient, but it also makes food safer for consumers by making sure that food products meet the required standards. Molecular spectroscopy also has a lot of potential for monitoring and evaluating the environment. It is an important method for looking at pollutants and judging how healthy ecosystems are here. Researchers can find and measure dangerous substances in air, water, and soil samples by using different types of spectroscopies. These analyses help create plans for protecting the environment and making it last longer. They also show how molecular spectroscopy can help us understand how human activity affects natural resources. Furthermore, technological advancements, exemplified by reports from the Boston University Photonics Center, demonstrate how interdisciplinary research in spectroscopy can yield innovative solutions to urgent environmental issues (Anonymous, 2013). So, even though molecular spectroscopy has many uses, its role in environmental science shows how important it is for promoting global health and ecological sustainability (Figure 1.10).

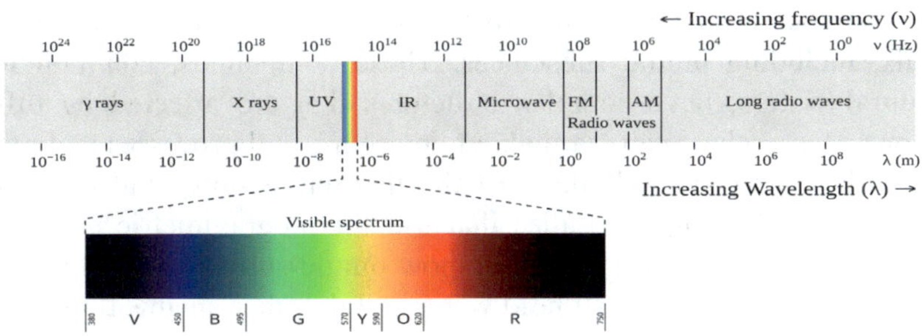

Figure 1.10. *Applications of molecular spectroscopy.*

Source: https://cms-media.bartleby.com/wp-content/uploads/sites/2/2021/10/08091027/EM.png.

1.5.1. Role in Chemical Analysis

Molecular spectroscopy is a key method in chemical analysis because it can give us information about how molecules are structured and how they move. This technique uses light's absorption, emission, and scattering to show what compounds are like at the molecular level. Researchers can find out what substances are in a sample and how much of each one there is by looking at how light and matter interact. Additionally, molecular spectroscopy helps find molecular conformations and keep an eye on chemical reactions as they happen. As the field progresses, sophisticated spectroscopic methods like terahertz time-domain spectroscopy have developed, showcasing the ability to investigate molecular dynamics with unparalleled resolution and precision, thereby improving the dependability of chemical analyses across multiple sectors, including pharmaceuticals and materials science (Anonymous, 2016). The use of terahertz time-domain spectroscopy in chemical analysis is a good example of how molecular spectroscopy has changed. This non-contact, non-destructive method has been especially helpful for studying amorphous solids, which are hard to study with traditional analytical methods. To make amorphous drugs more bioavailable, it's important to understand how their molecules move and how they crystallize. Researchers can make smart choices about formulation and stability, which are important parts of drug delivery systems, because they can figure out when and how strong molecular motion is. The incorporation of advanced spectroscopic techniques into chemical analysis facilitates enhanced characterization, underscoring their critical importance in the pharmaceutical sector, where accuracy is essential (Anonymous, 2016). Moreover, molecular spectroscopy occupies a distinctive role in the examination of intricate organic materials, including humic substances. These compounds, common in soil and natural waters, have complex structures that are affected by different metal ions. Researchers can learn about environmental processes and material properties by using methods that test the thermal stability and reactivity of humic acids. For example, studies that use thermogravimetric analysis and differential scanning calorimetry show how humic substances and metal ions interact, showing how they can hold water and do other chemical things. This example shows that molecular spectroscopy not only helps us understand the basic properties of chemicals, but it also has real-world uses in environmental science and sustainability. This shows that it can be used in many different ways for chemical analysis.

1.5.2. Applications in Pharmaceuticals and Medicine

The growing use of molecular spectroscopy in medicine and pharmaceuticals has changed the way drugs are made and given to people. Researchers can describe the molecular structure of drug compounds with unmatched accuracy using methods like infrared (IR) and nuclear magnetic resonance (NMR) spectroscopy. This level of understanding is important for improving the effectiveness of drugs, maximizing their benefits, and reducing their possible side effects. Recent progress has shown how useful hybrid nanoparticles, especially Fe_3O_4-Au hybrid nanoparticles (HNPs), can be for making drug delivery systems that respond to heat. When these nanoparticles are added to smart scaffolds, they can be used to remotely control drug release by heating specific areas. This is a promising way to target therapy, especially in cancer treatment, where controlling where drugs work is very important (Hoskins *et al.*, 2013). Moreover, the investigation of natural products as essential elements in drug development highlights the importance of molecular spectroscopy. A large number of new drugs come from natural sources, so it is more important than ever to understand the molecular parts of traditional medicines. Countries with a lot of different kinds of plants and animals, like Bhutan, are important places for bioprospecting, where traditional medicine could help modern medicine.

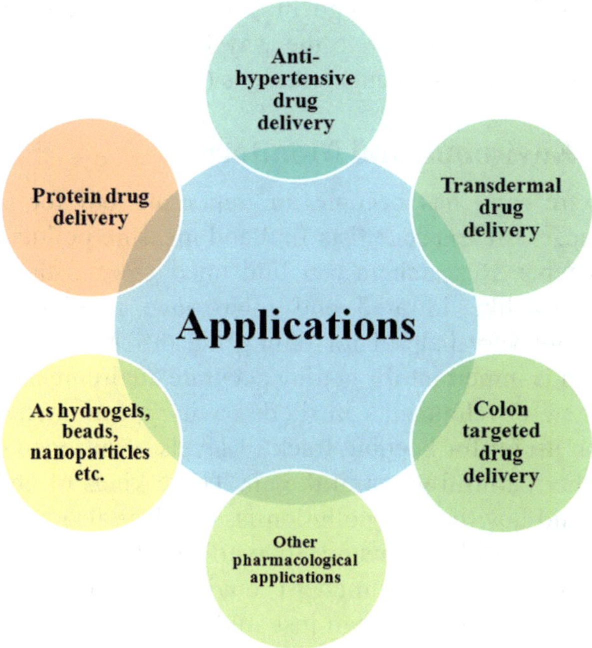

Figure 1.11. *Schematic representation of pharmaceutical application.*

Source: https://www.researchgate.net/profile/D-Gowda/publication/301560508/figure/fig3/AS:353634941456387@1461324548328/Schematic-representation-of-pharmaceutical-application-of-GG.png.

Molecular spectroscopy's ability to clarify complicated plant-based compounds opens up new ways to improve current treatments and find new drugs. Traditional medicines are becoming more popular because they often offer holistic treatment options with fewer side effects. This makes them a good focus for modern drug discovery programs (Wangchuk *et al.,* 2008). Additionally, molecular spectroscopy is useful for more than just drug characterization and development; it is also very important for quality control and therapeutic monitoring. The pharmaceutical industry is putting more and more pressure on drug companies to meet strict standards for drug safety and effectiveness. Molecular techniques make it possible to keep checking drugs throughout their lives. For quantitative analysis, such as figuring out how much active pharmaceutical ingredients are in a formulation, techniques like UV-Vis spectroscopy are very important. Furthermore, advancements in spectroscopy have practical applications in personalized medicine, facilitating the customization of therapeutic strategies according to an individual's distinct biological composition. This change in personalized treatment shows how molecular spectroscopy can change the way healthcare works by improving patient outcomes and creating better solutions (Figure 1.11).

1.5.3. Use in Environmental Monitoring

Molecular spectroscopy has become an important tool for monitoring the environment because it lets scientists find and measure pollutants in different types of ecosystems. Researchers can find trace gases with high sensitivity by using methods like infrared and microwave rotational spectroscopy. These methods are very helpful for telling the difference between different pollutants, which is important for getting accurate environmental assessments. For example, new developments in diode-cavity spectrometers have shown that they can be useful for keeping track of levels of nitrogen dioxide, sulfur dioxide, and other harmful gases in the air. These kinds of abilities not only help us understand how bad air pollution is, but they also help policymakers make decisions that will lessen its negative effects. This shows how important molecular spectroscopy is to modern environmental science. Molecular spectroscopy is useful for more than just studying the atmosphere; it can also be used to monitor water and soil. Researchers can find harmful substances like heavy metals and other toxic compounds by carefully looking at the molecular makeup of water samples, especially when they are contaminated. This is especially important when checking the health of ecosystems and making sure the water is safe to drink. Many spectroscopic methods are non-destructive, which means that samples can be measured over and over again without changing them. This makes longitudinal studies more reliable. The extensive data produced offer a more precise understanding of environmental health, enabling focused remediation strategies for contaminated locations. The use of molecular spectroscopy in different areas of environmental monitoring shows how useful it is for protecting the health of ecosystems. The educational value of molecular spectroscopy is also important because it helps train the next generation of scientists with the analytical skills they need. The implementation of related programs, such as those outlined by Anonymous (2010), promotes advanced diploma qualifications in scientific fields, ensuring that students gain hands-on experience with modern spectroscopic methods. Institutions can help with ongoing efforts to monitor the environment by training a new generation of scientists in these advanced analytical methods. Moreover, this transfer of knowledge is essential for improving research projects that aim to stop climate change and environmental damage. In conclusion, molecular spectroscopy has many uses in environmental monitoring. It helps scientists learn more and also helps build the educational framework that will be needed to keep these efforts going in the future.

1.6. CHALLENGES AND LIMITATIONS

Molecular spectroscopy is an important way to learn about the structure and behavior of different chemical systems, but it does have its problems and limits. One major problem is that molecular spectroscopic techniques are very sensitive to outside factors like temperature and pressure, which can change the results. Changes in these parameters can cause differences in spectral data, which can make it harder to understand how molecules interact and how they are shaped. Additionally, the inherent noise present in spectroscopic measurements can obscure subtle spectral features, complicating the accurate analysis of complex mixtures. For instance, sum-frequency generation (SFG) spectroscopy can provide extensive information about molecules at interfaces, but it performs poorly in real-world samples due to highly variable environmental conditions. Studies of water–air interfaces by Backus et al. (2020) illustrate this limitation. So, it's important to understand and reduce these outside factors in order to make molecular spectroscopic analyses more reliable. Another limitation of molecular spectroscopy is that it is hard to understand and model the data. It can be hard to assign specific molecular features to different parts of a mixture because the spectra often show overlapping peaks. Also, the fact that advanced computer techniques are needed to simulate and compare experimental spectra makes things even more complicated. While the integration of SFG with simulation techniques can improve comprehension of molecular-level structures, it necessitates considerable computational resources and advanced modeling methodologies (Backus *et al.*, 2020). This duality makes it hard to get accurate results and shows that we need to use both experimental and theoretical methods to solve these problems. When spectroscopists and computational chemists work together, they can come up with better ways to interpret data and learn more about complicated systems. Lastly, the technological limits of different spectroscopic methods make it even harder to analyze molecular systems well. For example, Fourier transform infrared (FTIR) and Raman spectroscopy are two types of vibrational spectroscopy that have been used in a lot of different situations, but each one has its own set of problems. FTIR, for instance, can be affected by water vapor, which can make analysis harder. Also, the spatial resolution of spectroscopic methods often makes them less useful, especially when looking at heterogeneous materials or interfaces, like those in painted textiles. This makes it hard to see structural features at the nanoscale, which is important for understanding how things break down in conservation work. Ultimately, acknowledging these challenges and constraints is essential for the progression of molecular spectroscopy as a dependable analytical instrument, requiring ongoing innovation and adaptation in methodologies and interdisciplinary collaborations (Figure 1.12).

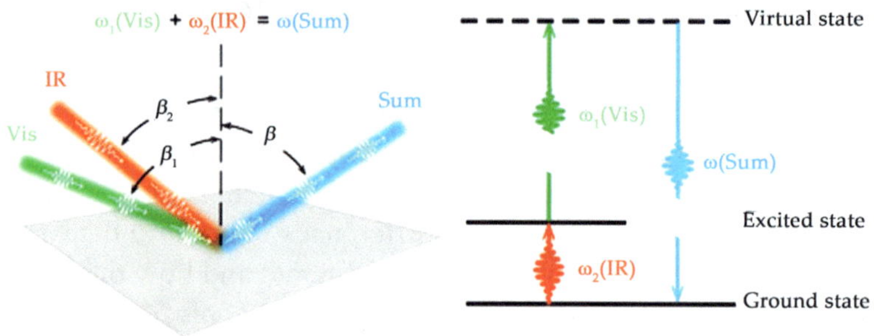

Figure 1.12. *Sum-frequency generation.*

Source: https://ars.els-cdn.com/content/image/3-s2.0-B978012409547213954X-f13954-01-9780081019832.jpg.

1.6.1. Sensitivity and Detection Limits

In molecular spectroscopy, sensitivity and detection limits are basic parameters that determine how well analytical methods work. Sensitivity is the ability of an instrument to pick up on small changes in a signal compared to a baseline noise. This is important for finding low-concentration analytes in complex samples. A high sensitivity enables the distinction of markedly low concentrations of substances, thereby enhancing sophisticated analytical applications such as environmental monitoring, pharmaceutical analysis, and biochemical diagnostics. When choosing a molecular spectroscopic method, researchers should know how sensitive the technique is so they can choose the right instruments for their analytical needs. So, making sensitivity better not only makes the data more trustworthy, but it also opens up new uses for molecular spectroscopy in many areas of science. Detection limits, on the other hand, are the lowest amount of an analyte that can be reliably found in a sample matrix but not necessarily measured. Instrument calibration, sample preparation, and the molecules' own properties, like their absorption cross-sections, all affect detection limits. For example, improvements in techniques like molecular contrast agents in optical coherence tomography have greatly increased detection limits, making these technologies much more useful in clinical settings (Yang *et al.,* 2004). So, improving detection limits is very important for getting accurate measurements of biomolecules, which will lead to new discoveries in medical diagnostics and therapeutic monitoring. Furthermore, novel methodologies like cavity-enhanced laser-induced fluorescence (CELIF) exhibit considerable enhancements in both sensitivity and detection thresholds. Researchers have created a system that combines laser-induced fluorescence with cavity ring-down methods. This system boosts the signal while reducing noise, which greatly increases the range of detection (Berden *et al.,* 2013). This enables the precise assessment of diverse molecular attributes, including absolute absorption coefficients and fluorescence quantum yields, within a singular measurement framework. So, constantly improving methods to make them more sensitive and lower detection limits is very important for the progress of molecular spectroscopy. This way, researchers can be sure that they can analyze samples that are getting more complicated with confidence in their results (Figure 1.13).

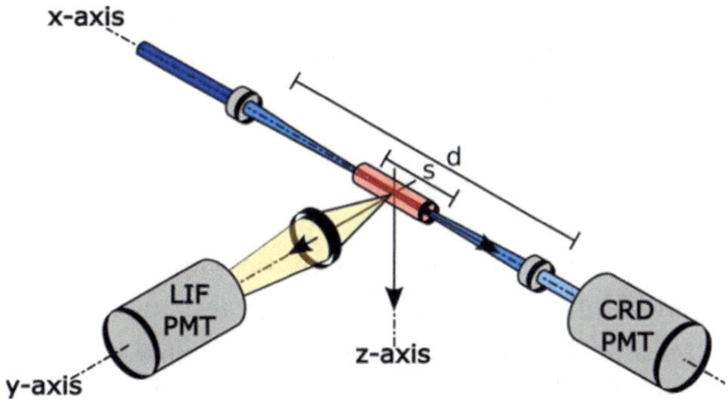

Figure 1.13. *Schematic of the cavity-enhanced laser-induced fluorescence (CE-LIF).*

Source: https://www.researchgate.net/publication/361358664/figure/fig8/AS:1167810050965510@1655439021617/A-schematic-of-the-cavity-enhanced-laser-induced-fluorescence-CELIF-apparatus-This.jpg.

1.6.2. Interference from Other Substances

Interference from other substances is a big problem in molecular spectroscopy. Spectroscopic techniques depend on the detection of distinct molecular signatures within a sample; however, these signatures may be concealed by overlapping spectra from coexisting compounds. For example, when testing the air quality inside a building, finding formaldehyde has shown that it overlaps a lot with other chemicals that are often found in those kinds of places. A study found 11 substances that made it harder to find formaldehyde in the UV region. This shows how hard it is to accurately analyze air samples. The competitive nature of these spectral features requires the creation of new detection methods to reduce interference, which will improve the accuracy of measuring target molecules (Davenport *et al.,* 2015). This overlap of spectra and interference shows how important it is to carefully develop methods in molecular spectroscopy. Additionally, interference can affect detection in different ways, such as by causing fluorescence quenching. For instance, the use of chitosan-capped ZnS quantum dots (QDs) to find bisphenol A (BPA) showed that they could greatly lessen the effects of other substances. This method clearly showed that BPA quenched fluorescence, but the presence of most substances that could have interfered with the analysis did not have a big effect on the results. This shows how important it is to be selective when designing sensors. The linear relationship between fluorescence response and BPA concentration over a wide range also shows how well this new method works to get around possible spectral interferences. By making sure that most chemicals don't get in the way, this method shows how promising eco-friendly materials are for improving molecular detection technologies (Bie *et al.,* 2014). In the end, dealing with interference from other substances is very important for moving molecular spectroscopy forward. Comprehending the diverse mechanisms by which external compounds influence spectroscopic measurements is crucial for the advancement of sensitive and precise analytical methodologies. Researchers must be vigilant in identifying potential interferences and optimizing methods accordingly. Researchers can make spectroscopic analyses more reliable by using targeted methods, like using reference wavelengths where interference is low. The main problem is making systems that can tell the difference between real signals and noise, which will improve the accuracy of molecular analysis in many different fields. So, more research into ways to reduce interference is necessary for the future of molecular spectroscopy and its use in many fields.

1.6.3. Limitations of Current Technologies

Molecular spectroscopy is a useful tool for learning about chemical structures and interactions, but it has some technological problems that make it less useful and less efficient in general. One major problem is that existing spectroscopic systems, especially Raman spectroscopy, are very expensive and hard to use. Current Raman systems are too expensive for many research settings, even though they are very good at finding chemical compounds and making biomedical images without labels. This financial barrier makes it hard for people to get important tools that could greatly improve molecular analysis. Moreover, the intricate nature of the technology frequently requires specialized training for operators, leading to discrepancies in data acquisition and interpretation across various laboratory settings (Eibl *et al.,* 2014). Another major problem is that molecular spectroscopic techniques are not very fast or have a high temporal resolution. Because of their slow acquisition times, traditional Raman spectrometers often have trouble providing real-time analysis. This can be especially problematic in dynamic systems where things change quickly. For example, in drug discovery, getting quick feedback on how molecules interact is important for making decisions about experiments. The creation of new techniques, like the time-encoded Raman (TICO-Raman) setup, aims to solve these problems by providing faster speeds and a wider range of spectral coverage. But these kinds of new technologies aren't very common yet, so many labs still use older technologies that aren't efficient enough for modern research needs (Eibl *et al.,* 2014). Also, current technologies can't give complete spectral information, which is necessary for accurate chemical characterization. The resolution and completeness of the spectral data collected can have a big effect on the analysis in many spectroscopic applications. Current techniques frequently produce incomplete or noisy spectra, complicating the identification of subtle variations in molecular characteristics. For instance, improvements in atomic force microscopy (AFM) force spectroscopy have made it easier to find protein-protein interactions by improving the signal-to-noise ratio. However, similar improvements in spectral resolution for molecular spectroscopy are still hard to find. These limitations can make findings less reliable and slow down progress in fields like drug discovery and material science as scientists try to fully use the power of molecular spectroscopy.

SUMMARY

The study of molecular spectroscopy gives us a deep understanding of how light and matter interact, which is important for many scientific uses. Researchers can break down the complicated structures of molecules and figure out how they work by using different methods, like UV-Vis, IR, and NMR spectroscopy. This analytical method not only helps us understand how molecules behave better, but it also helps chemistry, biology, and materials science make big strides. Being able to accurately describe molecular species has many important effects, such as drug development and monitoring the environment. As we keep improving these methods, we can expect to see more analytical tools, which will help us connect what we know about theory with what we can do in practice. In addition, to understand molecular spectra, you need to know how a molecule's electronic, vibrational, and rotational transitions are related to each other. This knowledge is essential for analyzing experimental outcomes and their application to practical situations. For example, the binding dynamics of nitric oxide (NO) to heme-nitric oxide/oxygen binding (H-NOX) proteins demonstrate how molecular spectroscopy can elucidate the intricacies of biochemical interactions. These interactions have significant ramifications for cellular signaling pathways and underscore the critical role of complex formation in biological systems. The intricate interaction between NO and heme compounds exemplifies how molecular spectroscopy functions as both an analytical instrument and a substantial reservoir of information for elucidating dynamic biological processes, thereby reaffirming its essential role in modern scientific inquiry (Britt *et al.,* 2017). In the end, the ongoing improvements in molecular spectroscopy will probably lead to new discoveries and innovations. This is because technological and theoretical progress are coming together. As scientists learn more about how molecules interact with each other, we expect new spectroscopic methods to come out that improve sensitivity and resolution. The dedication to establishing stringent criteria and educational frameworks, exemplified by ANONYMOUS (2010), will enhance the discipline by cultivating a new generation of scientists endowed with both theoretical understanding and practical competencies. When we think about how important molecular spectroscopy is in science, it becomes clear that its contributions go far beyond just being able to analyze things. They open the door to groundbreaking research that can help us solve important global problems and deepen our understanding of the molecular world.

Molecular spectroscopy is a very important tool for figuring out what different substances are made of and how they act on a molecular level. This field includes a lot of different methods, such as infrared (IR) spectroscopy, nuclear magnetic resonance (NMR), and Raman spectroscopy. Each of these methods gives us different information about how molecules are structured and how they move. Researchers can learn important things about molecular interactions, chemical bonding, and functional groups in a sample by being able to find and study certain molecular vibrations and transitions. Because of this, molecular spectroscopy is not only important for basic research, but it is also very important in areas like pharmacology, environmental science, and materials development. This all-encompassing approach shows how important the field is, with uses in everything from sensor technology to quality control in many fields. In the quest to improve analytical techniques, molecular spectroscopy has made great strides in recent years. For example, Fourier-transform infrared (FTIR) and Raman spectroscopy have changed the way we look at complex mixtures by making it easier to find the chemical parts. Research examining the nectar composition of *Leptospermum scoparium* has illustrated the effectiveness of these methodologies in quantifying specific constituents, including dihydroxyacetone (DHA) and sugars such as fructose and glucose. The incorporation of predictive modeling in this study is especially significant, as it elucidates the determinants affecting nectar production. These improvements show how molecular spectroscopy is still changing and becoming more useful in both academic and business settings, especially in fields that depend on accurate chemical analysis (Nickless *et al.,* 2015). Recent advancements in time-resolved methodologies have facilitated novel approaches to comprehending ultrafast phenomena in molecular systems. Time-resolved coincidence imaging, demonstrated through investigations on 1,3-butadiene, elucidates complex dynamics occurring on femtosecond timescales, providing detailed insights into excited-state molecular behavior. The sensitivity of time-resolved photoelectron angular distributions (TRPADs) demonstrates that contemporary spectroscopic methods can greatly improve our understanding of molecular evolution following excitation. However, the higher number of dimensions in the data makes it harder to analyze, which shows how important it is for different fields to work together to connect experimental results with theoretical models. This convergence of experimental and theoretical work signifies a pivotal advancement in molecular spectroscopy, with the potential to elucidate the complexities inherent in molecular interactions and reactions (Hockett *et al.,* 2013).

REFERENCES

1. Adhikari, C. M., Dawes, R., Jentschura, U. D., Kolachevsky, V., et al. (2019). Pressure shifts in high-precision hydrogen spectroscopy: I. Long-range atom-atom and atom-molecule interactions. http://arxiv.org/abs/1904.03025 (accessed on 29th October 2024).
2. Agterberg, D., Agterberg, D., Alford, M., Als-Nielsen, J., Altman, E., Andreev, A., Andreev, A., et al. (2011). Quantum liquid-crystal order in resonant atomic gases. http://arxiv.org/abs/1112.0773 (accessed on 29th October 2024).
3. Alfano, R. R., Cheng, G., Liu, C. H., Shi, L., et al. (2019). Optical biopsy identification and grading of gliomas using label-free visible resonance Raman spectroscopy. https://core.ac.uk/download/323064657.pdf (accessed on 29th October 2024).
4. Allegrini, M., Aymar, M., Bouloufa, N., Chotia, A., et al. (2010). Photoionization spectroscopy of excited states of cold cesium dimers. http://arxiv.org/abs/1005.3283 (accessed on 29th October 2024).
5. Anderson de la Llana, S., Anderson, I. S., Capitán Vallvey, L. F., Cecchini, M., et al. (2014). A general perspective of the characterization and quantification of nanoparticles: Imaging, spectroscopic, and separation techniques. https://core.ac.uk/download/186617942.pdf (accessed on 29th October 2024).
6. Andrews, D., Andrews, D., Cialla, D., Das, R., Keller, J., Krafft, C., Long, D., et al. (2016). On the emergence of Raman signals characterizing multicenter nanoscale interactions. https://core.ac.uk/download/41993322.pdf (accessed on 29th October 2024).
7. Appleby, J. F. (2025). Workshop on Advanced Technologies for Planetary Instruments, Part 1. https://core.ac.uk/download/pdf/42806195.pdf (accessed on 29th October 2024).
8. Arnaud, N., Briand, F., Gascuel, O., Gimenez, F., et al. (2012). Final report of the ModSysC2020 Working Group–Data, models and theories for complex systems: New challenges and opportunities. https://core.ac.uk/download/52630003.pdf (accessed on 29th October 2024).
9. Backus, E. H. G., Bonn, M., Imoto, S., Mukamel, S., et al. (2020). Molecular structure and modeling of water-air and ice-air interfaces monitored by sum-frequency generation. https://core.ac.uk/download/323075765.pdf (accessed on 29th October 2024).
10. Baianu, I. C. (2011). Translational oncogenomic and human cancer interactome networks. https://core.ac.uk/download/pdf/290188.pdf (accessed on 29th October 2024).

11. Basson, M. D., Chen, L., Dong, L., Hou, C., et al. (2019). Optimization of protein-protein interaction measurements for drug discovery using AFM force spectroscopy. https://core.ac.uk/download/305123513.pdf (accessed on 29th October 2024).

12. Bauer, J., Bornmann, L., & Haunschild, R. (2019). Influential cited references in FEMS Microbiology Letters: Lessons from Reference Publication Year Spectroscopy (RPYS).

13. Bauke, S., Ctistis, G., Golibrzuch, K., Müller, T., et al. (2018). Quantitative measurement of combustion gases in harsh environments using NDIR spectroscopy. http://arxiv.org/abs/1810.09239 (accessed on 29th October 2024).

14. Beechem, J., Levenson, R., & McNamara, G. (2012). Spectral imaging in preclinical research and clinical pathology. https://core.ac.uk/download/323068082.pdf (accessed on 29th October 2024).

15. Bellatreccia, F., Della Ventura, G., & Marcelli, A. (2013). SR-FTIR microscopy and FTIR imaging in the earth sciences. http://arxiv.org/abs/1312.5750 (accessed on 29th October 2024).

16. Berden, G., Wrede, E., Gagliardi, G., Lehmann, K. K., Nahler, N. H., Willis, O. R., Sanders, S. E., et al. (2013). Absolute absorption and fluorescence measurements over a dynamic range of 10^6 with cavity-enhanced laser-induced fluorescence. https://core.ac.uk/download/162910697.pdf (accessed on 29th October 2024).

17. Bie, J., Cao, X., Guo, J., Liu, Y., et al. (2014). Facile synthesis of chitosan-capped ZnS quantum dots as an eco-friendly fluorescence sensor for rapid determination of bisphenol A in water and plastic samples. https://core.ac.uk/download/231848140.pdf (accessed on 29th October 2024).

18. Blake, G. A. (2009). Terahertz spectroscopy in the lab and at telescopes. https://core.ac.uk/download/4886017.pdf (accessed on 29th October 2024).

19. Britt, R. D., Guo, Y., Herzik, M. A., Iavarone, A. T., et al. (2017). Regulation of nitric oxide signaling by formation of a distal receptor-ligand complex. https://core.ac.uk/download/323081579.pdf (accessed on 29th October 2024).

20. Broadway, D. A., Hall, L. T., Hollenberg, L. C. L., Simpson, D. A., et al. (2017). Quantum probe hyperpolarization of molecular nuclear spins. https://core.ac.uk/download/162232040.pdf (accessed on 29th October 2024).

21. Cadorim, H. R., Del Campo Menoyo, J., Carvalho, G. S., Duarte, F. A., et al. (2017). Development of a fast-screening method for the direct

determination of chlorinated persistent organic pollutants in fish oil by high-resolution continuum source graphite furnace molecular absorption spectrometry. https://core.ac.uk/download/153371322.pdf (accessed on 29th October 2024).

22. Cai, J., Kost, M., & Plenio, M. B. (2015). Resolving single molecule structures with nitrogen-vacancy centers in diamond. https://core.ac.uk/download/77002999.pdf (accessed on 29th October 2024).

23. Caycedo-Soler, F., Hauer, J., Huelga, S. F., Lim, M., et al. (2015). Vibronic origin of long-lived coherence in an artificial molecular light harvester. http://arxiv.org/abs/1502.01717 (accessed on 29th October 2024).

24. Cestellos-Blanco, S., Kornienko, N., Lim, J., Liu, C., et al. (2018). Physical biology of the materials-microorganism interface. https://core.ac.uk/download/287623640.pdf (accessed on 29th October 2024).

25. Coy, S., Flannery, C., Foy, B., Hetzler, R., et al. (1990). Determination of molecular spectroscopic parameters and energy-transfer rates by double-resonance spectroscopy. https://core.ac.uk/download/pdf/10419419.pdf (accessed on 29th October 2024).

26. Crossfield, I. J. M. (2015). Observations of exoplanet atmospheres. http://arxiv.org/abs/1507.03966 (accessed on 29th October 2024).

27. Davenport, J., Hodgkinson, J., Saffell, J. R., Tatam, R. P., et al. (2015). A measurement strategy for non-dispersive ultra-violet detection of formaldehyde in indoor air: Spectral analysis and interferent gases. https://core.ac.uk/download/42144215.pdf (accessed on 29th October 2024).

28. Eibl, M., Huber, R., Karpf, S., Klein, T., et al. (2014). Time-encoded Raman: Fiber-based, hyperspectral, broadband stimulated Raman microscopy. https://core.ac.uk/download/211702172.pdf (accessed on 29th October 2024).

29. Flambaum, V. V., & Stadnik, Y. V. (2016). Improved limits on interactions of low-mass spin-0 dark matter from atomic clock spectroscopy. http://arxiv.org/abs/1605.04028 (accessed on 29th October 2024).

30. Gardiner, A. T., Britton, G., Britton, G., Britton, G., Uragami, C., Frank, H. A., Hashimoto, H., et al. (2018). Understanding/unravelling carotenoid excited singlet states. https://core.ac.uk/download/157582964.pdf (accessed on 29th October 2024).

31. Hockett, P., Ripani, E., Rytwinski, A., Stolow, A., et al. (2013). Probing ultrafast dynamics with time-resolved multi-dimensional coincidence imaging: Butadiene. http://arxiv.org/abs/1301.6928 (accessed on 29th October 2024).

32. Hoskins, C., Lees, M. R., McGarvey, D. J., Roach, P., et al. (2013). Remotely triggered scaffolds for controlled release of pharmaceuticals. https://core.ac.uk/download/16663137.pdf (accessed on 29th October 2024).

33. Hrubesh, L. W. (1973). Microwave rotational spectroscopy: A physical technique for specific pollutant monitoring. https://core.ac.uk/download/pdf/42898929.pdf (accessed on 29th October 2024).

34. Lennard, F., Smith, M., & Thompson, K. (2017). A literature review of analytical techniques for materials characterization of painted textiles–Part 2: Spectroscopic and chromatographic analytical instrumentation. https://core.ac.uk/download/84591645.pdf (accessed on 29th October 2024).

35. ANONYMOUS. (2003). College of Engineering. https://ecommons.cornell.edu/bitstream/handle/1813/38520/COS_v94_2002_2003_05.pdf?sequence=1&isAllowed=y (accessed on 29th October 2024).

36. N/A. (2013). The Boston University Photonics Center Annual Report 2013–2014. https://open.bu.edu/bitstream/2144/29752/1/Photonics-Center-Final-Annual-Report-2013-2014-v2.pdf (accessed on 29th October 2024).

37. Nickless, E. M. (2015). Influential Factors in Nectar Composition and Yield in Leptospermum scoparium: A Thesis Presented in Partial Fulfilment of The Requirements for The Degree of Doctor of Philosophy in Plant Science, Institute of Agriculture and the Environment, College of Sciences, Massey University, Palmerston North, New Zealand. https://core.ac.uk/download/159449396.pdf (accessed on 29th October 2024).

38. Nováčková, K. (2009). Stability of salts and complexes of humic acids. https://core.ac.uk/download/30278457.pdf (accessed on 29th October 2024).

39. Park, J. W., Will, S. A., & Zwierlein, M. W. (2015). Two-photon pathway to ultracold ground state molecules of ^{23}Na ^{40}K. http://arxiv.org/abs/1505.01835 (accessed on 29th October 2024).

40. Pronti, L., Perino, M., Cursi, M., Santarelli, A., et al. (2018). Characterization and digital restauration of XIV–XV centuries written parchments by means of non-destructive techniques: Three case studies. https://core.ac.uk/download/pdf/188822276.pdf (accessed on 29th October 2024).

41. Van Hooydonk, G. (2008). Ionic Kratzer bond theory and vibrational levels for achiral covalent bond HH. https://core.ac.uk/download/55784608.pdf (accessed on 29th October 2024).

42. Wangchuk, P. (2008). Health impacts of traditional medicines and bio-prospecting: A world scenario accentuating Bhutan's perspective. https://core.ac.uk/download/1323136.pdf (accessed on 29th October 2024).

43. Yang, C. (2004). Molecular contrast optical coherence tomography: A review. https://core.ac.uk/download/4869760.pdf (accessed on 29th October 2024).

CHAPTER 2

CHEMICAL ANALYSIS PROCESS

Contents

INTRODUCTION

The study of substances is the most important part of chemistry, both for science and for business. The process of chemical analysis includes many different methods that are used to find out what chemicals are made of, how they are put together, and what they are like. This complicated mix of methods is important not only for learning basic chemical concepts but also for solving real-world problems in many fields. From drugs to environmental science, being able to accurately and effectively analyze chemicals can lead to better product quality, compliance with regulations, and better safety measures. Consequently, a thorough understanding of the chemical analysis process is essential for both aspiring chemists and seasoned professionals addressing contemporary challenges. The development of advanced analytical methods has changed the way chemical analysis is done, allowing scientists to study the details of matter with more precision than ever before. Soft sensors have become well-known as useful tools in the process industry, especially for keeping an eye on and getting important process variables.

Figure 2.1. *Method validation in chemical analysis.*

Source: https://blogger.googleusercontent.com/img/b/R29vZ2xl/AVvXsEg4fzRI-ati1WMCwnwSKlNqggVuY-IScYOAhlj2V-Wy8pixqH7nGqB0vuUl3-8nHGVA9m-huga_kAMAScXmiVeJzTUcU18m6XvMskS7cfmtc5nOgAoxvhf1iDfs3nkJEBvx9x-uvRJ9-kKtf41vCO4WutXdCmJZhHSdyKCyPC0xNcLdYkDV7kRDrY9jaX5nufA/s1792/Method%20Validation%20in%20Chemical%20Analysis.jpeg.

Recent literature has emphasized that the advancement of data-driven soft sensors holds substantial promise for improving process control and monitoring across multiple sectors, including the chemical and bioprocess industries. Soft sensors can give insights that were previously impossible by making good use of the huge amounts of data generated in these settings. This improves processes and makes sure that industry standards are followed more closely (Abdi *et al.,* 2009). This evolution highlights the necessity for a flexible framework that integrates both conventional and contemporary analytical techniques in the chemical analysis process. The incorporation of stable isotope techniques further illustrates the extensive range of analytical methodologies present in chemical analysis. These methods, especially isotope ratio mass spectrometry (IRMS), let you measure isotopic ratios very accurately. This gives you important information about nitrogen cycling and other ecological processes. Recent advancements in methodologies, including the development of compound-specific nitrogen isotope analysis, underscore the potential for enhanced sensitivity and precision in quantifying compounds at low concentrations (Norlin *et al.,* 2005). As chemical analysis continues to evolve, understanding these methodologies will empower researchers to not only delve deeper into the intricacies of chemical substances but also to apply these analytical findings in practical contexts, ultimately fostering innovation and advancement within the field (Figure 2.1).

Definition of Chemical Analysis

Chemical analysis is an important part of chemistry that helps scientists find and measure the amounts of chemicals in a sample. This complex process includes a number of different methods, each of which is best suited for a certain type of material or analytical need. Chemical analysis can be divided into two main types: qualitative analysis, which finds out if certain parts are present, and quantitative analysis, which finds out how much of those parts are present. Chemical analysis gives us more information about materials and helps us use that information in real life in many fields, such as food safety, environmental monitoring, and pharmaceuticals. As businesses learn more about how important chemical analysis is, the need for strong analytical methods grows. This opens the door for new ideas in analytical techniques and tools. The rules that govern chemical analysis include a systematic way to prepare samples, choose the right analytical methods, and understand the results. This process usually starts with careful sampling to make sure that the sample is a good representation of the larger material it came from. After samples are obtained, different methods like spectroscopy, chromatography, and mass spectrometry can be used to reach the desired analytical goals. Every method has its own pros and cons, so you need to know how to choose the right one for the situation. Furthermore, the incorporation of methodologies like life cycle assessment (LCA), as highlighted in recent literature, offers prospects to improve the comprehensiveness of chemical analyses by integrating environmental factors into process optimization (Jacquemin *et al.,* 2011). This blending of analytical and environmental methods is a good example of how chemical analysis is changing. It makes the analytical process better while also promoting sustainability. In the realm of intricate chemical processes, comprehending the dynamics of energy transfers and irreversibility is essential for enhancing analytical techniques. For example, using advanced exergy analysis in established process simulation software shows where chemical production systems are not working as well as they could (Baudet *et al.,* 2012). This analytical method not only helps find energy losses, but it also helps plan ways to reduce these losses by changing the process. These methods are very important for making chemical industries more efficient and better for the environment. So, combining chemical analysis with energy assessments helps us learn more about chemical processes, which helps us in both theory and practice. So, the definition of chemical analysis includes not only figuring out what materials are and how much of them there are but also making processes better by making smart choices.

Importance of Chemical Analysis in Various Fields

Chemical analysis is important in many areas, but it is especially important in medicine, where accurate diagnoses are necessary for effective treatment. Healthcare professionals can figure out what chemicals are in biological samples by using different analytical methods, like chromatography and mass spectrometry. This level of accuracy makes it possible to find diseases early and keep an eye on how well treatments are working. For example, toxicology screenings can find harmful substances in patients, which can help doctors decide what to do right away. As analytical methods keep getting better, the possibility of personalized medicine grows. This means that treatments can be made to fit each person's unique biochemical profile. This not only improves patient outcomes, but it also helps the field of pharmacogenomics, where knowing how drugs interact at the molecular level can greatly improve their effectiveness and safety. Chemical analysis is an important part of environmental science because it helps keep an eye on pollutants and check the health of ecosystems. Researchers can find toxic substances and their levels by carefully testing soil, air, and water samples. This information can then be used to make decisions about public policy and clean-up efforts. For instance, looking at heavy metals in water can show where contamination comes from and what risks it poses to both human health and biodiversity. These methods also help businesses follow environmental laws, which means they have less of an impact on the environment. As scientists improve analytical methods and tools, they can find lower amounts of pollutants. This shows how chemical analysis is becoming more important for protecting public health and the environment. The process of chemical analysis is also very important in materials science, where figuring out what substances are made of and how they behave leads to new ideas and better materials. Scientists can use techniques like spectroscopy and microscopy to look at materials at the atomic and molecular levels. This gives them information about their mechanical, electrical, and thermal properties. These studies help scientists make new materials, like carbon nanotubes, which have been hard to understand how they are made and grow. Recent research has proposed a dynamic ferromagnetic mechanism to enhance the formation of carbon nanotubes, illustrating the significant influence of chemical analysis on technological progress. The ongoing investigation of material properties via chemical analysis not only advances scientific comprehension but also fosters the creation of innovative applications across diverse sectors, ranging from electronics to nanotechnology.

2.1. TYPES OF CHEMICAL ANALYSIS

There are many different ways to do chemical analysis, and each one is meant to help you understand the makeup and properties of substances. In general, these methods can be put into two groups: qualitative analysis and quantitative analysis. Qualitative analysis aims to determine the chemical constituents in a sample without measuring their concentrations, utilizing methods like chromatography and spectrophotometry. Quantitative analysis, on the other hand, tries to find out the exact amounts of each part, usually by using gravimetric and titrimetric methods. These analytical methods are very important in fields like environmental science and pharmaceuticals because they give researchers and regulators basic information that helps them make decisions. To accurately assess chemicals and come up with new ideas in different fields of study, you need to know how these methods are different and how they can be used. Instrumental methods of chemical analysis have changed the way we do analyses in a big way. Mass spectrometry and nuclear magnetic resonance (NMR) spectroscopy are two methods that help scientists learn about the structure and mass of molecules. These are important for figuring out what complex substances are. These methods use advanced instruments to get results that are more sensitive and faster than those from traditional methods. For instance, mass spectrometry can find even the smallest amounts of contaminants in a sample.

This is very important for quality control and assurance in many fields, especially in pharmaceuticals and food safety. As technology gets better, these instrumental techniques are not only getting better, but they are also becoming easier to get, which is changing how chemical analysis is done in many fields. The introduction of soft sensors in recent years has added a new and useful tool to traditional chemical analysis methods. Soft sensors use data-driven models to check and keep an eye on process variables in real time. This makes them very useful in dynamic and complicated industrial settings, like the chemical and bioprocess industries. These sensors can use data that is already there to make predictions about what will happen, which improves process control and efficiency. The use of soft sensors in chemical analysis is a new strategy that focuses on automation and data analytics. Their ongoing development will likely bridge gaps in existing methodologies, leading to more robust and reliable analyses that can address contemporary challenges within the chemical landscape (Abdi *et al.,* 2009; Avni *et al.,* 2025) (Figure 2.2).

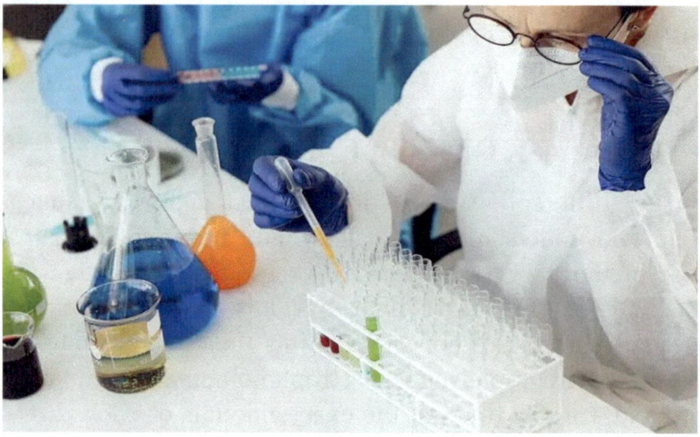

Figure 2.2. *Types of chemical analysis.*

Source: https://www.projectpractical.com/wp-content/uploads/2022/11/Types-of-Chemical-Analysis.jpg.

2.1.1. Qualitative Analysis

Qualitative analysis is an important part of chemical analysis that helps scientists figure out what a chemical substance is and what its properties are. This method goes beyond just measuring things; it focuses on figuring out the qualitative properties of materials using methods like spectroscopy, chromatography, and microscopy. Researchers can use these methods to study the makeup of complex mixtures and find out not only what each part is but also how they fit together. Being able to show this kind of information is very important in many fields, such as drugs, environmental monitoring, and materials science. Qualitative analysis is therefore essential for fostering innovation and maintaining safety across diverse industries, highlighting the necessity of thorough chemical analysis methodologies. Furthermore, the amalgamation of qualitative analysis with risk assessment frameworks enhances our comprehension of engineered nanomaterials (ENMs) and their prospective environmental ramifications. Because these materials are so complicated, qualitative methods can be used to look into the interactions and behaviors that could affect the environment. For example, risk assessments need a deep understanding of how ENMs might act when they are put into different ecosystems. Current literature, however, emphasizes a disparity between the requirements of decision-makers for thorough risk assessment frameworks and the limitations of existing tools to meet those requirements (Bilal *et al.,* 2017). By improving qualitative analyses, stakeholders can be better prepared to make smart choices about how to safely use nanotechnology, which will ultimately encourage sustainable practices. Finally, qualitative analysis can help chemical processes get better all the time by finding the root causes of problems. Researchers can find problems that are causing chemical operations to not work as well as they could by looking at both traditional process measurements and extra sources of information, like utility and mechanical data. Using these different data sources helps us get a better overall picture of how processes work, which makes it possible to make targeted changes that improve safety and efficiency. In this context, the function of qualitative analysis in discerning causal factors signifies substantial ramifications for industries dependent on chemical processes. Recent studies have indicated that the adoption of a semi-automated methodology could markedly improve the precision of root-cause analysis, thereby facilitating continuous operational enhancement (Bauer *et al.,* 2012). Qualitative analysis is not only a way to explain composition, but it is also an important part of improving the whole chemical analysis process.

2.1.2. Quantitative Analysis

Quantitative analysis is a very important part of chemical analysis because it gives exact measurements of the chemical substances in a material. This part of analytical chemistry tries to figure out how much of each chemical is in a sample and what its concentration is. This is important for many things, such as quality control, environmental monitoring, and process optimization. To get very accurate and reliable results, people often use gravimetric analysis, titrations, and instrumental methods like spectrophotometry and chromatography. Chemists can use these methods to make smart choices based on quantitative data, which helps both research and industry. Quantitative analysis is the basis of good chemical characterization because it connects substance concentration to physical and chemical properties. This helps us understand how complex mixtures behave in different situations. In the field of material sciences, quantitative analysis has important effects on fields like metallurgy and wood processing. The difficulties in analyzing wood-based materials, owing to their opaque characteristics, have prompted the creation of novel analytical methodologies. In this regard, photoacoustic and transient infrared spectroscopies have become useful tools for quickly identifying chemicals. By combining these spectral techniques with chemometric methods, researchers can now quickly find out the chemical and physical properties of wood chips. This makes the process monitoring and quality assurance phases of production easier. These kinds of improvements not only make real-time chemical analysis more efficient, but they also cut down on the time it takes to do things the old-fashioned way. This makes the industry as a whole more productive (Bajic *et al.,* 1997). The examination of vanadium microalloyed steels exemplifies the importance of quantitative analysis in metallurgy. Quantitative methods for characterizing microstructure help us learn more about how elements like vanadium affect the properties of steel, especially during processes like tempering and controlled rolling. The advancement of sophisticated techniques like electron energy loss spectroscopy has facilitated comprehensive analyses of the chemical composition of precipitates, thereby enhancing the understanding of material behavior under varying temperatures and processing conditions. Researchers are still interested in the nucleation sequences and precipitation mechanisms, so it is becoming more and more important to use quantitative analysis in metallurgical studies to help make sense of things and guide the development of new materials, as shown by Baker et al. (2009) (Figure 2.3).

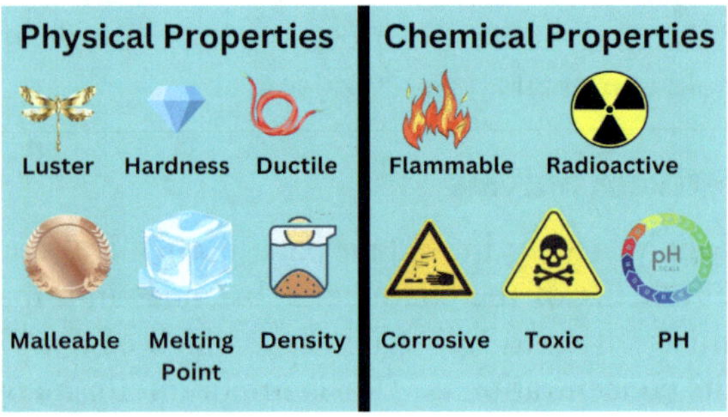

Figure 2.3. *Physical and chemical properties.*

Source: https://i.ytimg.com/vi/8uAca9uKvPk/maxresdefault.jpg.

2.1.3. Instrumental Analysis

Instrumental analysis is a key part of chemical analysis because it gives us advanced tools that help us learn more about the composition and properties of materials. Instrumental methods are more sensitive and specific than classical methods, which often don't show how different parts of a sample interact with each other. For instance, infrared spectroscopy and nuclear magnetic resonance (NMR) are two methods that can accurately identify functional groups and molecular structures. This makes it easier to study how chemicals change. By using instrumental methods, researchers can learn more about how processes work, like how organic waste is composted, than they can with traditional methods. This gives them important information about the stability and quality of the end products, as shown in studies of winery and distillery wastes (Bustamante *et al.,* 2015). Instrumental analysis also combines different methods to get complete results that help support sustainable practices in many areas. In the field of textile treatment, the use of non-thermal plasma technologies is an example of this integration. Researchers can measure changes to surfaces very accurately by using advanced methods like scanning electron microscopy and X-ray photoelectron spectroscopy. This assessment not only illustrates alterations in wettability and adhesion characteristics but also highlights the feasibility of environmentally sustainable alternatives to conventional chemical methodologies (Carneiro *et al.,* 2012). These improvements show how flexible instrumental analysis can be and how important it is for making and processing things in a way that is good for the environment and efficient. Instrumental analysis does more than just find and describe things; it is an important tool for improving and creating new chemical processes. It helps scientists make new materials and processes that can greatly improve the performance of products and their impact on the environment by letting them look closely at how materials interact with each other and their properties. In composting studies, the integration of traditional analytical methods with sophisticated instrumental techniques facilitates a comprehensive understanding of organic matter turnover, thereby enabling the optimization of compost quality (Bustamante *et al.,* 2015). In the same way, improvements in textile treatment using instruments have led to better functions, like better surface adhesion for functionalizing materials (Carneiro *et al.,* 2012). The combination of these analytical methods is what drives the field of chemical analysis forward, making sure that it keeps changing to meet new challenges and take advantage of new opportunities.

2.2. SAMPLE PREPARATION

Sample preparation is a very important part of chemical analysis as a whole. It connects the collection of raw materials with the actual measurement of the analysis. This step is often what decides how good and reliable the analytical results are, since bad or incomplete preparation can lead to wrong results. For example, in the wood and paper industry, traditional methods need a lot of sample preparation, which can slow down the quantitative analysis that is needed for monitoring and control. The shortcomings of traditional methods—especially concerning the opacity and reflectivity of wood materials—frequently obstruct efficient analysis. Researchers can quickly characterize complex materials by improving sample preparation methods, such as by combining them with new spectroscopic methods. This makes sure that analysis can be done quickly and accurately, which helps quality assurance processes in industrial settings (Bajic *et al.,* 1997). Along with traditional methods, new technologies, especially in microfluidics and nanotechnology, have changed how samples are prepared in many fields of study. These new technologies have made it possible to do quick and accurate analyses, which are very important in fields like managing infectious diseases and making sure food is safe. Microfluidic systems are made to use as little reagents and samples as possible while maximizing the speed and sensitivity of the analysis. These kinds of improvements not only cut costs, but they also make workflows more efficient, which makes sample preparation easier for real-time use. The addition of cutting-edge microfluidic platforms has greatly improved the ability to prepare samples. This has led to progress in molecular diagnostics that are easy to use and can be used in many different situations (Abate *et al.,* 2017). Effective sample preparation is also very important for improving the accuracy and reliability of data from analytical processes. Analysts can get rid of or lessen possible interferences and background noise by carefully preparing samples. These are often the causes of analytical errors. This careful process also makes the analytical method more specific, which makes it easier to understand the results. For instance, when looking at the chemical makeup of wood-based materials, using advanced sample preparation methods can give you important information that can help you better understand the materials. This kind of improvement not only helps with decisions about processing, but it also helps with research and development projects that aim to make the most of wood products. The interaction between sample preparation and chemical analysis is essential for maintaining the integrity and significance of analytical results across diverse scientific fields (Figure 2.4).

Figure 2.4. *Wood and paper industry.*

Source: https://www.dorst-technologies.com/fileadmin/Industries/Industries-square/wood_and_paper_industry.jpg.

2.2.1. Collection of Samples

The collection of samples is an important first step in the chemical analysis process because it directly affects the quality and reliability of the results that follow. Good sampling methods make sure that the samples taken are a good representation of the materials or residues being studied. In the context of forensic tire analysis, different ways of collecting data were tested to see how well they worked on different types of roads. Research demonstrated that while picking was ideal for concrete roads with gravel aggregates, a fingerprint lift was more effective on bitumen surfaces (Sarkissian *et al.,* 2009). This study underscores the significance of context-specific sampling strategies, as the selection of methodology can profoundly influence the analytical effectiveness and precision of the outcomes. Sampling errors frequently occur due to improper collection methods, resulting in contamination or misrepresentation of the sample. This false representation makes the analytical data collected during the chemical analysis process less reliable. For example, the study found that fingerprint lifts on bitumen surfaces contaminated the adhesive, which made the pyrolysis-gas chromatography/mass spectrometry (PyGCMS) analysis less accurate. These kinds of problems show how important it is to think about both the surface material and the specific features of the evidence when planning how to collect samples (Sarkissian *et al.,* 2009). Forensic scientists must deal with these problems because they need to make sure that their sample collection methods are both reliable and accurate in order to draw valid conclusions. Along with the physical method of collection, the chemical differences in the samples must also be carefully taken into account. Different things can cause changes in chemical composition, such as how the materials were made and how old they are. The tire samples and their residues showed a 5–10% difference in the amount of polymer they contained, which changed depending on the tire model and the production batch (Sarkissian *et al.,* 2009). Recognizing these differences makes the chemical analysis process even more complicated and shows how important it is to use strong and standardized sampling methods to make sure that all the data is analyzed, which will help forensic investigations find the right sources more accurately.

2.2.2. Preservation and Storage of Samples

Maintaining and storing samples is very important for making sure that chemical analyses are accurate. A key part of this process is getting samples that accurately show the conditions in the field at the time of analysis. For instance, in water quality monitoring, the time between collecting a sample and analyzing it in a lab can cause big changes in the chemical makeup of the sample because reactions are still happening in it. These kinds of changes can make analytical results less reliable, which means that conclusions based on the data may not be correct. Freezing and other effective preservation methods can lower these risks by keeping the sample's composition stable until it can be analyzed. This makes the chemical analysis process more accurate and makes sure that the empirical data shows the true characteristics of the original sample (Patocka *et al.,* 1972). Freezing is a well-known method for keeping samples intact, but the way that freezing is done can have a big effect on the quality of the preservation. Studies have demonstrated that conventional isobaric freezing techniques can result in adverse outcomes, including weight reduction and structural impairment in food and biological specimens.

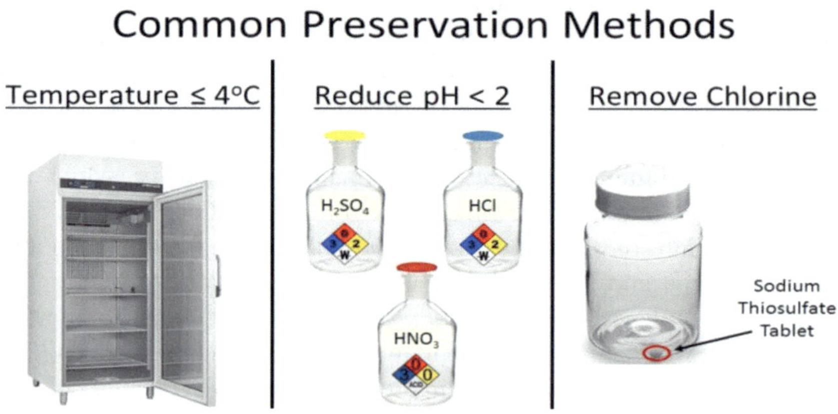

Figure 2.5. *Sample preservation.*

Source: https://store.americanwatercollege.org/wp-content/uploads/2017/01/Sample-Preservation-768x432.jpg.

Alternatively, isochoric freezing systems may offer a superior preservation method by reducing mechanical damage caused by ice crystals and preserving osmotic equilibrium within the samples. Preliminary studies suggest that isochoric freezing results in reduced enzymatic degradation, thereby maintaining the overall quality and integrity of samples (Lyu *et al.,* 2017). Therefore, laboratories aiming to improve their sample preservation methods during chemical analysis must consider incorporating advanced freezing technologies. The benefits of good sample preservation go beyond just making sure that the analysis is correct; they also help with the bigger goals of sustainability and resource management in science. Keeping samples in their original condition cuts down on waste and the need for repeated tests, which can take a lot of time and money. As environmental policies place greater emphasis on the significance of precise data for the sustainable management of natural resources, the methodologies utilized in sample preservation become increasingly vital. As a result, laboratories that do chemical analysis need to have clear rules about how to freeze samples correctly and how to check their stability. These practices not only maintain the scientific integrity of the collected data but also promote the responsible management of valuable environmental resources (Patocka *et al.,* 1972) (Figure 2.5).

2.2.3. Preparation Techniques for Analysis

To get reliable and accurate results, it's important to use the right preparation methods for chemical analysis. The quality of analytical results is mostly determined by how well the samples are prepared before they are analyzed. For example, in industries like forest products, where wood and paper materials are tested, the limits of traditional methods can make it hard to get results quickly. The difficulty comes from the fact that these materials are hard to see through and reflect light, which makes it hard to measure them accurately. New methods, such as photoacoustic and transient infrared spectroscopies, have been developed to solve these problems. They make it easier to characterize wood chips without the need for a lot of sample preparation. These techniques show how improvements in technology can make the analysis process faster, which can lead to better quality control and process monitoring in industrial settings (Bajic *et al.,* 1997). Other preparation methods have been looked into to improve material analysis, especially for hybrid materials, in addition to spectroscopy. The development of hybrid organic-inorganic membranes signifies a notable progression in material science, highlighting the significance of efficient preparation techniques. For example, mixing a sulfonated polyamideimide resin with silica filler shows how the right sol-gel processes can affect the properties of the material. Researchers can make silica nanoparticles that are well dispersed in a polymer matrix by using tetraethoxysilane (TEOS) as a silica precursor and the coupling agent aminopropyltriethoxysilane (APTrEOS) to make them more compatible. This careful preparation not only makes the membranes work better, but it also makes the tests needed to evaluate them better (Hill *et al.,* 2010). In general, the importance of preparation techniques in chemical analysis cannot be overstated. These methods are very important to the overall analytical process because they help adapt spectroscopic methods to deal with the unique properties of different materials or improve the performance of hybrid materials through careful preparation. As analytical technologies keep getting better, using new preparation methods will probably make assessment tools more advanced and accurate. When these methods are used successfully, they help with both basic chemistry research and real-world improvements in many fields. This leads to better understanding of how materials work and what they do. Focusing on the study and improvement of preparation methods will help make sure that the chemical analysis process stays strong and can handle the difficulties of modern analytical problems.

2.3. ANALYTICAL TECHNIQUES

Analytical techniques are very important for chemical analysis because they let scientists find and measure chemical compounds with great accuracy. Isotope ratio mass spectrometry (IRMS) is one of these methods that has become a very useful tool, especially in studies of how nitrogen moves through different ecosystems. IRMS is very useful for understanding how nitrogen moves around because it can measure the heavy (^{15}N) to light (^{14}N) isotopic ratios very accurately. Researchers can do compound-specific analyses that are very important for environmental science by using gas chromatography with combustion IRMS (GC-C-IRMS). This analytical method can also be used to get reliable isotopic ratios for ammonium and glycine in samples, showing how useful it is in different situations (Norlin *et al.,* 2005).

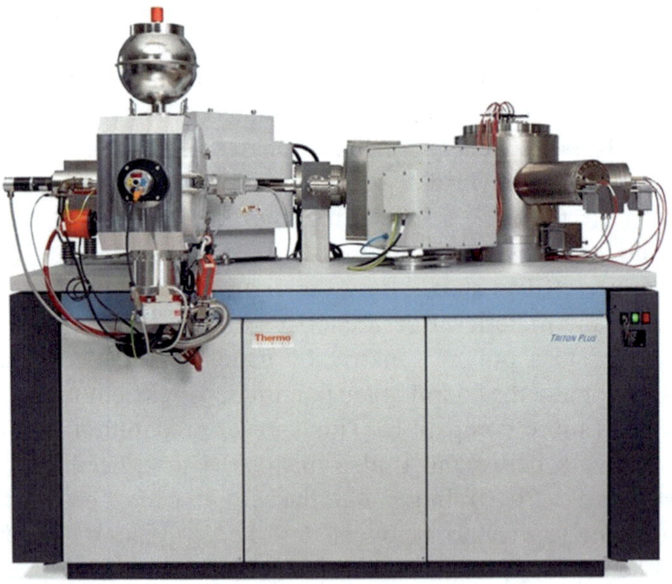

Figure 2.6. *Isotope ratio mass spectrometry (IRMS).*

Source: https://www.thermofisher.com/us/en/home/industrial/mass-spectrometry/isotope-ratio-mass-spectrometry-irms/jcr:content/MainParsys/textimage_9172/image.img.480.medium.jpg/1670864484574.jpg.

Furthermore, improvements in analytical methods have resulted in the creation of multiple complementary techniques for the isolation and analysis of nitrogen compounds. For example, headspace analysis, solid phase microextraction (SPME), and the purge and trap (P&T) method have all been shown to work well for pre-concentration of ammonium before isotopic analysis. Each method has its pros and cons. For instance, the SPME method makes things much more sensitive than equilibrium headspace analysis, but it might not be as accurate in some situations. Also, the P&T method has shown promise for measuring lower concentrations, but it has some problems with thermal desorption. As researchers keep improving these methods, they will be able to use them in more and more places. This shows how important new ways of analyzing chemicals are (Norlin *et al.,* 2005). The exploration and application of these analytical techniques not only strengthen environmental research but also deepen our comprehension of biochemical processes at a fundamental level. The capacity to quantify isotopic ratios in low-concentration samples, exemplified by glycine in soil extracts, underscores the continuous advancements in chromatography and mass spectrometry techniques. These new ideas, along with improvements in technique sensitivity and precision, help scientists study where nitrogen comes from and how it changes in ecosystems. Ultimately, the ongoing advancement of analytical techniques in the chemical analysis process empowers scientists to discover novel insights and tackle ecological challenges, highlighting their indispensable role across diverse fields of study (Norlin *et al.,* 2005) (Figure 2.6).

2.3.1. Spectroscopy

Spectroscopy is an important method for analyzing chemicals that helps scientists learn about the properties and structures of different substances. Spectroscopy reveals molecular composition and dynamics by studying how electromagnetic radiation interacts with matter. Infrared (IR), nuclear magnetic resonance (NMR), and mass spectrometry are all types of spectroscopies that use specific parts of the electromagnetic spectrum to learn about molecular vibrations, chemical environments, and mass-to-charge ratios. For example, IR spectroscopy is very good at finding functional groups in organic compounds, which helps us better understand how to describe complex mixtures. Using these methods, scientists can learn more about how materials behave and interact with each other. This opens up possibilities for everything from drug development to materials science. Spectroscopy is useful for improving many different technological processes because it is so flexible. In the making of D2B IR detectors, several surface treatment methods have been used to improve product quality. These include X-ray photoelectron spectroscopy

(XPS) and atomic force microscopy (AFM). These methods let us look closely at the surface chemistry and roughness of different treatments, which helps the detector work better. Recent studies have shown that using critical etching and annealing methods to improve these processes has not only made surfaces flatter, but it has also cut down on dark current leakage in detectors.

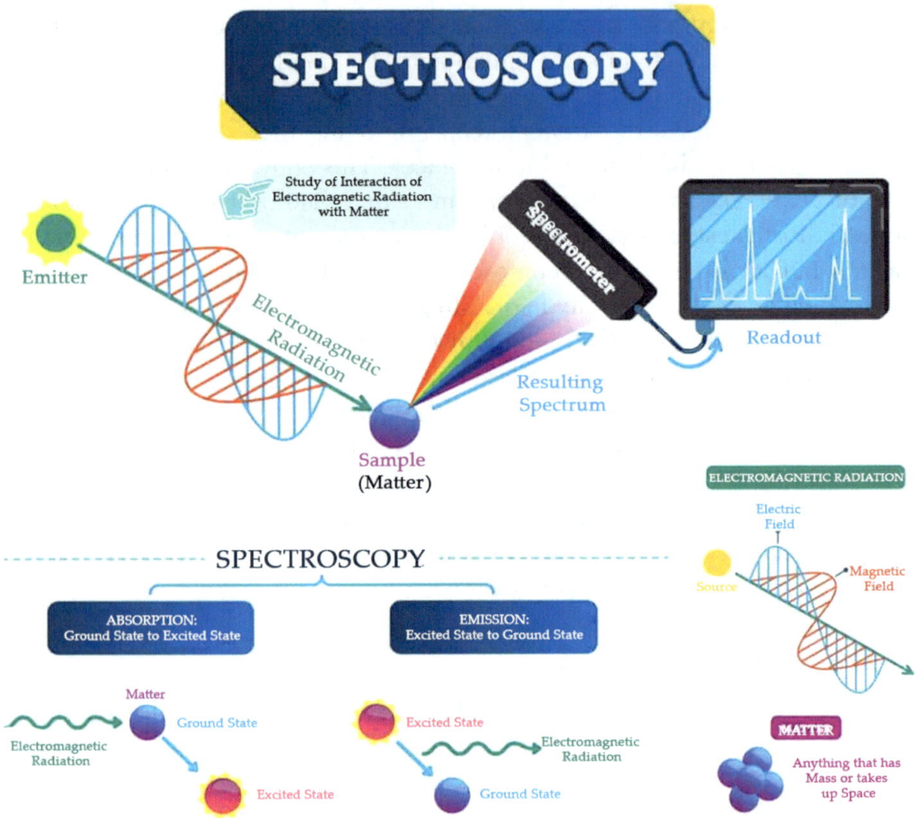

Figure 2.7. *Spectroscopy analytical technique.*

Source: https://www.chemicals.co.uk/wp-content/uploads/2021/09/spectroscopy-fact-sheet-640x586.jpg.

Consequently, spectroscopy plays an instrumental role in refining manufacturing practices, thus benefiting industries reliant on precision and accuracy in electronic components (Chase *et al.,* 1934). New technologies in spectroscopy are also changing how we think about and use amorphous solids, especially in the pharmaceutical field. For instance, terahertz time-domain spectroscopy (THz-TDS) lets you look at molecular dynamics in these materials without damaging them. This method has been very helpful in figuring out the physical properties of amorphous drug formulations, which are known to make active pharmaceutical ingredients more available in the body. Researchers can figure out the molecular mobility that causes crystallization processes by doing careful analysis. This lets them better control drug delivery systems. The capacity of THz-TDS to clarify the fundamental mechanisms in amorphous solids represents a revolutionary advancement in pharmaceutical analytics, highlighting the significance of spectroscopy in propelling research and development in the domain (ANONYMOUS, 2016) (Figure 2.7).

2.3.2. Chromatography

Chromatography is an important part of chemical analysis because it can be used to separate and analyze complicated mixtures. The basic idea behind chromatography is to separate parts between a stationary phase and a mobile phase. This process makes it possible to separate compounds based on how much they like each of the two phases. For example, in liquid chromatography, the choice of solvents and stationary phases can have a big effect on how well analytes are separated and how clear they are. These complexities highlight the necessity of optimizing chromatographic conditions for diverse applications, including environmental monitoring and pharmaceutical analysis. Chromatography is an important part of chemical analysis because it can give precise and repeatable results. This is especially important in labs where accurate chemical profiling is needed. Recent improvements in chromatography methods, especially in liquid chromatography coupled with mass spectrometry (LC-MS), have made chemical analysis even better. LC-MS combines the separation power of chromatography with the ability of mass spectrometry to figure out the structure of compounds. This makes it easier to analyze compounds at lower concentrations and with more accuracy. LC-MS has played a key role in drug development by helping researchers find and describe impurities and metabolites in drugs. The use of different mass spectrometric methods, like electrospray ionization, adds to the high-performance liquid chromatography (HPLC) used in this study, helping us better understand how compounds behave in biological systems (Lim *et al.*, 2002). This combination of advanced detection methods and chromatographic techniques shows how chemical analysis methods are always changing to meet the needs of science and industry. Moreover, the incorporation of numerical methods in the analysis of chromatographic data underscores the intricate nature of contemporary chemical analysis. Numerous chromatographic systems functions under conditions characterized by nonlinear differential equations, complicating empirical observations and data interpretation. To deal with this complexity, scientists have come up with ways to use math and computer simulations to guess the values of unknown parameters in these systems. For instance, parameter estimation techniques utilized on biochemical data can greatly improve comprehension of chemical interactions and reaction kinetics (Goulet *et al.,* 2015). These methods not only make chromatography results more reliable, but they also help plan future experiments. Chromatography's role in chemical analysis is growing as it connects theoretical frameworks with real-world uses. This helps fill in the gaps between empirical data and predictive modeling, which is important for progress in many scientific fields (Figure 2.8).

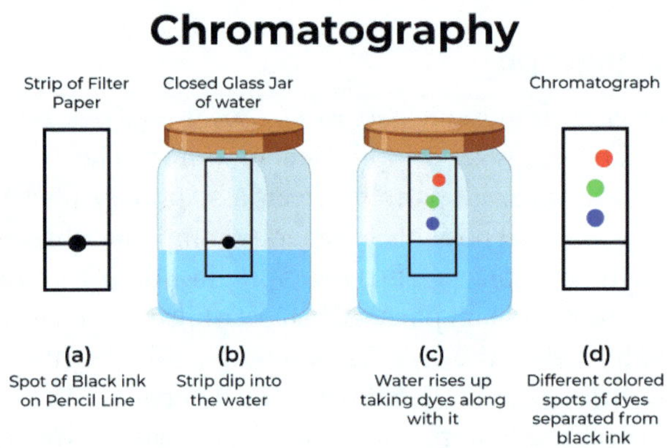

Figure 2.8. *Chromatography.*

Source: https://media.geeksforgeeks.org/wp-content/uploads/20230814185417/Chromatography-1.png.

2.3.3. Titration

Titration is a basic analytical method that chemists use a lot to find out how concentrated an unknown solution is. In this process, a titrant, which is a solution with a known concentration, is usually added slowly to the analyte, which is the substance whose concentration needs to be measured. When the titrant reacts with the analyte, the reaction reaches its endpoint at a certain point. This is often shown by a change in color, especially with pH indicators. This process is very important for quantitative analysis in many areas, such as food quality control, pharmaceuticals, and environmental monitoring. Researchers can figure out the concentration of the substance they are studying by carefully measuring how much titrant they used to reach the endpoint. This gives them important information about the substance's chemical properties and uses. The accuracy and dependability of titration results are greatly influenced by several factors, such as the selection of titrant, the nature of the reaction, and the technique employed to ascertain the endpoint. Acid-base titrations are a common type of titration in which a strong acid and a strong base react with each other to make a neutral solution. Redox titrations, on the other hand, involve the transfer of electrons between the analyte and the titrant. This gives us another way to find the concentration. Good lab practices, like calibrating equipment correctly and choosing the right indicators that give clear and distinct endpoint signals, are often what make titration work. Also, improvements in instruments, such as automated titration systems, have made these measurements more accurate and repeatable. These new ideas show how titration methods are still changing, which is an important part of the chemical analysis process. Titration is also very important for understanding complicated biochemical interactions, especially when it comes to studying how proteins and ligands bind to each other. This aspect is exemplified by the complex interplay between protein conformation and ligand affinity, wherein titration can clarify binding constants and conformational alterations linked to ligand interactions.

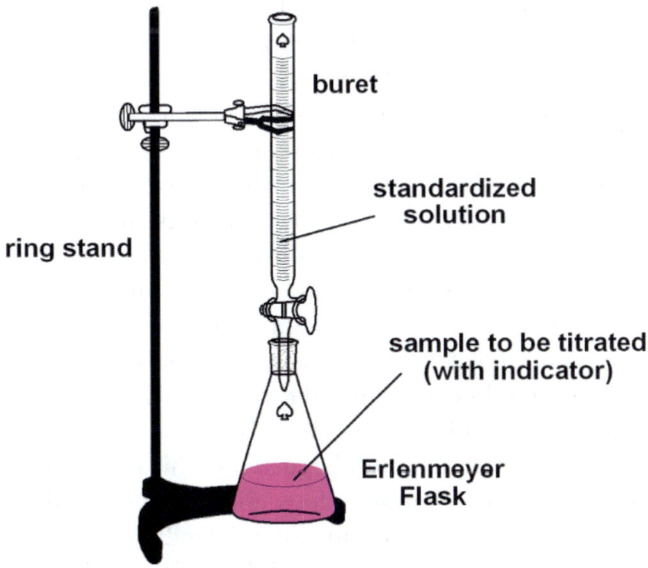

Figure 2.9. *Titrations.*

Source: http://sachiacidbase.weebly.com/uploads/2/7/9/1/27911357/570918_orig.png.

The dynamic nature of these interactions can complicate analyses, as minor variations in concentration can alter the binding kinetics observed through methods like NMR spectroscopy. These challenges have been mitigated through computational modeling to accurately interpret titration data, thereby identifying patterns crucial for understanding multistate systems (Kovrigin *et al.,* 2012). The importance of strong titration methods goes beyond basic chemistry; they can also be used to learn more about biological systems. This connects analytical chemistry and biochemistry in the larger context of the chemical analysis process (Cadorim *et al.,* 2017) (Figure 2.9).

2.4. DATA COLLECTION AND INTERPRETATION

In the field of chemical analysis, collecting and interpreting data is very important, especially when it comes to figuring out what materials are and what they are made of. The systematic gathering of information about samples through different analytical methods is the basis of effective data collection. In this case, non-destructive methods like Fourier transform infrared spectroscopy (FTIR) with attenuated total reflection (ATR) have become very useful. These methods allow scientists to determine the chemical composition of materials such as barkcloths without damaging them. Using principal component analysis (PCA) on the spectral data collected, it is possible to create distinct groups of old fabrics based on their molecular properties, such as the presence of tannins and lignins. This advanced data analysis helps both conservators and art historians by giving them important information about where these textile artifacts came from and how to keep them safe (Holmes-Smith *et al.,* 2019). Going beyond textiles, understanding data is just as important in the study of geomorphological processes, as shown by the study of coastal formations. Information about the rates of cliff erosion and the formation of notches helps scientists understand how the environment and geology change over time. Scientists can very carefully measure the shape of coastal notches using advanced methods like terrestrial light detection and ranging (LiDAR). The data gathered not only shows how erosion happens, but it also gives us a way to understand past sea levels and how erosion is still happening. This kind of analysis is very important for figuring out notch morphology and telling the difference between U-shaped and V-shaped formations in coastal landscapes. Moreover, these results enhance the comprehension of coastal dynamics, facilitating predictions of future alterations and informing effective conservation initiatives (Antunes *et al.,* 2018). The convergence of data collection and interpretation in chemical analysis highlights the interdisciplinary essence of scientific investigation. Every method used, such as FTIR for old textiles or LiDAR for coastal erosion, shows a dedication to accuracy and rigor in gathering data. The complexities of data interpretation enable researchers to formulate significant conclusions and identify correlations among ostensibly unrelated data sets. In both instances, the capacity to categorize and correlate findings augments the comprehension of material characteristics and environmental transformations, integrating historical context with modern analysis. Ultimately, strong frameworks for collecting and interpreting data not only move chemical analysis forward, but they also lay the groundwork for future research, which will let us learn more about materials and the places they live (Figure 2.10).

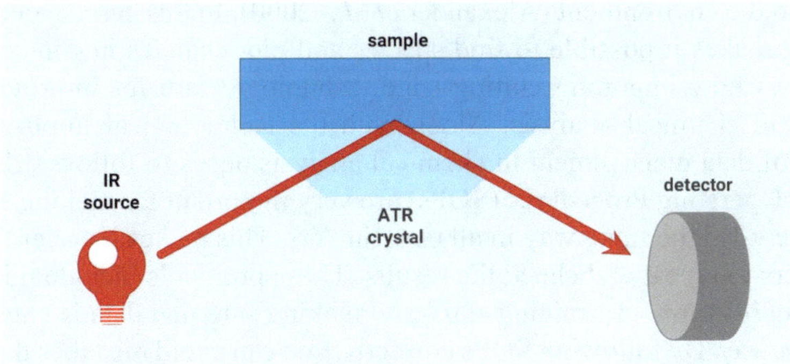

Figure 2.10. *Attenuated total reflectance (ATR).*

Source: https://www.bruker.com/content/bruker/int/en/products-and-solutions/infrared-and-raman/ft-ir-routine-spectrometer/what-is-ft-ir-spectroscopy/atr-attenuated-total-reflectance/jcr%3acontent/root/sections/section_copy_copy/sectionpar/twocolumns/contentpar-1/image.coreimg.jpeg/1733913483273/atr-ft-ir-images-1.jpeg.

2.4.1. Recording Data Accurately

The accuracy of data recording is very important in chemical analysis because it is what makes experimental results trustworthy. Accurate data management stops mistakes from happening that could make the results less reliable. In numerous chemical experiments, especially those related to nanotechnology or intricate material properties, minor measurement inaccuracies can result in substantial variations in outcomes. For instance, the process of documenting different physical properties of materials, as detailed in the review of nanofabrication technologies, necessitates a careful classification of patterning techniques and their uses. Not accurately recording these differences can lead to bad analyses and failed uses of these materials, like in ultra-high density recording technologies or magnetic logic devices (Lodder *et al.,* 2006). So, it is important to set strict rules for how data is recorded so that research can come to valid conclusions. Also, new recording technologies have changed the way scientists collect and analyze data. The use of new tools like holographic imaging systems in chemical and biological studies shows how much progress has been made in making data more accurate. The HoloCam system created as part of the HoloMar project shows how useful pulsed laser holography can be for taking three-dimensional pictures of marine organisms and particles in their natural environment. This non-destructive method lets researchers look at a lot of data at once, which gives them rich datasets while causing the least amount of harm to the environment (Alexander *et al.,* 2000). In this case, accurate data recording makes it possible to find species and plot changes in concentration. This shows how important cutting-edge technologies are for improving data accuracy in chemical analysis. Along with the technological improvements, the field of data management in chemical analysis needs to follow strict rules. Standard Operating Procedures (SOPs) are very important for making sure that data is recorded the same way in all experiments. This makes it easier for other researchers to repeat and check the results. These protocols include things like calibrating instruments, training staff, and making sure that data is entered in a consistent way. By following SOPs correctly, you can avoid mistakes that might happen because of people or equipment problems. Following these steps is especially important when looking at complicated materials, because the small details of how they act can have a big impact on the results of an experiment. Researchers can make their findings more credible and help chemical science move forward by putting accurate data recording first and using structured methods.

2.4.2. Statistical Methods for Data Analysis

For accurate interpretation and decision-making, it is very important to use statistical methods for data analysis in the context of the chemical analysis process. In the field of Life Cycle Assessment (LCA), the use of statistical methods can improve the reliability of the data by dealing with the uncertainties that are common in the chemical industry. Statistical techniques, including goodness-of-fit tests and maximum likelihood estimation, facilitate the characterization of data distributions, thereby ensuring that the parameters employed in analyses accurately reflect the intrinsic characteristics of the underlying processes. The results presented by Fukushima et al. (2018) demonstrate that using structured statistical methods to incorporate uncertainty information is important, as it enhances the reliability of the findings. This method not only helps us look at different chemical processes, but it also makes sure that the choices we make based on these analyses are based on sound statistical methods. The rise of soft sensors in the chemical industry is another sign of how important data-driven statistical methods are becoming. Soft sensors look at different types of process data to give real-time monitoring and control, which makes chemical processes work better. These new methods use statistical models to get useful information from datasets that might be too complicated for older methods. By concentrating on data-driven soft sensors, professionals can adjust to changing process conditions and keep the best operational parameters, as described by Abdi et al. (2009). This is especially important in industries like bioprocessing and steel, where getting exact measurements of important variables has a direct effect on the quality of the product and the efficiency of the process. The implications of such advancements indicate a shift towards more robust statistical frameworks in data analysis. Finally, the use of advanced statistical methods and soft sensor technology together helps people make smart choices during the chemical analysis process. By carefully analyzing statistical data, stakeholders can systematically look at the variability and uncertainty that are part of chemical processes. This reduces the risks that come with developing new products and running operations inefficiently. The suggested incorporation of statistical methodologies in industry-oriented Life Cycle Inventory databases, as detailed by Fukushima et al. (2018), illustrates how proficient statistical analysis can uncover the interrelations of inventory flows while maintaining data confidentiality. This combination of strict statistics and new technologies helps industries improve their processes and get a better picture of how their actions affect the environment. This will eventually lead to more environmentally friendly ways of making and using chemicals (Figure 2.11).

Figure 2.11. *Statistical methods for chemical analysis.*

Source: https://www.innovatechlabs.com/wp-content/uploads/2014/11/chemical-analysis.jpg.

2.4.3. Interpretation of Results

Interpreting the results of a chemical analysis is an important step in turning raw data into useful and understandable information. This step is very important for figuring out how important the results are in relation to the hypotheses that were made at the start of the analysis. To interpret data correctly, you need to know how the analysis was done and what the data can't tell you. When chemical analysis shows that certain compounds or pollutants are present, it is important to put this information in the context of health or environmental issues. Consequently, the interpretation process must not only be meticulous in its calculations but also reflective of the broader implications these results may yield, especially within the scope of public policy or environmental standards. Also, the strictness of data interpretation means that possible sources of error must be carefully looked at. This includes differences that can happen because of sample contamination, instrument calibration, and human factors in doing and analyzing experiments. A thorough comprehension of these variables can profoundly influence the analysis's results. By encouraging educational practices that get students involved, they can learn more about these important processes. For example, changing how labs are run so that students help write protocols can help them understand the interpretation process better. As shown by Chiva et al. (2018), this method gets students to think about both the how and the why of each experiment, which leads to more meaningful interpretations of the results. In addition, being able to correctly interpret results can lead to progress in both scientific and practical uses of chemistry. Teachers can help students develop analytical skills that will help them make smart choices in school and in the real world by getting them to think deeply about what they find. Well-analyzed data can help make recommendations about safety standards, environmental rules, and public health programs. In this context, the critical analysis and interpretation of chemical results are essential elements of scientific literacy. They enable future chemists to make knowledgeable contributions to sustainable practices and policies, guaranteeing that their discoveries are comprehended and applied to instigate societal change, as highlighted by Pinto et al. (2017).

2.5. QUALITY CONTROL IN CHEMICAL ANALYSIS

Quality control in chemical analysis is necessary to make sure that the results are correct and trustworthy. It includes a methodical way to keep an eye on and judge the whole analytical process, from getting the samples ready to understanding the results. By putting quality control measures in place, labs can find possible sources of variation and error, which makes their results more trustworthy. For example, using standard reference materials, regularly calibrating instruments, and validating methods are all important parts of quality control that help make chemical analyses more reliable. By using these methods to keep checking, labs can find trends, make sure they follow the rules, and keep the trust of people who depend on their analytical results. As chemical analysis gets more complicated, quality control needs to use more advanced methods. Conventional single-method strategies may prove inadequate, especially in varied contexts such as the examination of insecticides in durable mosquito nets. Recent advancements underscore the necessity for multi-residue methodologies capable of effectively addressing complex samples. In a study examining the extraction of deltamethrin and alpha-cypermethrin from treated fabrics, researchers established robust methodologies enabling simultaneous detection via gas chromatography with electron capture detection (GC-mu ECD). This new way of doing things shows how important quality control is when making methods that shorten extraction times while increasing the recovery and accuracy of the active ingredients (Ouattara *et al.,* 2013). Also, quality control is very important for the validation steps in chemical analysis. The accuracy and precision of any newly developed method depend on thorough validation, which includes things like specificity, limit of detection, and recovery rates. Recent studies have shown that the validated multi-residue method for finding insecticides has mean recovery rates of over 86% with very low relative standard deviations. This shows that the method is reliable. This shows that thorough quality control can lead to big improvements in analytical methods, making sure they meet both scientific and regulatory standards. These kinds of changes are very important, especially in areas that directly affect the health and safety of the public. Therefore, incorporating thorough quality control in chemical analysis is essential for promoting progress and preserving analytical integrity (Pinto *et al.,* 2017) (Figure 2.12).

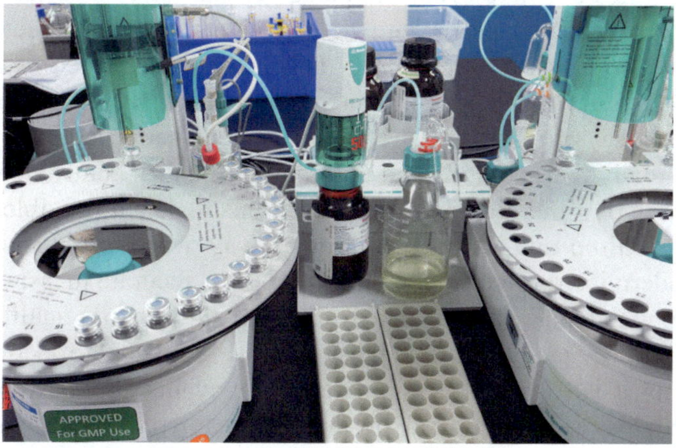

Figure 2.12. *Quality control in chemical analysis.*

Source: https://assets-global.website-files.com/6477b4213b69665fa49f89bf/65983bca7c02ff2fdb4762d5_2023_1013_Seran_QC_Lab.10.jpg.

2.5.1. Importance of Quality Control

Quality control is a necessary part of the chemical analysis process. It makes sure that the materials made are safe and effective. In a laboratory environment, particularly in pharmaceutical compounding, quality control mechanisms can substantially reduce the risks linked to incorrect outputs. For example, a study with second-year pharmacy students showed how adding an analytical assessment to a compounding lab exercise helped them learn about common mistakes, like weighing things wrong and contamination. The results showed that compounded products had an average error rate of 23.7%, which shows how important it is to have strong quality control (Almoazen *et al.,* 2014). The majority of students said they had a better understanding of how important it is to keep quality in their preparations, which will probably affect how they work in the future. Quality control protects public health and also teaches future chemical professionals the right way to do things and how to think about things. The significance of quality control transcends individual laboratory practices, serving as a critical pillar in the broader chemical manufacturing and analysis landscape. To follow the rules and keep the trust of customers, businesses need to have good quality control systems. For example, NASA's workshops on life support systems recognized the need to set evaluation standards and combine different levels of modeling to reach process control excellence (Crabb *et al.,* 2025). Industries can make their products more reliable and safer by constantly improving these standards. This is especially important in fields like pharmaceuticals, where even small changes can have bad effects. So, strict quality control measures make sure that products not only meet regulatory standards but also follow best practices. This leads to a culture of excellence in chemical analysis. Also, the intersection of quality control and new technologies opens up new ways to improve the chemical analysis process. With the rise of advanced analytical methods and automation, quality control can now be a part of every step of the production process. This makes it possible to monitor quality in real time and take action right away when problems arise. For instance, modern instruments make it easier to take accurate measurements and analyze data, which makes it possible to find deviations from quality standards early on. In the previously mentioned lab exercise, students' reflections showed how important it is to know how to use these advanced tools to make fewer mistakes and make compounded products better overall (Almoazen *et al.,* 2014). By using technology as part of their quality control systems, labs and manufacturers can not only make their products more accurate and consistent, but they can also improve their reputation in a market that is getting more and more competitive. So, a strong commitment to quality control is necessary for the chemical analysis process to keep getting better.

2.5.2. Standard Operating Procedures (SOPs)

Standard operating procedures (SOPs) are an important part of the chemical analysis process. They make sure that all steps are done the same way and correctly. SOPs are detailed instructions that tell you what to do when you do experiments or analyses. This helps keep the results honest and trustworthy. For example, in environmental science and oceanographic research, SOPs are very important for measuring nutrients at sea. The Global Ocean Ship-based Hydrographic Investigations Program (GO-SHIP) has put together a full manual that talks about the best ways to measure nutrients, focusing on methods like gas-segmented continuous flow analysis (CFA), sample collection, and data reporting (Aminot *et al.,* 2010). So, SOPs not only help people follow established methods, but they also improve the overall quality of the data collected, which is very important for later analysis and interpretation. Implementing SOPs is more than just following the rules; it also creates an atmosphere of quality control and constant improvement.

Figure 2.13. *Global ocean ship-based hydrographic investigations program (GO-SHIP).*

Source: https://tos.org/oceanography/assets/images/content/oceanObs2021-roem-mich-f4.jpg.

In labs, detailed SOPs cover all parts of chemical analyses, from making calibration solutions to keeping track of reference materials. Quality control mechanisms, such as using reference materials for nutrients in seawater to check how well a system works, show how important accuracy is in science. The GO-SHIP manual shows how these methods improve the quality of each analysis and make the data collected from different cruises and research projects more reliable (Aminot *et al.,* 2010). So, setting up clear, well-documented processes is very important for making scientific collaboration easier and building trust in the data that comes from it. SOPs also help keep the workplace safe and make sure that rules are followed, especially in labs where dangerous materials are often used. In the public sector, unions put health and safety first to keep workers safe from injuries that could happen on the job. The Public Employees Federation stresses the need to be aware of dangers at work and follow safety rules. This is similar to what happens in labs where chemical tests are done. By including safety measures in their standard operating procedures (SOPs), labs can lower risks and make sure that chemical analysis results are still reliable. In general, the connection between SOPs in chemical analysis not only makes science more rigorous, but it also protects the health and safety of everyone involved in the research process (Figure 2.13).

2.5.3. Calibration and Validation of Instruments

In order to make sure that chemical analysis processes are reliable and accurate, it is very important to calibrate and validate instruments. Calibration is the process of comparing the readings from an instrument to a known standard. This lets you change the instrument's output to match the true value. This is important when using advanced methods like Fourier Transform Near-Infrared Spectroscopy (FT-NIRS), which need to be very precise in order to analyze complex mixtures correctly. For example, the method created for FT-NIRS analysis of soybean-based foods shows how important calibration is for getting accurate measurements of the composition of soybean products, such as proteins, oils, and carbohydrates (Baianu *et al.,* 2011). Analysts can reduce systematic errors and improve the accuracy of their measurements by creating a calibration curve using known standards. This makes it easier to interpret the data. Validation, on the other hand, is the systematic checking of the calibration process and how well the instrument works as a whole. This step makes sure that the instrument gives accurate results in controlled conditions and is also reliable with different types of samples and matrices. When it comes to spectroscopic techniques, validation might mean testing the instruments' responses with different sample mixtures to see how consistent they are and how well they can handle sample variability. For instance, the intrinsic difficulties arising from physical discrepancies in sample preparation, including variations in particle size and packing, can profoundly influence spectral measurements (Chen *et al.,* 2011). Researchers can ensure that the instrumental methodology is reliable and reproducible in heterogeneous samples by validating it through rigorous experimental designs. This confirms the integrity of the chemical analysis process. Furthermore, the relationship between calibration and validation underscores the necessity for proficient model development to alleviate prevalent analytical difficulties. As shown by the new ways to analyze spectral data, it is very important to deal with the multiplicative effects that light scattering causes in order to get accurate quantitative results in heterogeneous mixtures. The suggested method for separating spectral contributions from physical variations shows how calibration and validation methods are changing over time. Chemists can better handle the complexities of spectroscopic data by using advanced modeling techniques to improve the estimation of multiplicative parameters (Chen *et al.*, 2011). This dual focus on strict calibration and strong validation is important for keeping the standards of analytical chemistry high. This will lead to more reliable chemical analysis results and better scientific progress.

2.6. APPLICATIONS OF CHEMICAL ANALYSIS

Chemical analysis is used in many fields, but it is most common in environmental monitoring, pharmaceuticals, and materials science. Chemical analysis is very important for finding pollutants and checking the health of ecosystems in environmental settings. For example, gas chromatography mass spectrometry (GC-MS) can accurately find and measure dangerous chemicals in water and soil samples. These kinds of tools help regulatory bodies make sure that environmental standards are followed and make it easier to clean up polluted areas. Chemical analysis also helps figure out how bioavailable toxic substances are, which helps with risk assessment and management plans. This integration shows how important chemical analysis is for protecting the environment and how it affects society as a whole. Chemical analysis is very important in the pharmaceutical industry, not only for making new drugs but also for making sure that therapeutic agents are safe and work. High-performance liquid chromatography (HPLC) and other techniques give important information at different stages of drug formulation, such as testing for identity, checking for impurities, and studying stability. The analytical results help formulation scientists improve product designs and fix stability problems. These methods are also very important during the process of getting regulatory approval to make sure that drugs meet strict safety standards. The pharmaceutical industry would have a lot of trouble managing public health without reliable chemical analysis. This shows how important it is to connect lab research with real-world use. Advances in chemical analysis, especially in the characterization of new compounds, have greatly helped the growing field of materials science. For instance, when researchers look into metal-organic frameworks (MOFs), they use computational chemistry to guess and describe their structures and properties. Recent literature delineates various techniques that assist researchers in predicting crystal structures and evaluating mechanical properties, thereby facilitating the development of innovative materials for applications in gas storage and separation technologies (Coudert *et al.,* 2015). The rise of data-driven soft sensors is another example of how chemical analysis and modern engineering are coming together. These sensors improve process monitoring in industries like chemicals and bioprocessing, showing how chemical analysis can change as technology changes (Abdi *et al.,* 2009). The link between chemical analysis and new ideas shows how important it is for both scientific research and business use (Figure 2.14).

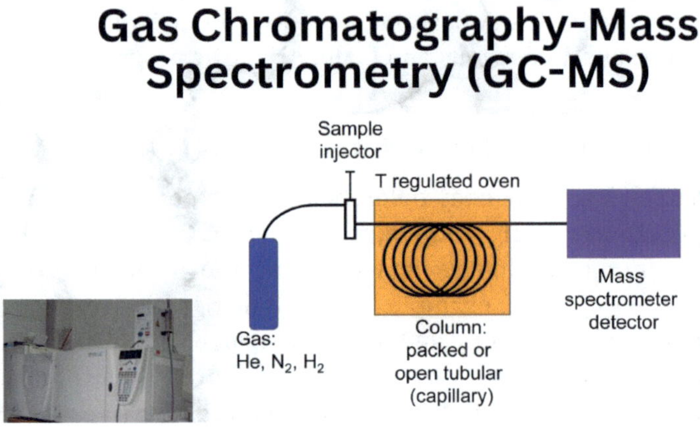

Figure 2.14. *Gas chromatography-mass spectrometry (GC-MS).*

Source: https://scienceinfo.com/wp-content/uploads/2023/02/Gas-Chromatography-Mass-Spectrometry-GC-MS.jpg.

2.6.1. Environmental Monitoring

Environmental monitoring is an important part of the chemical analysis process, especially when it comes to figuring out how human activities affect natural ecosystems. Environmental monitoring helps measure pollutants like heavy metals, organic compounds, and microplastics in soil, water, and air by using a variety of analytical methods in a systematic way. This measurement is very important for following the rules and keeping people healthy. Effective monitoring also helps stakeholders make smart choices, which lets them come up with ways to reduce pollution while encouraging sustainable practices. In addition, using more advanced chemical analysis techniques makes it easier to find contaminants, which gives us a better picture of environmental health. These findings have both scientific and social consequences. They can affect policy, raise awareness in the community, and lead to efforts to protect the environment. This shows how chemical analysis and environmental stewardship are linked. Technology has changed the way chemical analysis is done in environmental monitoring. New technologies like autonomous sensor networks have become important tools for gathering data in real time in a wide range of settings. These networks make it possible to monitor environmental parameters over a wide area and period of time, which improves both the accuracy and speed of data collection. As discussed in ongoing projects, these technologies are a great resource for both researchers and policymakers because they let us find and study pollutants on a scale never seen before. These sensors can give us high-resolution temporal data that can help us make better choices about environmental policies and public health programs. As the need for more thorough monitoring strategies grows, so does the reliance on these cutting-edge technological solutions. This brings the fields of chemical analysis and environmental monitoring even closer together and shows how important they are together in solving environmental problems. Also, environmental monitoring is very important for making sure that environmental rules and standards are followed. Agencies that enforce these rules often use data from chemical analysis to find out if there are any contaminants near industrial sites, waste disposal sites, and cities and how they might affect them. For example, reports from the Hazardous Waste Research and Information Center stress the need for constant monitoring to find dangerous chemicals that could harm ecosystems and human health (Gatz *et al.,* 1985). Such constant watchfulness shows how important it is to use strict analytical methods to keep the environment healthy. Also, using these monitoring frameworks not only protects ecosystems, but it also builds public trust in how the environment is managed. Environmental monitoring is an important link between scientific research and public policy. It does this by using clear processes and good communication of results. This encourages collaboration that is good for both community health and environmental sustainability, which shows how important the chemical analysis process is in today's world.

2.6.2. Pharmaceutical Development

The process of developing drugs is closely tied to the process of chemical analysis, which is an important part of making sure drugs are safe and work. At first, finding possible drug candidates depends a lot on thorough chemical analysis to make clear molecular properties and biological interactions. As pharmaceutical research progresses, entities are increasingly cognizant that conventional methodologies are inadequate for tackling the swift transformations in drug discovery. Scientists and analysts in the industry have said that the lack of new drugs being made shows that we need more collaborative models, even though funding for research and development is still going up. Improving collaboration between different sectors not only makes it easier to share information, but it also helps researchers get through the "valley of death," where promising research often fails to turn into viable drug candidates. This shows how important good chemical analysis is during the development process (Crossman *et al.,* 2008). In the field of pharmaceutical manufacturing, the use of cutting-edge technologies like hot-melt extrusion (HME) shows how important chemical analysis is for keeping product quality high. HME is a continuous manufacturing process that makes things more efficient and allows for more production. However, it needs to be carefully watched to make sure that the extrudates meet the specifications that were set. In this case, it is very important to use different process analytical techniques. These methods not only keep an eye on important factors like temperature and pressure, but they also help us learn more about how materials behave during extrusion. Recent evaluations have shown that using advanced sensors to keep an eye on different important quality attributes improves both process control and quality assurance for the final product. This ongoing supervision is in line with the main goals of drug development, which are to make sure that safe and effective drugs get to patients as quickly as possible (Beer *et al.,* 2013). The changing world of drug development is always pushing for new ways to analyze chemicals so that they can keep up with the complexities of making and designing drugs today. As the industry moves toward more collaborative frameworks, it's important to change the way research institutions, regulatory agencies, and pharmaceutical companies work together. The field can use the collective knowledge of its members to solve problems more effectively by encouraging openness and sharing resources. Also, the ongoing use of analytical technologies in the drug development cycle not only makes things easier, but it also makes sure that all rules are followed. This thorough and varied method of drug development shows how important chemical analysis is for getting good results, which will eventually lead to the development of effective treatments that can have a big effect on public health (Beer *et al.,* 2013; Crossman *et al.,* 2008).

2.6.3. Food Safety Testing

Food safety testing is an important part of the chemical analysis process. Its goal is to make sure that food products are safe to eat and don't contain any harmful substances. This testing is necessary to deal with the many problems that come with modern food production, which has many factors to consider, such as the use of pesticides and the possibility of foodborne pathogens. Labs can find out how much contamination is present in samples by looking for chemical residues. This can help them find levels that could be harmful to people's health. Also, as more people become aware of the safety of food products, there is a greater need for openness and responsibility in the food supply chain. Adding chemical analysis makes it easier to understand these elements by giving us real-world data that can help educate consumers and guide regulatory decisions, which will make sure that food products on the market are safe. In recent years, the methods used to test food safety have changed a lot to account for the many different types of contaminants that could be present. Mass spectrometry and chromatography are two advanced chemical analysis methods that have made it easier to find pesticide residues and other harmful substances in food. As was said, more than 400 pesticides are now allowed to be used in food production. To protect public health, strict testing procedures are needed. The chemical analysis process not only looks for these chemical residues, but it also looks at the trust issues that consumers have with food choices. For example, a study showed how important it is for people to trust the credibility of food safety measures, since trust had a big effect on their preferences for products that follow strict safety testing protocols (Hartmann *et al.,* 2020).

Figure 2.15. *Food safety testing in microbiology.*

Source: https://img.freepik.com/premium-photo/food-safety-testing-microbiology-laboratory_798657-3439.jpg.

So, the strength of food safety testing is linked to how honest consumers think it is. In the bigger picture of food safety testing, it is important to deal with the tensions between food safety, chemical use, and what people want. As the use of pesticides has gone up to meet the needs of farmers, so have worries about public health risks. This shows how complicated the relationship is that all parties involved must deal with. Proponents of pesticide application contend that it is essential for maintaining food security, while detractors emphasize the potential health risks linked to these substances (Graham *et al.*, 2019). This split shows how important it is to have thorough chemical analysis processes that can both make sure safety and help consumers make decisions. As a result, the ongoing evolution of food safety testing not only protects public health but also helps the food industry meet the growing demand for safer, more transparent, and sustainably produced goods. Food safety will change all the time in the future as people talk to each other and make changes to meet changing consumer needs and government rules (Figure 2.15).

SUMMARY

The chemical analysis process shows how important it is to correctly evaluate and understand chemical data. This process is the basis for making smart choices in many areas, including environmental science, pharmacology, and materials engineering. A thorough understanding of chemical analysis not only improves the ability to identify substances and measure their properties, but it also helps to create safer and more effective solutions to important problems. As the field of chemical assessment changes, it becomes more and more important to deal with the uncertainties and complexities that come with it in order to make sure it is reliable and useful in real life. It is important to know the limits of current models because recent research shows that it is too early to fully adapt risk-assessment policy frameworks to the problems posed by modern toxicological studies (Elliott *et al.,* 2000). Furthermore, it is essential to involve the wider community impacted by the results of chemical analysis. For example, in environmental remediation efforts, scientific findings must be combined with local concerns, which means that all parties involved, including tribal communities, need to talk to each other openly. Including community interests in the evaluation matrix for environmental technologies helps us understand how they might affect cultural, environmental, and health aspects in more detail (Jeanne *et al.,* 2003). Therefore, it is important to encourage scientists, policymakers, and affected communities to work together to come up with plans that not only meet legal requirements but also build public trust and encourage people to be involved in decision-making. This kind of involvement is necessary to meet the needs of different groups of people who are often more affected by chemical hazards than others. In conclusion, the complex nature of chemical analysis shows how important it is to keep doing research and talking about it. Future assessments must take into account the increasingly intricate environmental, health, and regulatory factors that affect chemical safety and effectiveness. The scientific community can better inform risk-management frameworks and make sure that new technologies are in line with social values and priorities by putting an integrative approach to chemical assessment first. As we work toward more sustainable ways of doing things, the information we get from working together on assessments can help us find solutions to chemical-related problems in society that are fair, effective, and respectful of different cultures. Having thoughtful conversations about these issues will eventually lead to a future with better safety and resilience in chemical management.

To fully understand the chemical analysis process, you need to look at a few key parts that make it work. To get reliable results, the first step is to choose the right methods. This means carefully looking at the chemical properties of the things being studied, which can help you decide whether to do a qualitative or quantitative analysis. Also, proper sample preparation and handling are very important for making sure that the analysis is accurate. Contamination or degradation can have a negative effect on the results. The process also uses advanced tools and methods that can find a wide range of chemical compounds with great accuracy and sensitivity. These basic parts show how important it is to plan and carry out the chemical analysis process very carefully. This is what makes accurate and repeatable results possible. However, chemical analysis has effects that go beyond just technical details; it also has effects on safety and regulations. The Environmental Protection Agency (EPA) and the Food and Drug Administration (FDA) are two examples of public health agencies that depend on strong chemical analysis to make decisions and protect consumers. This dependence is apparent in the examination of chemicals in diverse contexts, especially those presumed to possess carcinogenic characteristics. The Office of Advocacy's report on scientific evaluations of dangerous substances was very revealing. It showed how scientific integrity and the interests of big chemical companies can be at odds (Greenhaw *et al.,* 2013). This intersection of science and regulatory oversight increases the need for clear methods and objective analyses, making sure that public health stays a top priority even when businesses are putting pressure on it. Finally, the interplay between scientific evaluation and external factors necessitates an examination of the ethical implications inherent in the chemical analysis process. Advocacy groups and lobbying efforts make the stories about chemical safety even more complicated. For example, trade groups spend a lot of money on lobbying to show how far industries will go to get the rules they want (Colten *et al.,* 1990). This has serious consequences for the credibility of chemical assessments, especially when groups with a stake in the outcome put pressure on regulatory bodies. To restore public trust in scientific methods and make sure that decisions based on chemical analysis really do reflect the goal of health and safety for everyone involved, we need to deal with these ethical issues. So, the ongoing discussion about chemical analysis shows not only how technically difficult it is, but also how it affects society as a whole.

REFERENCES

1. Abate, A., Adley, T., Aebersold, R., Ahlford, A., Almassian, R., Almoguera, C., Asiello, P., et al. (2017). Advances in microfluidics and lab-on-a-chip technologies. http://arxiv.org/abs/1703.08583 (accessed on 29th October 2024).
2. Abdi, H., Alhoniemi, E., Angelov, P., Angelov, I., Angelov, D., Arazo-Bravo, M. J., Atkeson, C. G., et al. (2009). Data-driven soft sensors in the process industry. https://core.ac.uk/download/75803.pdf (accessed on 29th October 2024).
3. Albiol Chiva, J., Carda-Broch, S., Doménech-Carbó, M. T., Esteve-Amorós, M., et al. (2018). Design of the laboratory script by the students in chemistry practical's: Analysis of water. https://core.ac.uk/download/304372108.pdf (accessed on 29th October 2024).
4. Alexander, S., Anderson, S., Craig, G., Hendry, R., et al. (2000). HoloCam: A subsea holographic camera for recording marine organisms and particles. https://core.ac.uk/download/335466.pdf (accessed on 29th October 2024).
5. Almoazen, H., Birnie, C., Eley, J., Hinkle, J., Kadi, A., McGill, C., Newton, C., et al. (2014). Analysis of compounded pharmaceutical products to teach the importance of quality in an applied pharmaceutics laboratory course. https://core.ac.uk/download/287923243.pdf (accessed on 29th October 2024).
6. Aminot, A., Aoyama, M., Bakker, K., Becker, S., et al. (2010). Recommendations for the determination of nutrients in seawater to high levels of precision and inter-comparability using continuous flow analyzers. https://core.ac.uk/download/11900585.pdf (accessed on 29th October 2024).
7. Antunes, P., Antunes, I., Costa, A., Ferreira, M., Griggs, J., Hobbs, S., Moura, A., et al. (2018). A high-precision LiDAR-based method for surveying and classifying coastal notches. https://core.ac.uk/download/186632082.pdf (accessed on 29th October 2024).
8. Avni, R., & Venugopalan, M. (2025). Analysis of glow discharges for understanding the process of film formation. https://core.ac.uk/download/pdf/42847544.pdf (accessed on 29th October 2024).
9. Baianu, I., & Guo, J. (2011). Development of novel calibrations for FT-NIR analysis of protein, oil, carbohydrates and isoflavones in foods. https://core.ac.uk/download/pdf/290259.pdf (accessed on 29th October 2024).

10. Bajic, S., Stanley, H., Hames, B., Jones, R., Jones, T., et al. (1997). Rapid analysis of wood using transient infrared spectroscopy and photoacoustic spectroscopy with PLS regression. https://core.ac.uk/download/38932127.pdf (accessed on 29th October 2024).
11. Bakel, B. van, & Postma, G. (1997). Challenges for chemical information extraction and text retrieval. https://core.ac.uk/download/pdf/11454209.pdf (accessed on 29th October 2024).
12. Baker, T. N. (2009). Processes, Microstructure and Properties of Vanadium Micro Alloyed Steels. https://core.ac.uk/download/9022903.pdf (accessed on 29th October 2024).
13. Baudet, P., Baudouin, O., Floquet, P., Ghannadzadeh, A., et al. (2012). General methodology for exergy balance in ProSimPlus® process simulator. https://core.ac.uk/download/18540207.pdf (accessed on 29th October 2024).
14. Bauer, C., Chiang, M., Choudhury, A., Fiske, S., Jiang, L., Maurya, R., et al. (2012). Process performance analysis in large-scale systems integrating different sources of information.
15. Bilal, M., Cohen, Y., Godwin, H. A., Romero-Franco, M., et al. (2017). Needs and challenges for assessing the environmental impacts of engineered nanomaterials (ENMs). https://core.ac.uk/download/323076183.pdf (accessed on 29th October 2024).
16. Biris, A. R., Biris, A. S., Baker, R. T. K., Baker, C., Berger, M., Biris, A., Couteau, C., et al. (2007). On the dynamical ferromagnetic, quantum Hall, and relativistic effects on the carbon nanotubes nucleation and growth mechanism. http://arxiv.org/abs/0705.0407 (accessed on 29th October 2024).
17. Branco Pinto, D., Constales, D., Marin, G., Yablonsky, G., et al. (2017). Kinetic intersections as a source of information on rate coefficients. https://core.ac.uk/download/91288153.pdf (accessed on 29th October 2024).
18. Bustamante, M. A., Gomis, P., Marhuenda Egea, F. C., Martín-Mata, J., et al. (2015). Chemical, thermal and spectroscopic methods to assess biodegradation of winery-distillery wastes during composting. https://core.ac.uk/download/32324967.pdf (accessed on 29th October 2024).
19. Cadorim, H. R., Del Campo Menoyo, J., Carvalho, G. S., Duarte, F. A., et al. (2017). Development of a fast-screening method for the direct determination of chlorinated persistent organic pollutants in fish oil by high-resolution continuum source graphite furnace molecular absorption spectrometry. https://core.ac.uk/download/153371322.pdf (accessed on 29th October 2024).

20. Carneiro, N., Fernandes, M., Oliveira, F. R., Souto, A. P., et al. (2012). Functionalization of textile materials by double barrier discharge plasma. https://core.ac.uk/download/55637260.pdf (accessed on 29th October 2024).
21. Chase, R. K. (1934). The drafting of the housing legislation. https://core.ac.uk/download/62558940.pdf (accessed on 29th October 2024).
22. Chen, Z. P., Jin, J. W., Li, L. M., Steponavicius, T., et al. (2011). Quantitative spectroscopic analysis of heterogeneous mixtures: The correction of multiplicative effects caused by variations in physical properties of samples. https://core.ac.uk/download/9037183.pdf (accessed on 29th October 2024).
23. Cleary, J., Collins, F., Diamond, D., Fay, C., et al. (2013). Distributed environmental monitoring. https://core.ac.uk/download/11311498.pdf (accessed on 29th October 2024).
24. Colten, C. E., & Samsel, T. B. (1990). The Calumet Area Hazardous Substance Data Base: A User's Guide with Documentation. https://core.ac.uk/download/4812361.pdf (accessed on 29th October 2024).
25. Coudert, F. X., & Fuchs, A. H. (2015). Computational characterization and prediction of metal-organic framework properties. http://arxiv.org/abs/1506.08219 (accessed on 29th October 2024).
26. Crabb, T. M., Evanich, P. L., & Gartrell, C. F. (2025). 1992 NASA Life Support Systems Analysis Workshop. https://core.ac.uk/download/pdf/42806684.pdf (accessed on 29th October 2024).
27. Crossman, C., Rai, A. K., Reichman, J. H., Uhlir, P. F., et al. (2008). Pathways across the valley of death: Novel intellectual property strategies for accelerated drug discovery. https://core.ac.uk/download/62565976.pdf (accessed on 29th October 2024).
28. De Beer, T., Remon, J. P., Saerens, L., Vervaet, C., et al. (2013). Process monitoring and visualization solutions for hot-melt extrusion: A review. https://core.ac.uk/download/55761837.pdf (accessed on 29th October 2024).
29. Elliott, K. C. (2000). A case for caution: An evaluation of Calabrese and Baldwin's studies of chemical hormesis. https://core.ac.uk/download/72056494.pdf (accessed on 29th October 2024).
30. Fukushima, Y., Hellweg, S., Hirao, M., Hungerbühler, K., et al. (2018). Using standard statistics to consider uncertainty in industry-based life cycle inventory databases (7 pp). https://core.ac.uk/download/159146996.pdf (accessed on 29th October 2024).

31. Gatz, D. F., & Sweet, C. W. (1985). Atmospheric research and monitoring study of hazardous substances. https://core.ac.uk/download/4812338.pdf (accessed on 29th October 2024).

32. Goulet, D. (2015). Modeling, simulating, and parameter fitting of biochemical kinetic experiments. http://arxiv.org/abs/1508.05359 (accessed on 29th October 2024).

33. Graham, K. Z. (2019). Federal regulation of pesticide residues: A brief history and analysis. https://core.ac.uk/download/232208315.pdf (accessed on 29th October 2024).

34. Greenhaw, K., Weatherford, K., & Rabinowitz, R. (2013). Small businesses, public health, and scientific integrity. https://core.ac.uk/download/71361208.pdf (accessed on 29th October 2024).

35. Hartmann, M., Langen, N., & Yeh, C. H. (2020). The role of trust in explaining food choice: Combining choice experiment and attribute best–worst scaling. https://core.ac.uk/download/322793605.pdf (accessed on 29th October 2024).

36. Hill, A., Dao, B. N., Dao, B., Sanchez, C., Yang, C., Powell, C. E., Cornelius, C. J., et al. (2010). Synthesis and characterization of hybrid organic-inorganic materials based on sulphonated polyamideimide and silica. https://core.ac.uk/download/10831334.pdf (accessed on 29th October 2024).

37. Holmes-Smith, A. S., Lennard, F., & Smith, M. J. (2019). Development of non-destructive methodology using ATR-FTIR with PCA to differentiate between historical Pacific barkcloth. https://core.ac.uk/download/293865553.pdf (accessed on 29th October 2024).

38. Jacquemin, L., Pontalier, P. Y., & Sablayrolles, C. (2011). Life cycle assessment (LCA) applied to the process industry: A review. https://core.ac.uk/download/12043304.pdf (accessed on 29th October 2024).

39. Kovrigin, E. (2012). NMR line shapes and multi-state binding equilibria. https://core.ac.uk/download/213086317.pdf (accessed on 29th October 2024).

40. Lim, C. K., & Lord, G. A. (2002). Current developments in LC-MS for pharmaceutical analysis. https://core.ac.uk/download/7439.pdf (accessed on 29th October 2024).

41. Lodder, J. C. (2006). Patterned nanomagnetic films. https://core.ac.uk/download/pdf/11470244.pdf (accessed on 29th October 2024).

42. Lyu, C., Nastase, G., Rubinsky, B., Serban, C., et al. (2017). A comparison of freezing-damage during isochoric and isobaric freezing of the potato.

https://core.ac.uk/download/323081574.pdf (accessed on 29th October 2024).

43. ANONYMOUS. (2003). PESH/OSHA Standards: Information for Workers. https://core.ac.uk/download/19958797.pdf (accessed on 29th October 2024).
44. Norlin, E. (2005). Nitrogen isotope analysis of ammonium and glycine. https://core.ac.uk/download/11694902.pdf (accessed on 29th October 2024).
45. Ouattara, J. P. N., Pigeon, O., & Spanoghe, P. (2013). Validation of a multi-residue method to determine deltamethrin and alpha-cypermethrin in mosquito nets by gas chromatography with electron capture detection (GC-μECD). https://core.ac.uk/download/55757436.pdf (accessed on 29th October 2024).
46. Patocka, R. R. (1972). Influence of freezing water samples prior to chemical analysis. https://core.ac.uk/download/345017867.pdf (accessed on 29th October 2024).
47. Rubin, J. M., Denham, K. C., & Tano, M. L. (2003). Identifying the burdens and opportunities for tribes and communities in federal facility cleanup activities: Environmental remediation technology assessment matrix for tribal and community decision-makers. https://core.ac.uk/download/71342300.pdf (accessed on 29th October 2024).
48. Sarkissian, G. (2009). Tyre profiling: Development and evolution of forensic methodology. https://opus.lib.uts.edu.au/bitstream/10453/29871/1/01Front.pdf (accessed on 29th October 2024).
49. Shimizu, R., & Tanaka, H. (2015). A novel coarsening mechanism of droplets in immiscible fluid mixtures. http://arxiv.org/abs/1509.03385 (accessed on 29th October 2024).

CHAPTER 3

MICROWAVE SPECTROSCOPY

Contents

INTRODUCTION

The introduction to microwave spectroscopy is an important first step in learning about both its theoretical basis and how it can be used in real life. This method works in the electromagnetic spectrum mostly between 1 GHz and 300 GHz. It looks at the rotational and vibrational transitions of molecules, which gives us important information about how molecules are structured and how they move. Microwave spectroscopy is important for more than just basic scientific research; it is also changing the way chemistry, biology, and materials science work. Researchers have improved the sensitivity and specificity of microwave spectroscopic techniques by using new detection technologies. This has made it possible to study how molecules behave under different conditions in great detail. This introduction will provide an overview of the historical context, the basic principles of microwave spectroscopy, and its importance in modern research. This will set the stage for more in-depth discussions about its uses and possible future directions. Microwave spectroscopy is also important because it has made a big difference in submillimeter astronomy, where its advancements have changed how we can observe things. The field of submillimeter astronomy is still very difficult, even though it has come a long way in the last 30 years. This is because current detection technology and the way the atmosphere transmits light make it hard to see.

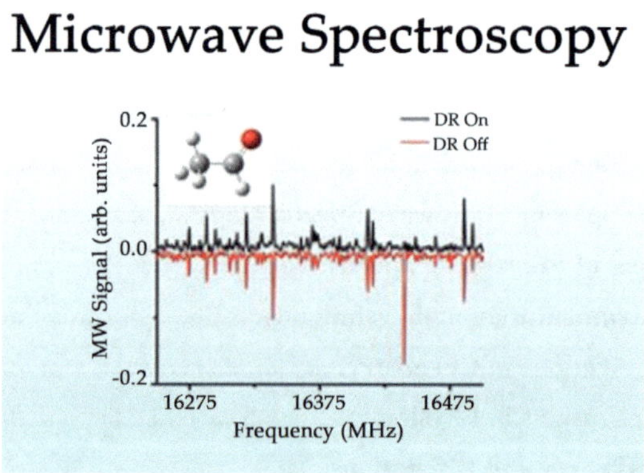

Figure 3.1. *Microwave spectroscopy.*

Source: https://external-content.duckduckgo.com/iu/?u=https%3A%2F%2Ftse2.mm.bing.net%2Fth%2Fid%2FOIP.czTB2czZp617qLIC5Kw2iQHa-EK%3Fpid%3DApi&f=1&ipt=d7e8be5c2e50a07a9c8ffb3301c4b1f3bd16e220ce340fd557c66d78290ded9b.

However, astronomers have been able to get around some of these problems by carefully building submillimeter telescopes in the best places, like Mauna Kea and Antarctica. These kinds of efforts led to the development of important tools like the ALMA interferometer and the ESA/NASA Herschel Space Observatory, which use non-invasive microwave spectroscopy methods to make progress. This interaction between microwave spectroscopy and submillimeter astronomy shows how important this method is in modern astrophysics and how it helps us learn more about the universe. Microwave spectroscopy is also very useful in the biological sciences, especially through the use of new biosensors.

Recent research has shown that microwave techniques, such as the microstrip ring resonator (MRR), can be used to study biological tissues at microwave frequencies. Using equivalent lumped elements to model biological tissues lets scientists look into important factors like overlay thickness and resonance frequency. This advancement represents a significant progression in biosensor technology, enhancing the applicability of microwave spectroscopy in assessing food quality and in medical diagnostics. The capacity to simulate and quantify outcomes establishes a solid foundation for validation in these contexts, thereby enhancing the utility of the technique. These kinds of improvements that stress working together across fields not only make the research more reliable, but they also put microwave spectroscopy at the cutting edge of new technologies in biomedicine (Figure 3.1).

Definition of Microwave Spectroscopy

Microwave spectroscopy is a key method in molecular analysis that lets scientists look into how molecules rotate. This technique takes advantage of the fact that microwave radiation interacts with the dipole moments of molecules. This lets scientists see changes between quantized rotational energy levels. Microwave spectroscopy is very useful for studying gases in different states of matter, such as in the atmosphere and in the lab, because it is very sensitive to molecular environments. Researchers can learn a lot about molecular structure, dynamics, and interactions by looking at how microwave radiation is absorbed at certain frequencies. Microwave spectroscopy not only reveals the complex nuances of molecular behavior but also holds significance in diverse domains, including astrophysics and atmospheric science, where comprehension of molecular interactions can substantially enhance research initiatives. The basic idea behind microwave spectroscopy is that matter is made up of quantum mechanics, which means that energy levels are quantized, and rotational transitions are distinct. Microwave radiation can give a molecule energy, which can cause it to move between these rotational levels. The moment of inertia and the charge distribution in the molecule are directly related to the frequency at which absorption happens. Because of this, each type of molecule has a different spectral signature in the microwave range, which makes it easy to tell them apart. Methods like Fourier-transform microwave spectroscopy have made this method even more sensitive and precise, allowing researchers to see complicated spectral patterns. This improved resolution helps us understand molecular complexities better, which makes microwave spectroscopy useful in many areas, including chemical analysis, medical diagnostics, and environmental monitoring. In light of the progress made in technologies and research methods, it is important to know what microwave spectroscopy means so that it can be used in modern scientific work. For example, the growing popularity of optical atomic clocks shows how molecular spectroscopy and precision measurement techniques work together. This shows how improvements in microwave spectroscopy can lead to better timekeeping and frequency standards. Microwave spectroscopy also helps atmospheric physics by looking at the molecular parts of the atmosphere. This shows how useful the methods are for understanding changes in the climate and how pollution moves around, as shown by the progress made in atmospheric physics research (Beckley *et al.,* 2025). This shows how important it is to keep improving and expanding the uses of microwave spectroscopy, which will lead to new ideas that will change both basic research and practical uses.

Historical Development of the Technique

The historical progression of microwave spectroscopy exemplifies the advancement of analytical methodologies in physical chemistry. Microwave spectroscopy was first based on quantum mechanics from the early 1900s. It became more popular in the 1930s when radiofrequency technology became available. Researchers began to understand that they could learn about molecular structures and interactions by looking at the resonance frequencies of molecules in the microwave range of the electromagnetic spectrum. The introduction of dielectric materials and sensitive detection methods led to important progress, allowing for more detailed observations of molecular vibrations and rotations. While scientists worked on the necessary technology, early research showed that microwave spectroscopy was an important tool for studying molecular dynamics. This opened the door for future advancements that could help many fields, such as chemistry and materials science. The middle of the 20th century was a time of great progress in technology, which led to the creation of more advanced microwave spectrometers. Researchers could look at a wider range of molecular interactions and transitions thanks to these tools. Fourier Transform Microwave Spectroscopy (FTMW) was first used in the 1980s, and it changed the field by making it much faster and more sensitive to collect data. This new technology made it possible for scientists to study transient species that were hard to see before and analyze complicated gas-phase reactions. Also, better computer technology made it possible to work with large datasets, which helped scientists understand spectral data better. Such advancements not only improved the precision of molecular characterization but also expanded the utility of microwave spectroscopy in fields like astrophysics and atmospheric sciences, where comprehensive molecular data is essential for comprehending environmental phenomena. Microwave spectroscopy is still making big strides in modern research, thanks in part to collaborations between different fields that use the latest technologies. Combining microwave spectroscopy with other methods like mass spectrometry and ambient ionization has made it possible to use it in real time for chemical analysis and environmental monitoring. This has made measurements more accurate and useful. MeteoMet-Metrology for Meteorology is one of many projects that show how important reliable data is for solving global problems like climate change (Antonsen *et al.*, 2013). Scientists are improving the way they analyze complex datasets by adding metrological points of view to microwave measurements. Moreover, advancements in instrumentation and calibration propel the investigation into novel fields, including nanotechnology and biomedicine, highlighting microwave spectroscopy's crucial contribution to contemporary scientific progress (Fanale *et al.,* 2025).

3.1. PRINCIPLES OF MICROWAVE SPECTROSCOPY

Microwave spectroscopy is based on the interaction between electromagnetic radiation and the energy states of molecules. The core of this method is that molecules can absorb microwave radiation, which causes them to move between different rotational energy levels. These changes happen when the frequency of the incoming radiation is the same as the energy difference between the quantized rotational states. Because of this, the absorption spectrum can give us useful information about the molecular structure and how rotational motion works. Microwave spectroscopy is especially useful for studying small and medium-sized molecules because their rotational transitions are often clear and distinct, which makes it possible to measure molecular parameters very accurately. Advanced instruments are very important for making microwave spectroscopy better. Modern spectrometers use designs like the Fabry-Perot resonator to make measurements much more accurate and sensitive. This high-resolution method works well with other methods, such as pulsed supersonic nozzles, which help cool samples without changing their temperature, which reduces the effects of Doppler broadening. Researchers can get detailed rotational spectra from these spectrometers because they work better. These spectra are very important for understanding complex molecules. The design, performance, and operation of high-resolution microwave spectrometers are thoroughly described, with many experimental examples showing how they can be used for molecular analysis (Cerny *et al.,* 2013). The intricate nature of spectral patterns, especially in larger polyatomic molecules, presents distinct challenges in microwave spectroscopy. Because these molecules have many vibrational and rotational states, the overlapping of energy levels creates complex spectral features that make it hard to understand the data. This complexity is especially clear in higher-lying overtone and combination vibrational levels, where the density of states goes up a lot. To tackle these issues, the utilization of double-resonance spectroscopy techniques, including time-resolved pump-probe measurements, has become imperative. These techniques not only enable the investigation of excited states but also yield significant information regarding inelastic energy transfer rates, pressure-broadening parameters, and other collision processes essential for modeling the atmospheric behaviors of various species, such as ozone and ammonia (Coy *et al.,* 1990). Consequently, comprehending the principles and complexities of microwave spectroscopy is essential for the progression of both theoretical and practical molecular spectroscopy (Figure 3.2).

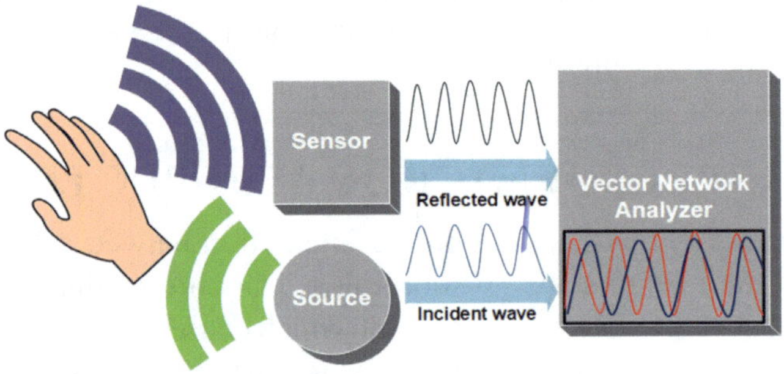

Figure 3.2. *Microwave spectroscopy working principle.*

Source: https://external-content.duckduckgo.com/iu/?u=https%3A%2F%2Ffigures.academia-assets.com%2F118954688%2Ffigure_004.jpg%3Fresponse-content-disposition%3Dattachment%253B%2Bfilename%253Dfigure-4-microwave-spectroscopy-working-principle&f=1&nofb=1&ipt=620a59343083fb78673e72ce2442216d935bf4a17896dc88d7d41035cfe75347.

3.1.1. Electromagnetic Spectrum Overview

There are many different wavelengths in the electromagnetic spectrum, and each one corresponds to a different type of electromagnetic radiation. This range includes radio waves, microwaves, infrared radiation, visible light, ultraviolet radiation, X-rays, and gamma rays. Each part of the spectrum has its own uses and properties. For example, microwaves, which have a frequency range of about 1 GHz to 300 GHz, are very important for communication technologies and spectroscopic analysis. Microwave spectroscopy, in particular, looks at how these waves interact with matter. This gives us information that is very important for understanding how molecules are structured and how they move. By looking at the absorption and emission profiles of microwave radiation, scientists can learn a lot about molecular arrangements, which is important for progress in many areas of science, from chemistry to materials science. The fundamental principles of microwave spectroscopy are intrinsically linked to the properties of the electromagnetic spectrum. This method takes advantage of the fact that molecules resonate at certain microwave frequencies that match their rotational and vibrational transitions. As a result, this makes it possible to find molecular species and look into how molecules interact with each other. High-resolution microwave spectrometers, which are built using advanced technologies, are very sensitive and accurate when measuring these transitions. This makes it easier to find very small spectral features. A Fabry-Perot resonator design, for instance, can greatly improve the accuracy of measurements, allowing scientists to look closely at how molecules interact with each other (Cerny *et al.,* 2013).

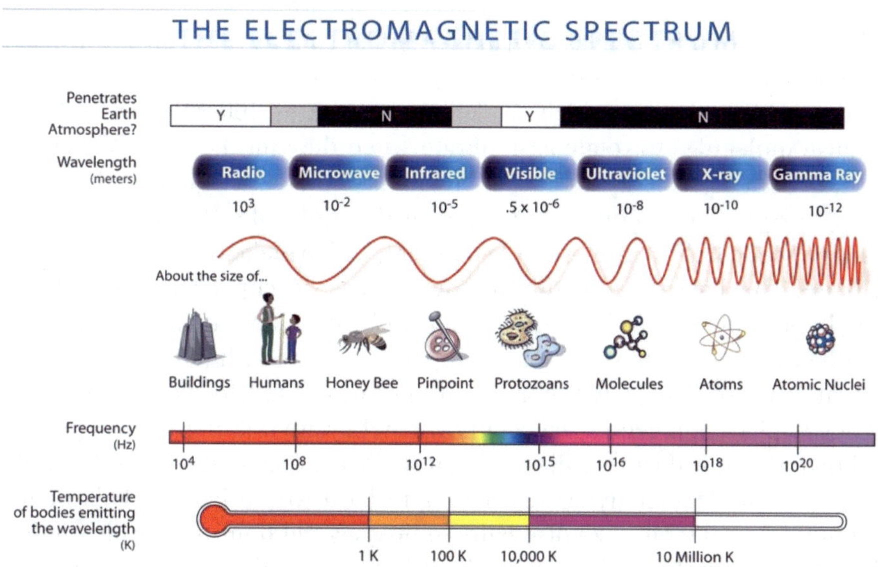

Figure 3.3. *Electromagnetic spectrum.*

Source: https://cdn1.byjus.com/wp-content/uploads/2018/11/physics/wp-content/uploads/2015/12/electromagnetic-spectrum.jpg.

The interaction between microwave radiation and molecular systems is essential for the progress of microwave spectroscopy, expanding the limits of our comprehension of chemical and physical phenomena. Additionally, microwave spectroscopy is being used in new ways in material science and telecommunications that go beyond its traditional uses. The emergence of innovative metamaterials and their interaction with microwave signals exemplifies the expanding potential in this domain. Research has demonstrated the efficacy of customized metamaterial absorbers in diverse applications, including imaging and stealth technology, leveraging their distinctive electromagnetic characteristics (Astorino *et al.,* 2018). Microwave spectroscopy is still changing, and as it does, it not only gives us basic information about how molecules behave, but it also opens up new possibilities for technological progress in areas like materials engineering and remote sensing. Microwave spectroscopy enhances the comprehension of the electromagnetic spectrum and its numerous applications by amalgamating emerging technologies with conventional techniques (Figure 3.3).

3.1.2. Interaction of Microwaves with Matter

The interaction of microwaves with matter fundamentally relies on the principles of electromagnetic radiation, especially their capacity to induce molecular transitions. Microwaves, which work in the range of 300 MHz to 300 GHz, can cause molecules to rotate and vibrate when they hit them. This excitation happens when the energy of the microwave photons is equal to the energy difference between two quantized states of the molecule. Because of this, the electromagnetic field creates a dipole moment, which makes the molecules rotate or vibrate. This can be seen with microwave spectroscopy. This method gives researchers important information about molecular structures and dynamics, which lets them look into the basic properties of gases, liquids, and solids in different states of matter. A major use of microwave spectroscopy is that it can look at gas-phase molecules, especially when they are rotating. Recent progress has shown that ultrafast two-dimensional terahertz spectroscopy works well to see complex molecular dynamics that were hard to see before. This method uses time-delayed terahertz pulse pairs to link several light-matter interactions, which shows complex information about how molecular dipoles are oriented and how they relax. As a result, scientists can turn nonlinear time-domain orientation signals into two-dimensional rotational spectra by mapping them into the frequency domain. This ability to see rotational transitions directly greatly improves our understanding of molecular dynamics and how they affect the properties and interactions of materials (Fleischer *et al.,* 2016). Additionally, the interaction of microwaves with matter transcends gas-phase molecules to include diverse quantum systems, such as quantum dots. Cavity quantum electrodynamics provides novel methodologies for manipulating light-matter interactions, illustrating the capacity for sophisticated control over quantum systems. In these configurations, quantum dots can interact via cavity photons, even when they are far apart, allowing for unique electrostatic and tunnel-free couplings. This framework makes it possible to scale quantum bit architectures and simulate electron interactions on a chip, which could change the way quantum computing and communications work. Microwave interactions are not limited to spectroscopy; they are also essential for the progress of quantum technologies, connecting fundamental science with practical applications in engineering and information technology (Bruhat *et al.,* 2012) (Figure 3.4).

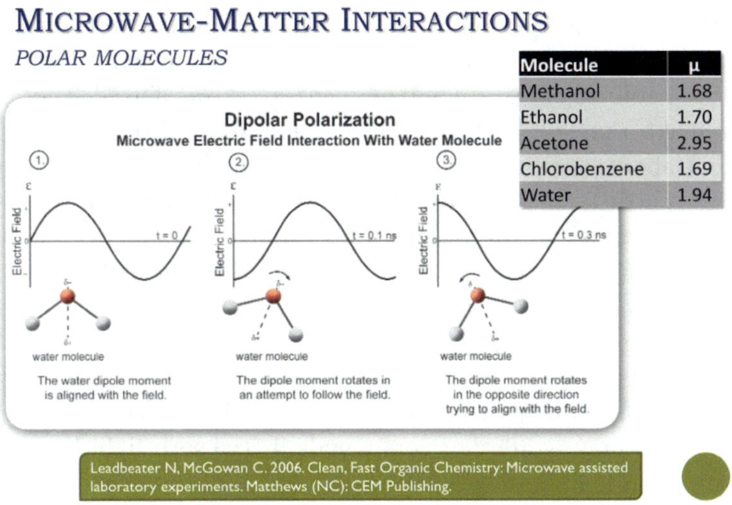

Figure 3.4. *Interaction of microwaves with matter.*

Source: https://image.slideserve.com/441648/microwave-matter-interactions-polar-molecules-l.jpg.

3.1.3. Energy Levels and Molecular Transitions

Understanding microwave spectroscopy, which looks at the rotational and vibrational transitions of molecules, starts with studying energy levels and molecular transitions. In a molecular system, energy levels represent quantized states that a molecule can occupy, determined by its structural and electronic characteristics. When a molecule interacts with electromagnetic radiation, it can move between these energy levels. Each transition has its own frequency, which gives information about the molecule's structure. For example, the two-step laser excitation method used to measure high-lying even parity levels of neutral Yb shows how accurate modern methods are for studying molecular transitions, which helps us learn more about molecular energy states related to microwave spectroscopy (Cheinet *et al.,* 2018). This information is very important for fields like astrophysics and atmospheric science, where understanding molecular transitions is the basis for finding and identifying different compounds. Furthermore, the complexity of energy transitions in molecular systems increases with the presence of polyatomic molecules, as vibrational and electronic interactions give rise to intricate spectral patterns. In these systems, overtone and combination levels that are higher up often show up, making it harder to understand spectral data. Advanced techniques like double-resonance spectroscopy can help make this complexity less of a problem. This technique uses time-resolved measurements to better explain how inelastic collisions happen and how energy is transferred. These methods not only make the energy levels clearer, but they also give us important information about transition moments and propensity rules that control how molecules act in different situations (Coy *et al.,* 1990). The ability of microwave spectroscopy to break down these complex transitions makes it an important tool for advancing molecular science and improving our understanding of the chemical environment. Understanding energy levels and molecular transitions has effects that go beyond just being interesting to scientists; they affect how things work in many different fields of science. For instance, what we learn from spectroscopic studies helps us make new materials and energy technologies. One example is how we design compounds that absorb light better for solar energy conversion. Researchers can design molecular systems that make the most of energy absorption and conversion processes by figuring out how certain energy levels interact with radiation. Also, looking at the collision broadening parameters from microwave spectroscopy helps us figure out what the atmosphere is made of, which is important for studying the environment and the climate. The complex interplay among energy levels, molecular transitions, and their practical applications underscores the vital significance of microwave spectroscopy in both theoretical and applied chemistry.

3.2. TYPES OF MICROWAVE SPECTROSCOPY

Microwave spectroscopy includes a number of methods, each designed to look at certain molecular properties by how microwaves interact with matter. Rotational spectroscopy is one of the most important methods because it looks at the rotational states of polar molecules. This method lets scientists get exact information about molecular structures, bond lengths, and angles, which gives them a better understanding of the chemical identity and behavior of substances in different states. It is well known that rotational spectroscopy can help us understand how molecules move. It gives us information about the quantum nature of molecular rotations, which are strongly affected by temperature and the environment of the molecules. Scientists can learn more about both basic and practical chemical processes by studying how microwave radiation is absorbed at certain frequencies that match these transitions. This helps us understand how complex molecular systems work (McDonnell *et al.,* 2016). Electron spin resonance (ESR) is another important part of microwave spectroscopy. It lets you look closely at unpaired electrons in chemical species. ESR directly measures the magnetic properties of these electrons, which is different from other methods. This makes it possible to find free radicals, transition metal complexes, and defects in solids. ESR is a great way to study reaction mechanisms in organic and inorganic chemistry because it is very sensitive to the electronic environment around unpaired electrons. Furthermore, progress in this field has led to the creation of methods like pulsed ESR, which allow for time-resolved measurements of dynamic processes. Researchers are discovering new ways to use ESR to learn more about how electrons move and how molecules interact in different systems. This shows how important ESR is for both basic and applied science (Fleischer *et al.,* 2016). Microwave spectroscopy also includes techniques like microwave photolysis, which uses both microwave and photochemical methods to figure out what happens to short-lived species during chemical reactions. This method makes it easier to create short-lived species, which are usually hard to analyze with regular spectroscopy. In this case, using microwaves lets you selectively excite molecular vibrations and rotations, which makes it easier to find transient intermediates. Consequently, microwave photolysis is essential in mechanistic investigations of reactive intermediates, providing understanding of their synthesis and degradation. The synergistic application of microwave energy enhances the efficacy of excitation mechanisms and deepens our comprehension of intricate molecular transformations, representing a notable progression in the domain of microwave spectroscopy (McDonnell *et al.,* 2016).

3.2.1. Continuous Wave (CW) Spectroscopy

Continuous Wave (CW) Spectroscopy is an important method in microwave spectroscopy that uses the constant emission of electromagnetic radiation to study molecular and atomic systems. This method is different because it keeps a constant frequency, which lets you measure energy transitions in the target substances very accurately. The ability to continuously emit microwaves makes it possible to excite different electronic or rotation-vibration states in molecules, which is very important for accurate spectral analysis. Researchers can study a wide range of phenomena with such precise control over frequency, such as characterizing materials and watching chemical reactions happen in real time. In the end, CW spectroscopy is a strong way to get high-resolution data that can be used in many different scientific studies. This shows how useful it is in both basic research and industrial settings. Combining CW techniques with more advanced methods, like Electron Paramagnetic Resonance (EPR) spectroscopy, makes it much easier to understand reactive oxygen and nitrogen species (ROS and RNS) in living things. Recent studies have shown that EPR spectroscopy gives us important information about physical-chemical oxidation processes and how free radicals behave.

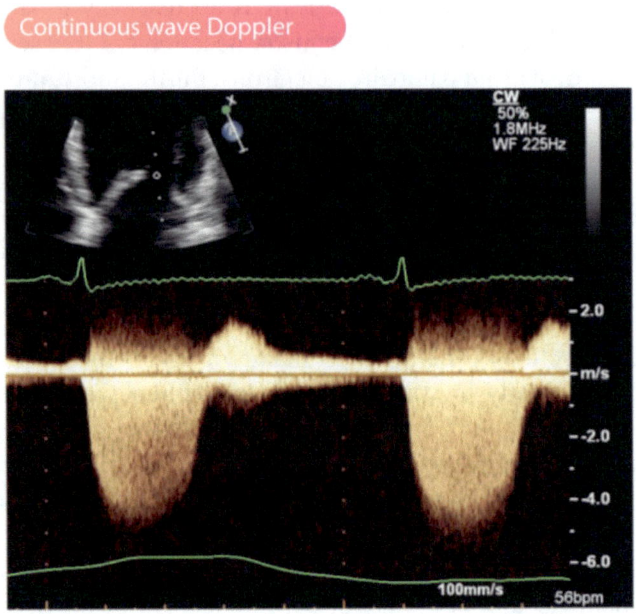

Figure 3.5. *Continuous wave doppler (CW) ultrasound echocardiography.*

Source: https://ecgwaves.com/wp-content/uploads/2019/11/continuous-wave-doppler-cw-ultrasound-echocardiography.jpg.

This makes CW spectroscopy even more useful. The integration of these techniques allows scientists to utilize methodological approaches that can comprehensively examine oxidative stress, an essential contributor to numerous diseases. CW spectroscopy is usually very good at studying how molecules respond, but when it is combined with EPR, it can look at transient radicals and their roles in biochemical pathways in a more complete way. This opens up new research possibilities that can help with urgent health issues (Gurer-Orhan *et al.*, 2017). The emergence of optical frequency combs, especially those produced by high-Q micro-resonators, illustrates the advancement of continuous wave (CW) spectroscopy towards enhanced precision and efficiency. These micro-combs help improve spectroscopy by giving it wide optical bandwidths, which are necessary for high-resolution measurements. Their small size and low power use make them especially useful for RF photonics applications, which go beyond the limits of other spectroscopic methods. As micro-comb technology improves, it opens up new possibilities for CW spectroscopy, making it possible to add more features and connect to more complicated systems. Micro-combs can easily fit into current experimental setups, which opens up a lot of possibilities for practical use. This adds to the field of microwave spectroscopy and makes it more important in both academia and industry (Chu *et al.,* 2018) (Figure 3.5).

3.2.2. Pulsed Microwave Spectroscopy

Pulsed microwave spectroscopy has become an essential method in molecular analysis, especially for examining intricate systems at the molecular scale. This method uses short bursts of microwave radiation to let scientists see fast-moving processes inside molecules. This makes it easier to study structural and interaction properties that are often hidden in traditional spectroscopy. Pulsed microwave spectroscopy has the unique ability to capture transient states that help us understand how molecules behave in different environments. Improvements in pulse shaping and timing strategies have made the data even more precise and detailed, allowing researchers to look into a wide range of macromolecular interactions. This is especially important for figuring out complicated details in biological macromolecules, which are very important for fields like pharmacology and materials science. In the realm of pulsed microwave spectroscopy, the incorporation of cutting-edge methodologies like light-induced triplet-nitroxide DEER (LiDEER) has markedly enhanced the ability to quantify nanometric distances within intricate molecular assemblies. The comparative efficacy of LiDEER and laser-induced magnetic dipole spectroscopy (LaserIMD) underscores the continuous enhancement of experimental methodologies in this field. For example, research shows that using chirp inversion pulses in the LiDEER method can greatly improve measurement accuracy by making the modulation depth of signal acquisition deeper. This ability is very important because it lets us better understand how molecules move and interact with each other by giving us more accurate measurements of the distances between atoms that make up molecular structures. As research continues to improve these methods, they have more and more important effects on our understanding of biological processes, which helps us understand basic molecular biology better. The use of pulsed microwave spectroscopy goes beyond just looking at the structure of molecules; it also gives us important information about how energy moves and collides between molecules. Recent advancements have shown that combining double-resonance spectroscopy with time-resolved methods can help us learn more about inelastic collision processes, which are important for modeling how energy moves around in different places. This method has been essential for studying a variety of species important to atmospheric chemistry and environmental science, including ozone and ammonia. By clarifying parameters associated with pressure broadening and transition moments, researchers can proficiently model intricate atmospheric processes. These insights enhance our comprehension of molecular behavior in isolated environments and possess substantial implications for environmental and planetary science, underscoring the significance and adaptability of pulsed microwave spectroscopy within the scientific community (Figure 3.6).

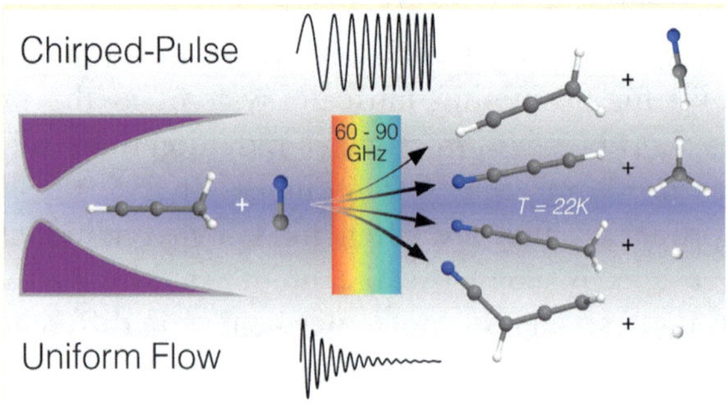

Figure 3.6. *Chirped pulse in uniform supersonic flow.*

Source: https://perso.univ-rennes1.fr/ian.sims/spectro/CPUF.gif.

3.2.3. Fourier Transform Microwave Spectroscopy (FTMW)

Fourier transform microwave spectroscopy (FTMW) is a big step forward in the field of microwave spectroscopy. It has better sensitivity and resolution than older methods. This method uses the principles of Fourier transform to get multiple spectral data points at the same time. As a result, FTMW can quickly gather complete spectra, which is important when looking at complicated molecular systems. FTMW uses microwave radiation to study the rotational transitions of molecules, which gives us a lot of information about their structure, dynamics, and interactions. This method has become more popular because it can study small samples, which is especially useful in fields like atmospheric analysis and astrophysics, where only small amounts of substances may be available. The technological advancements that have made FTMW possible have made it useful in many different fields of science. The use of optical frequency combs, for example, has changed how precise measurements are made in spectroscopy. These micro-combs create stable and wide optical bandwidths while staying small and energy-efficient, which greatly improves FTMW capabilities. Recent reviews have shown that micro-combs have a lot of potential in radio-frequency photonics because they can do more things and take up less space, which is a big advantage over older systems (Chu *et al.,* 2018). This change shows how FTMW is always changing and adding new technologies to give us more detailed information about molecules. In addition, the development of FTMW has been helped by big improvements in both theory and practice in spectroscopic methods. Recent advancements in instrumentation have enhanced the overall efficacy and accessibility of FTMW, promoting its incorporation into diverse research domains. The investigation of synchrotron radiation has facilitated high-resolution time-resolved studies in mineralogy and environmental science (Bellatreccia *et al.,* 2013). This move toward more advanced, high-resolution methods shows how important FTMW is for learning about the properties and behaviors of molecules. As these methods keep getting better, FTMW is ready to solve new problems in spectroscopy, which will open the door to new ideas and uses in many different fields of science.

3.3. INSTRUMENTATION AND TECHNIQUES

Microwave spectroscopy depends a lot on advanced tools and methods that make it easier to study the structures and movements of molecules. Heterodyne receivers are at the heart of this field, and they have changed a lot in the last few decades. These tools make it possible to measure spectral lines very accurately, which is important for figuring out what molecules are in different places. The introduction of technologies like bolometric detectors in space-based submillimeter wave telescopes marks a major turning point. These tools let scientists study atmospheres on Earth and in space. For example, projects like SWAS and MIRO use these advanced methods to make lists of light gases and map how molecules are spread out in interstellar dust. This shows how instrumentation development and astronomical research can work together, as noted by Siegel et al. (2007). Microwave spectroscopy technologies have also found use in more practical fields, like environmental monitoring, in addition to astrophysical research. Advanced instruments that were first made for aerospace use have been successfully adapted for use in industry, especially for monitoring air pollution. Microwave spectroscopy devices that can find very small amounts of pollutants help public health and regulatory efforts a lot by giving real-time information about air quality. This transfer of technology from one area to other shows how useful and important advanced instruments are in both scientific research and everyday life, as shown by Lieberman et al. (2025). It highlights the possibility of collaboration between different fields that can improve measurement methods and make microwave spectroscopy more useful overall. The combination of microwave spectroscopy instruments with data analysis methods makes them even more useful in many scientific fields.

Fourier Transform Microwave Spectroscopy (FTMW) is a good example of a big step forward that lets you take high-resolution spectral measurements. This method has been very useful for learning about the rotational transitions of molecules, which has helped us learn more about their properties and how they act. Scientists are working on making these methods better, such as by adding machine learning algorithms to help them understand data. This should make analyses much more accurate and efficient. These advancements not only enhance our comprehension of chemical processes but also catalyze innovations in diverse applications, from pharmaceuticals to climate science, underscoring the increasing importance of microwave spectroscopy as a fundamental instrument in contemporary research (Figure 3.7).

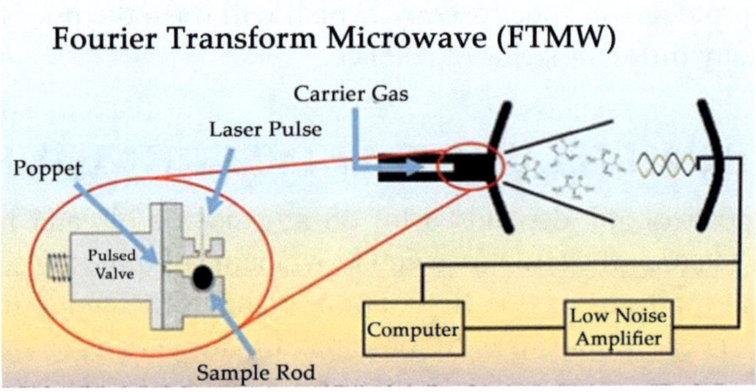

Figure 3.7. *Fourier transform microwave spectroscopy.*

Source: https://external-content.duckduckgo.com/iu/?u=https%3A%2F%2Ftse3.mm.bing.net%2Fth%2Fid%2FOIP.M5YifM361WTHUoKlF2o17gHaEK%3Fpid%3DApi&f=1&ipt=11fc1690903e7961f8653548d29c7ace6eb37bddf1c6b6373fd047482e3754ef.

3.3.1. Microwave Sources and Detectors

The development of microwave sources and detectors is very important for microwave spectroscopy. It has led to big advances in both research and real-world uses. Solid-state sources like Gunn diodes and IMPATT diodes are at the top of the list. They have changed the way microwave frequencies are made. These devices produce continuous wave output, which makes it easier to take high-resolution spectroscopic measurements. The creation of high-power microwave systems has also made it easier to find molecular signatures with more accuracy and sensitivity. Traditional sources have their limits, but new technologies have made it possible to find better ones that can work in a wider range of frequencies and bandwidths in spectroscopy. Researchers are still looking into new materials and configurations, and the efficiency and effectiveness of microwave sources will definitely improve their ability to study complex molecular systems. It is also very important to be able to pick up microwave signals, since this is what makes spectroscopic analyses work. Bolometers and superconducting detectors are two examples of modern microwave detectors that are very sensitive and can pick up small changes in spectral lines. For example, bolometric detectors can measure the strength of microwave radiation by looking at how the resistance changes when the temperature changes. These improvements have direct uses in many areas, including astrophysics and atmospheric science, where they let scientists see molecular interactions in a wide range of settings. The link between detection technology and source innovation is still important. This is because better detectors need better source capabilities to work at their best. The field of microwave spectroscopy will benefit greatly from the ongoing integration of these technologies, both in terms of resolution and versatility. The practical use of microwave sources and detectors is shown even more clearly by how they are used on Earth and in space. Recent efforts have shown how useful these technologies are for studying celestial events, as shown by astronomy instruments made for submillimeter-wave observations. For example, improvements in low-temperature detectors have helped researchers study the cosmic microwave background (CMB), which has given us useful information about how the universe began. These detectors are being used more and more in missions that involve large-scale atmospheric surveys. This greatly improves our understanding of how molecules are made up and where they are found in both planetary and interstellar environments. As a result, ongoing advancements in microwave spectroscopy and its associated technologies are not only revolutionizing scientific understanding but also providing practical applications across multiple fields. This is evidenced by the significant increase in satellite missions employing microwave technologies, which offer improved capabilities for detecting molecular signals in space (Miller *et al.,* 2009; Siegel *et al.,* 2007).

3.3.2. Sample Preparation Methods

In microwave spectroscopy, good sample preparation is very important because it directly affects how accurate and reliable the spectroscopic data is. Because different matrices can be very complicated, it is often necessary to use special preparatory methods to separate the analytes of interest from other substances. Solid-phase extraction (SPE) and liquid-liquid extraction (LLE) are two common ways to clean samples. However, microwave-assisted extraction (MAE) has shown to be much faster and more efficient. Microwave-assisted extraction (MAE) uses microwaves to speed up the extraction process, which often leads to higher recovery rates and less use of solvents. Being able to quickly prepare samples can be very helpful in analyses that need to be done quickly, since even small differences in how samples are treated can cause differences in spectral data that could change how the data is interpreted and what conclusions are drawn.

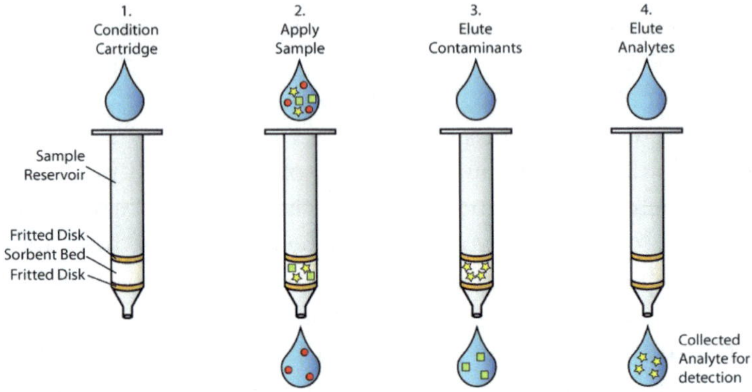

Figure 3.8. *Solid phase extraction principles.*

Source: https://www.specartridge.com/wp-content/uploads/2018/09/Solid-Phase-Extraction-Principles.jpg.

It is also interesting that microwave-assisted synthesis can be used to make samples for microwave spectroscopy. Recent advancements suggest that microwave synthesis can facilitate the preparation of intricate molecules, thereby enhancing yield and purity. For example, Boc-protection chemistry on Kaiser oxime resin has been successfully used to make polyamides with high yields that are useful in spectroscopic studies. These new ways of preparing samples not only make experiments more efficient, but they also make it much easier to see spectral features. Researchers can get more accurate spectroscopic readings by making sure that samples are free of contaminants and that their structure stays the same. The capacity to swiftly synthesize compounds under regulated conditions may facilitate the investigation of previously unreachable analytical targets, thereby expanding the applications of microwave spectroscopy. In conclusion, it is impossible to overstate how important it is to have strong sample preparation methods in microwave spectroscopy. It includes a variety of methods, from old-school ones to new microwave-assisted ones, all of which help keep the sampled analytes intact. Citing advancements like the successful microwave synthesis of complex polymers, the interconnectedness of preparation methods and their implications for analytical precision becomes clearer. As the field progresses, continuous research and development are expected to produce increasingly innovative preparation strategies that will augment the capabilities and applications of microwave spectroscopy. This ongoing enhancement in sample preparation will not only refine current methodologies but also facilitate new applications, thereby reinforcing the role of microwave spectroscopy across diverse scientific fields (Bacsa *et al.,* 2012; Cadorim *et al.,* 2017) (Figure 3.8).

3.3.3. Data Acquisition and Analysis

In microwave spectroscopy, data collection and analysis are the most important steps in getting useful information from experimental observations. Accurate data collection is the basis of any strong analytical method, especially in fields where high-frequency microwave signals are used to study the details of molecular interactions. Techniques such as Electron Paramagnetic Resonance (EPR) spectroscopy have significantly advanced data acquisition methodologies, particularly with the recent introduction of novel light-inducible spin labels. These innovations enable improved resolution in quantifying distances and interactions within intricate macromolecular systems, as demonstrated in the study showcasing the efficacy of triplet-nitroxide DEER and LaserIMD techniques (Bieber *et al.,* 2018). So, the quality and accuracy of the data that is collected depend not only on the tools used but also on the methods used to analyze the results. Data analysis in microwave spectroscopy is complicated and has many parts, and it often takes advanced computer techniques to figure out what the spectral data means. The dependability of these analyses is affected by various factors, such as noise levels and the modulation depth attained during experiments. Recent research indicates that the modulation-depth efficacy of light-induced methodologies can be substantially improved through the application of chirp inversion pulses, thereby augmenting the signal-to-noise ratio and enhancing the clarity of the acquired data (Bieber *et al.,* 2018). These changes give researchers the chance to look into new ways of doing spectroscopy, which will eventually help them understand molecular structures and dynamics better. As the field moves forward, using these advanced methods will be very important for improving the way data is analyzed in microwave spectroscopy. New technologies that make it easier to collect digital data are opening up a new era for microwave spectroscopy. Digital cavities, which allow radiofrequency and microwave outputs to be changed, are a promising idea for improving the resolution of spectral measurements while keeping the Q-factors flexible. These new technologies not only improve the way data is collected, but they also improve the ability to sync data, which is important for many uses, such as high-resolution spectroscopy. As researchers make use of these advanced data acquisition systems, the ways they analyze data will change, which will help them understand more about how molecules interact with each other. The combination of better ways to collect data with more advanced analytical tools will surely make microwave spectroscopy better and open up more uses for it in science.

3.4. APPLICATIONS IN CHEMISTRY

Microwave spectroscopy has made great strides in chemistry, especially when it comes to studying the structures and dynamics of molecules. This powerful method makes use of the interaction between electromagnetic radiation and molecular species to let chemists' study how molecules rotate. Researchers can get exact information about bond lengths, angles, and molecular conformations by using high-resolution microwave spectrometers. The Fabry-Perot resonator, which is a key part of advanced spectrometers, makes these measurements more sensitive and precise, making it possible to study complicated chemical systems in great detail (Cerny *et al.,* 2013). This ability has far-reaching effects on many areas of chemistry, such as environmental studies, where knowing the rotational spectra of atmospheric constituents can help find pollutants and see how they react with other chemicals. Microwave spectroscopy is also very important for studying biochemical processes, especially when it comes to reactive species like free radicals. It is important to know how these temporary molecules act because they are important to many normal and abnormal processes in the body. Recent studies have employed microwave spectroscopy in conjunction with various analytical methods to enhance the understanding of oxidative stress mechanisms in living organisms. The technique's capacity to elucidate the behavior of reactive oxygen and nitrogen species augments our understanding of their functions as intermediaries in biochemical reactions (Gurer-Orhan *et al.,* 2017). Researchers are better able to study complex biological matrices thanks to ongoing improvements in methods and tools. This helps them understand how living systems work more clearly. Microwave spectroscopy is useful for more than just basic molecular analysis; it can also be used in real-world situations in chemistry. The fast and correct retrieval of information about molecular structures helps a lot of fields, such as making drugs and agrochemicals. The ability to study high-energy molecular states has grown a lot of thanks to improvements in spectrometer design, like the use of pulsed supersonic nozzles that cool samples adiabatically. These improvements let chemists' study new compounds with more accuracy and sensitivity, and they also give them important information about how reactive and stable they might be (Cerny *et al.,* 2013). Ongoing research and innovation in microwave spectroscopy are expected to enhance our fundamental comprehension of chemical interactions and accelerate the discovery and optimization of new chemical entities essential for scientific progress (Figure 3.9).

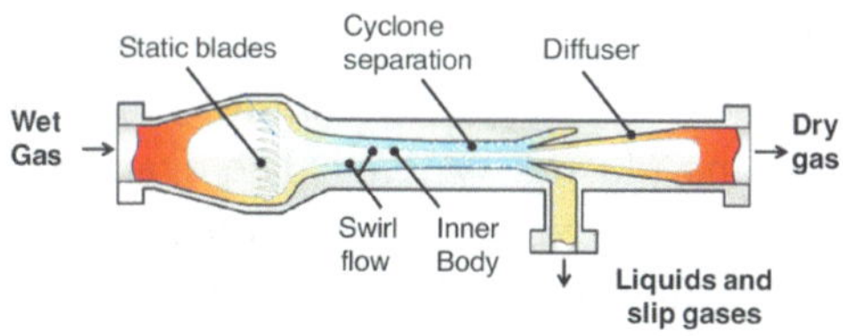

Figure 3.9. *Pulsed supersonic nozzles.*

Source: https://external-content.duckduckgo.com/iu/?u=https%3A%2F%2Fwww.researchgate.net%2Fprofile%2FPavel_Ditl%2Fpublication%2F279063450%2Ffigure%2Fdownload%2Ffig4%2FAS%3A391766588051456%401470415841100%2FDesign-of-a-supersonic-nozzle-for-NG-dehydration.png%3F_sg%3DQlqsrc1dEW2LXjvYPF_21rdjU3x-ulH1EwdicuJOlHQOM6MwX2bdOgO46379umkxCnVVx4jvQu8&f=1&nofb=1&ipt=e409faf853c7c1cb38135d1fb2ad77178dfa7c2c1c815aacdc398d44ee4165b1.

3.4.1. Molecular Structure Determination

Finding out what molecular structures are is very important for figuring out how different molecular species behave and interact with each other. Microwave spectroscopy is an invaluable tool in this effort because it can give very detailed information about rotational transitions. Scientists can use this method to look at how atoms are arranged in space by looking at how microwave radiation is absorbed or emitted. The rotational spectra generated can elucidate bond lengths and angles, facilitating a deeper comprehension of molecular geometry. This information is important for understanding the structures of both simple diatomic molecules and more complex polyatomic compounds. It helps us understand substances that are important in atmospheric chemistry and planetary science. The intricate structures of small to medium-sized polyatomic molecules require sophisticated techniques for precise analysis. These molecules have high-density energy levels that can make it hard to understand their microwave spectra, especially when overtone and combination vibrational transitions are involved. To deal with these problems, scientists often use double-resonance spectroscopy, which makes it easier to separate overlapping spectral lines and learn more about how energy moves from one place to another. For instance, examining the relationship between inelastic collisions and spectral line broadening can provide essential information about the molecular dynamics of significant atmospheric constituents, including ozone and ammonia. Consequently, the integration of microwave spectroscopy with various spectroscopic methods can enhance the comprehension of molecular structures and their corresponding thermodynamic characteristics, underscoring the importance of thorough analyses in molecular research (Coy *et al.,* 1990). Moreover, improvements in spectroscopic methods have made it possible to determine molecular structures in a wider range of settings, including plasma systems. The examination of the C_2 Swan bands in CO_2 microwave plasmas demonstrates the application of microwave spectroscopy in characterizing molecular states in high-energy contexts. Here, being able to accurately measure rotational and vibrational temperatures helps us understand how molecules interact with each other at higher temperatures. The relationship between gas temperature and the rotational temperature of molecules such as C_2 in microwave plasmas is especially interesting because it shows how molecules are made in these reactive environments. This empirical evidence, in conjunction with state-by-state fitting methods, illustrates the versatility and accuracy of microwave spectroscopy in clarifying both structural details and the dynamic behaviors of various molecular species under specific conditions (Carbone *et al.,* 2020).

3.4.2. Reaction Dynamics Studies

Understanding reaction dynamics is very important for learning more about chemical processes, especially in microwave spectroscopy. This method has a unique benefit because it lets scientists look into the temporary states of reactive species during chemical reactions. Scientists can use microwave radiation to make molecules switch between rotational and vibrational energy levels. This gives them important information about reaction pathways and intermediates. These kinds of insights are important for more than just figuring out how basic chemical processes work. They are also important for making catalysts and improving reactions in fields like materials science and biochemistry. Researchers are using microwave spectroscopy to look at how reactions happen, which means that accurate measurement methods and data interpretation are more important than ever. This is pushing the limits of what we know about how molecules behave in different situations. The interaction between reactive intermediates and free radicals underscores considerable challenges and prospects in the study of reaction dynamics. Reactive oxygen species (ROS) and reactive nitrogen species (RNS) are important parts of many biochemical reactions. They act as intermediates that can change the rates and results of reactions. Conventional techniques for identifying these species may be insufficient, frequently resulting in ambiguous outcomes. New microwave spectroscopy methods could make these measurements more accurate, which would let scientists look more closely at the oxidative processes that affect living things. This enhanced precision facilitates a more comprehensive understanding of the kinetics of reactions involving reactive oxygen species (ROS) and reactive nitrogen species (RNS), which is essential for evaluating oxidative stress and its ramifications in health and disease, thereby exerting a substantial influence in disciplines such as pharmacology and toxicology (Gurer-Orhan *et al.,* 2017). Furthermore, microwave spectroscopy can be used to study the conditions under which different types of molecules are made and react. For example, in high-pressure CO_2 microwave plasmas, C_2 Swan bands are a useful way to figure out the temperature of the gas and how reactive molecules move around in the plasma. The capacity to evaluate processes that result in the formation of particular molecules, such as C_2, offers essential insights into the evolution of molecular interactions in non-equilibrium environments. This ability not only helps us understand basic reaction mechanisms better, but it also helps us use them in real-world situations like plasma-assisted combustion and materials processing. As reaction dynamics studies increasingly incorporate advanced spectroscopic techniques, the capacity to obtain accurate temperature measurements and comprehend collisional phenomena will be essential for optimizing chemical reactions across diverse experimental conditions (Carbone *et al.,* 2020) (Figure 3.10).

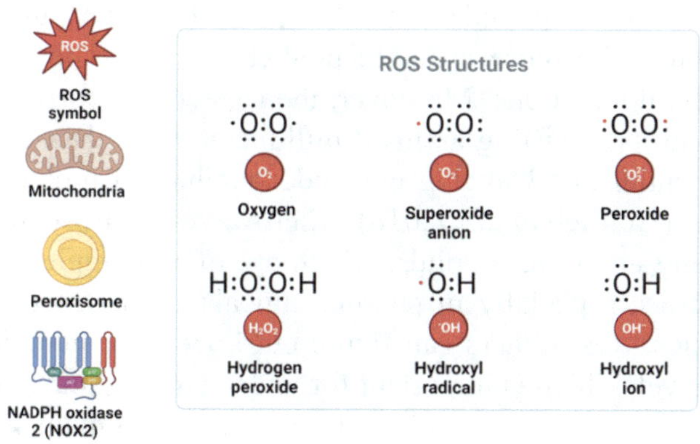

Figure 3.10. *Structure of reactive oxygen species and their sources.*

Source: https://www.researchgate.net/publication/370358957/figure/fig1/AS:11431281387325022@1745111960751/Structure-of-reactive-oxygen-species-and-their-sources-ROS-reactive-oxygen-species.tif.

3.4.3. Analysis of Complex Mixtures

Microwave spectroscopy has become a useful tool for analyzing complex mixtures because it can give a detailed spectral fingerprint of different compounds. This method uses the interaction of microwave radiation with molecular rotational transitions to find and describe many different types of molecules in complex samples. Researchers have successfully detected a wide range of chemical species in hydrocarbon/CS_2 discharge mixtures using advanced techniques like chirped-pulse and cavity Fourier-transform microwave spectroscopies. These techniques have revealed more than 60 different entities, along with their isotopic and vibrationally excited states. These results show the elemental and isotopic diversity that is naturally present in the mixtures. This is a first step toward a better understanding of their composition and behavior. In the end, the usefulness of microwave spectral taxonomy shows that complex mixture analysis has become much more advanced and accurate, opening up new areas for research and exploration in this field (Gottlieb *et al.,* 2018). It can be hard to understand how mixtures form and how their parts interact because they are so complicated. Microwave spectroscopy significantly enhances our understanding by enabling high-resolution measurements that reveal structural and dynamic information about molecular species. For example, studying the vibrationally excited states of carbon-sulfur chains has not only added to the list of molecular species, but it has also helped us understand how they form by using isotopic analysis. This fine-scale method lets researchers figure out where different compounds came from and how they grew over time. This helps them make more accurate predictions about how mixtures will behave in different situations. Moreover, the capacity to correlate microwave spectral data with pre-existing archival millimeter-wave data underscores the technique's potential in advancing our understanding of molecular dynamics and interactions (Gottlieb *et al.,* 2018). Microwave spectroscopy is important for characterizing amorphous solids, which are often complex mixtures used in many industries, especially in pharmaceuticals. It is also used to identify molecular species. Researchers can figure out how these materials behave at the molecular level, which is important for understanding things like how well they dissolve and how they crystallize. Terahertz time-domain spectroscopy, a complementary technique, improves this analysis by giving us non-destructive information about how the parts of the solids move and interact with each other. By examining the initiation and intensity of molecular mobility, researchers can acquire insights into crystallization processes crucial for drug delivery systems. Consequently, microwave spectroscopy and its associated methodologies significantly enhance the formulation of strategies designed to optimize the performance of amorphous solid formulations in the pharmaceutical sector, showcasing a remarkable synergy in the analysis of intricate mixtures.

3.5. APPLICATIONS IN PHYSICS

The combination of microwave spectroscopy with other physical applications shows how useful and important it is in modern research. One of the most important fields that microwave techniques have changed our understanding of is quantum optics and atomic physics. Microwave spectroscopy provides insights into the fundamental nature of photons and atoms by enabling the investigation of radiation-matter interactions. Recent progress in the field has resulted in innovative systems that employ circuit quantum electrodynamics (QED) to generate nonclassical propagating microwave fields. These kinds of new ideas are important not only for helping us understand quantum mechanics better, but also for making quantum information processing possible. For these uses, high-efficiency microwave photon detectors need to be used together. This is why studying microwave spectroscopy is so important for future research (Solano *et al.,* 2009).

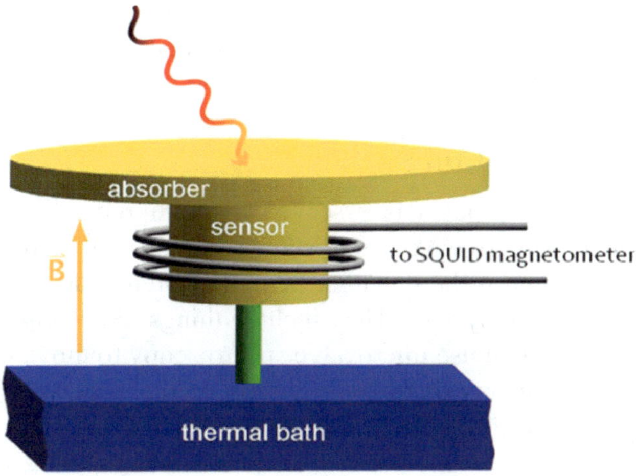

Figure 3.11. *Metallic magnetic calorimeters (MMCs).*

Source: https://www.researchgate.net/publication/284754816/figure/fig1/AS:1132477838049316@1647015165717/Schematic-principle-of-a-metallic-magnetic-calorimeter-consisting-of-an-x-ray-absorber.jpg.

Moreover, the progress in microwave spectroscopy has practical uses in many scientific fields, especially when it comes to designing advanced detection systems. Metallic magnetic calorimeters (MMCs) are an example of a new technology that is being developed in this field. They show how useful they can be in many different fields, such as nuclear physics and material science. As the need for more complicated experiments grows, so does the need for big detector arrays. The current work is on creating multiplexing methods, such as the microwave SQUID multiplexer, which works well in high-frequency areas. Adding these improvements to big detector systems makes them work better and makes them more useful. Using high Q resonators makes it possible to accurately measure microwave signals. This shows how microwave spectroscopy can help experimental physics move forward (Enss *et al.,* 2013). In conclusion, microwave spectroscopy is an important method in the field of physics that has led to big improvements in both theoretical understanding and real-world uses. Microwave techniques are being used for more and more things, from helping us learn more about quantum interactions to creating new ways to find things. As these fields change, the combination of microwave spectroscopy with new technologies like quantum computing and advanced materials opens up exciting new areas for research. The recent breakthroughs in efficient detection systems show how theoretical developments and practical use can work together. This shows how important it is to keep studying microwave spectroscopy and its many uses in physics. As researchers enhance their comprehension and improve their methodologies, the likelihood of groundbreaking discoveries persists significantly (Figure 3.11).

3.5.1. Study of Quantum States

Understanding quantum states is essential for comprehending the complex behaviors that dictate microwave spectroscopy, a potent method for investigating the properties of matter at the quantum level. Quantum states are the specific conditions of a quantum system. They include things like energy, momentum, and spin. Researchers can use microwave spectroscopy to move between these quantum states, which lets them look closely at the energy level structures of different systems. Being able to change quantum states has big effects on the growth of quantum technologies like quantum computing and precision measurement. Microwave spectroscopy is useful for observing quantum behaviors, which opens the door to new technologies that take advantage of the strange things that happen in quantum mechanics. These technologies will help us understand how materials behave and interact over a wide range of scales. Recent developments have highlighted the significance of quantifying quantum states through microwave-induced phenomena, especially through frameworks like *Landau–Zener–Stueckelberg* transitions. These transitions facilitate a wealth of quantum-coherent phenomena, observable through repeated interactions driven by microwave fields. Experimental studies, such as those outlined by Fragner et al. (2009), demonstrate the possibility of improving qubit state preparation and control techniques essential for quantum information science. You can change the coherence of these quantum states with great accuracy, which makes basic operations in quantum computing architectures possible. The study of quantum states clarifies the relationship between energy levels and state superpositions, demonstrating how microwave spectroscopy can reveal fundamental mechanisms of quantum state transitions and coherence retention, an area ready for investigation and improvement. Furthermore, the examination of defects in materials, as emphasized by Cole et al. (2019), offers a distinctive viewpoint on the analysis of quantum states. In quantum circuits, two-level defects are very important because they limit performance and give us a chance to study their dynamics in great detail. Microwave techniques now allow researchers to look at individual defects in amorphous solids, which shows how their behaviors change when the field strength and temperature change. This duality shows that defects in superconducting circuits can be both a problem and a useful tool for studying basic quantum processes. The study of quantum states through microwave spectroscopy helps us understand these complicated interactions and also helps make quantum technologies stronger so they can work well even when materials are limited.

3.5.2. Investigating Phase Transitions

The study of phase transitions, especially through methods such as microwave spectroscopy, offers a significant approach for comprehending molecular behavior under diverse conditions. Phase transitions entail modifications in the physical state of a substance, which are intricately connected to variations in molecular interactions. Researchers can see these changes in great detail by using microwave spectroscopy. This lets them figure out how rotational and vibrational spectra change with temperature and pressure. This method is very sensitive to the structure of molecules, which makes it a great tool for studying chiral molecules. These molecules can show different spectral lines depending on how they are oriented and how they interact with other substances. As the molecular structure changes, spectroscopy gives us important information about how these changes happen, which helps us understand the thermodynamic principles that govern phase transitions better. The importance of microwave spectroscopy in examining phase transitions is further clarified through examples such as the analysis of methyltrioxorhenium (MTO). As a test molecule, MTO aids in the examination of the weak interaction effects that are essential for comprehending parity violation and molecular behavior during phase transitions. Researchers can gather a lot of information by using different spectroscopic methods, as explained by Shelkovnikov et al. (2010). This lets them assign many rotational and rovibrational lines that give them clues about the molecular structure of MTO. These thorough analyses show how different experimental setups can be used to study molecular properties in different states. This combination of methods not only proves that the method works, but it also sets the stage for future studies of the basic factors that affect phase transitions in more complicated chiral systems. Moreover, the continuous research facilitated by microwave spectroscopy has enhanced our understanding of intermolecular interactions during phase transitions. Research on molecular dimers, exemplified by the

H_3N–HOH system, has clarified the functions of tunneling and internal rotation in phase behavior. The findings by Blake et al. (1992) illustrate the complexity of rotational coupling and its impact on transition states. It also shows how small changes in the way molecules are arranged can change the overall phase behavior. This deeper understanding is important for being able to predict and control phase transitions in different materials. This is important in many fields, from materials science to biochemistry. Microwave spectroscopy is still making new discoveries, and it has the potential to help us understand the complexities of phase transitions. This will give us a strong foundation for future research and applications in advanced material design (Figure 3.12).

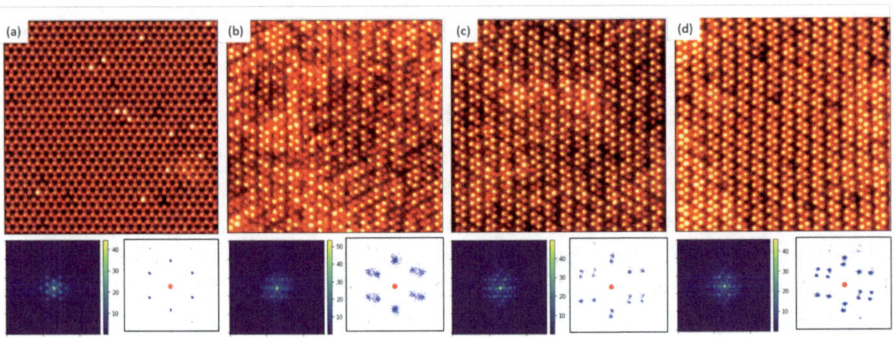

Figure 3.12. *Investigating phase transitions.*

Source: https://external-content.duckduckgo.com/iu/?u=https%3A%2F%2Fai2-s2-public.s3.amazonaws.com%2Ffigures%2F2017-08-08%2Fce3ee8a995a3499332c4f774b50e0d84160106c4%2F5-Figure1-1.png&f=1&nofb=1&ipt=f3a7309f4528d08a92c375a779baaf05fff1d41e5b5f08e2f183bab5c4d2282f.

3.5.3. Research in Plasma Physics

Research in plasma physics has increasingly emphasized the significance of microwave spectroscopy as a crucial instrument for analyzing plasma interactions and behaviors. Microwave spectroscopy enables the examination of molecular and atomic transitions in plasmas, yielding information about temperature, density, and composition. This method is especially useful for studies in space, where plasma phenomena can have a big effect on both cosmic radiation and planetary atmospheres. For example, it is important to understand the plasma that surrounds Earth in order to make models that can predict space weather events. This will help protect satellites and astronauts (Beckley *et al.*, 2025). Researchers are starting to connect their findings with technologies like microwave spectroscopy as they learn more about the complexities of plasma physics. This will help them improve their experimental and theoretical frameworks. The interaction between plasma physics and microwave spectroscopy has led to new discoveries in both fields, which have made it easier to study plasma environments. Improvements in the design and use of microwave spectroscopic methods have given scientists access to the dynamics of charged particles in plasma fields like never before. Researchers have started to see practical uses for microwave spectroscopy that can work in different plasma conditions by using hybrid superconducting circuits, for example. The creation of these hybrid systems not only improves the accuracy of measurements but also lets researchers make changes in real time to account for changes in plasma states. This is a big help for research in both labs and space (Akhmerov *et al.,* 2015). As the relationship between these two fields continues to grow, there is still a lot of room for new discoveries in plasma physics. Microwave spectroscopy is an important link between basic research and real-world uses in the larger field of plasma physics. It not only gives important information for theoretical models, but it can also be used in astrophysical studies where plasma is common. Researchers can study the composition and behavior of stellar formations and cosmic phenomena by using microwaves to look into the properties of astrophysical plasmas. Moreover, the results of this study may impact forthcoming technological initiatives, encompassing progress in fusion energy and space exploration missions (Beckley *et al.,* 2025). Researchers are combining microwave spectroscopy with plasma physics, which not only helps us understand the universe better but also opens up new possibilities for practical uses that could have a big impact on society.

3.6. CHALLENGES AND LIMITATIONS

Microwave spectroscopy is a very useful analytical tool, but it has some problems and limitations that could make it harder to use in other scientific fields. One major problem is that microwave spectroscopic methods are very sensitive to changes in the environment, such as temperature, pressure, and electromagnetic interference. Changes in these parameters can cause changes in spectral data, which makes it harder to analyze and understand. Also, the large amount of spectroscopic data can be too much for researchers to handle, so they need advanced computer tools to extract and understand it. Reliance on external factors and the difficulty of handling data can make the techniques less reliable in some situations. This shows how important it is to have strict standards and methodological improvements to improve overall performance and consistency in results. Microwave spectroscopy can also be limited by the technology that supports it, in addition to environmental sensitivities. Many modern microwave spectroscopic systems are costly and necessitate specialized apparatus, thereby restricting accessibility for researchers in smaller laboratories or emerging regions. The costs of running the business go beyond the initial investment. They also include maintenance and calibration, which require technical knowledge. This makes research capabilities unequal, which could limit the range of experimental studies. Also, the skill level required to use these systems properly can make it hard for new researchers to get started. To make microwave spectroscopy available to more people and get more scientists involved, it is important to deal with these financial and technical problems. Another major problem with microwave spectroscopy is that it can be hard to understand spectra because of the overlapping signals that can happen when there are a lot of different things in a mixture. Biological and chemical samples are often very complicated, which can make it hard to tell the difference between useful data and noise. Also, current microwave spectroscopic methods might not work well with samples that have very low concentrations because the signal-to-noise ratio can get too low. Integrated microwave and optical methods and other new technologies may help with these problems, but there is still a strong need for more research to improve spectral resolution and make data clearer. To get the most out of microwave spectroscopy for getting useful information from systems that are becoming more complicated, these problems need to be solved. This will make it more useful for scientific research (Chu *et al.,* 2018; Gurer-Orhan *et al.,* 2017).

3.6.1. Sensitivity and Resolution Issues

In microwave spectroscopy, sensitivity and resolution are two very important factors that have a big effect on how accurate and useful spectral measurements are. Sensitivity is the ability of a spectroscopic method to find weak signals in noise, while resolution is the ability of the instrument to tell the difference between closely spaced spectral features. These two things are linked because improving sensitivity can sometimes hurt resolution and vice versa. For example, new techniques like dual-comb interferometry have come about because of improvements in laser technologies, especially with frequency combs. This method has reached refresh rates that have never been seen before, which makes it possible to quickly collect data in a sequence. This directly improves both sensitivity and resolution in molecular spectroscopy. These kinds of changes can change the way we analyze things, making it possible to find spectral signatures that were previously too difficult to find. This opens up new possibilities for microwave spectroscopy in studying complex molecular systems. Furthermore, the significance of low-temperature detectors is accentuated when tackling sensitivity challenges in microwave spectroscopy. These detectors can pick up very weak microwave signals, which is useful for things like cosmic microwave background (CMB) and submillimeter astronomy. As was said in talks about improvements to detectors, sensitivity can be greatly improved by using better photon coupling and filtering architectures. These methods are directly related to how well you can get higher resolution in spectral measurements. Researchers can isolate and identify not only weak spectral lines but also their complex structures by making the detection process more sensitive. This could give them important information about molecular composition and dynamics. So, using the latest detector technologies is very important for making microwave spectroscopic methods work better overall. When looking at the subtle balance between sensitivity and resolution, it becomes clear that current research is pushing the limits of spectroscopic instrumentation. As shown by progress in electrooptic frequency comb technologies, the ability to measure complex spectra at refresh rates of 25 MHz is a big step forward in both sensitivity and resolution. This feature lets scientists quickly and reliably collects a lot of data, which makes it easier to study ultrabroadband waveforms that are useful for a variety of analytical uses. Moreover, as improvements in sensitivity yield more comprehensive spectral data, researchers must also concentrate on interpreting this information within a context that facilitates resolution. Connecting these parts together leads to a deeper understanding of how molecules interact and behave, making microwave spectroscopy an essential tool for modern scientific research (Duran *et al.,* 2015; Miller *et al.,* 2009) (Figure 3.13).

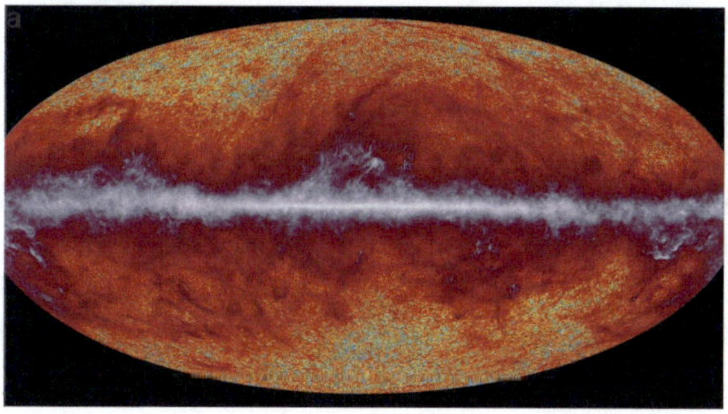

Figure 3.13. *Cosmic microwave background (CMB) radiation.*

Source: https://www.esa.int/var/esa/storage/images/esa_multimedia/videos/2018/07/the_legacy_of_planck/17602630-2-eng-GB/The_legacy_of_Planck_pillars.jpg.

3.6.2. Sample Size and State Constraints

In microwave spectroscopy, the size of the sample is very important for getting and understanding accurate data. The accuracy of spectroscopic measurements often depends on how big the sample is that is being looked at.

A bigger sample size can make the signal-to-noise ratio better, which makes the resulting spectra more reliable. However, this connection isn't straight; after a certain point, making the sample size bigger may not improve the quality of the data as much. Researchers must meticulously evaluate the trade-offs among sample size, measurement duration, and resolution, ensuring that their methodologies are congruent with the precise objectives of their spectroscopic analysis. The consequences go beyond just signal quality; a full understanding of sample size is necessary to confirm results and make sure that microwave spectroscopy studies can be repeated. State constraints are very important for determining whether microwave spectroscopic measurements are accurate. The physical conditions in which samples are analyzed, such as temperature, pressure, and electromagnetic field strength, cause these limits. Any of these things could change the molecular interactions being studied, which would affect the accuracy of the spectral data. A restrictive state environment may limit the availability of certain molecular states, thereby distorting results. The relationship between sample size and state constraints also means that data collection needs to be done in a careful way. For example, when only a few molecular states are available, increasing the sample size could lessen some of the negative effects of restrictive conditions. Consequently, researchers must maneuver through these limitations to obtain valuable spectral data while simultaneously upholding the principle of reproducibility in their results. The incorporation of rigorous experimental design, including sample size and the acknowledgment of state constraints, is essential for the progression of microwave spectroscopy research. Experimental studies aimed at elucidating intricate molecular interactions must emphasize the formulation of precisely defined parameters that encompass both dimensions. For instance, creating new methods that use large sample sizes and controlled state conditions can help us learn more about how dynamic molecular systems work. Researchers can now get around problems that were caused by state constraints thanks to new technologies like better detection methods and automated sample handling. In the end, a synergistic approach to dealing with these factors can greatly improve the quality and relevance of microwave spectroscopy findings, which will lead to more new ideas in both academic and practical research settings.

3.6.3. Interpretation of Spectral Data

Understanding the physical and chemical properties of different substances depends on how microwave spectroscopy interprets spectral data. By finding the unique signatures that molecules give off or take in at certain microwave frequencies, scientists can learn about molecular structures, dynamics, and interactions. Microwave radiation is especially good for probing rotational transitions in gaseous samples because its wavelengths are longer than those of other types of electromagnetic radiation. This helps researchers learn more about how molecules behave in different thermal conditions. Correctly interpreting this data can give us important information about how stable and reactive complex compounds are. This is important for a wide range of fields, from atmospheric science to astrophysics. Nevertheless, interpreting spectral data presents challenges, especially when external factors affect the outcomes. For instance, diffusion processes in a multi-component system can greatly change spectral profiles and cause wrong conclusions about how many molecules are present. Studies show that systematic errors, like those caused by wrong assumptions about the amount of helium, can lead to angular distance measurements from microwave observations of cosmic structures like galaxy clusters being off by as much as 25%. These errors become more pronounced at reduced intracluster medium (ICM) temperatures, emphasizing the necessity for precise models and data interpretation methodologies to accurately determine astrophysical parameters (Gilfanov *et al.,* 2014). These kinds of complexities make it necessary to keep improving both theoretical frameworks and experimental methods in order to make spectral data interpretation more reliable. Collaborative research efforts and funding initiatives, like those supported by NASA, also help make spectral data interpretation better. NASA-funded programs seek to improve methods in observational systems, such as microwave spectroscopy, which encourages new ways to collect and analyze data. The interaction between cutting-edge technology and thorough scientific research could help us improve the factors that affect spectral analyses. By using interdisciplinary methods, researchers can make sense of the complexities involved in interpreting spectral data, which will help us understand a wide range of phenomena in many different scientific fields (ANONYMOUS, 2025). These advancements not only offer improved understanding of current frameworks but also facilitate innovative applications that utilize the complex information contained within spectral data.

SUMMARY

When we look into microwave spectroscopy, we need to remember how much this method has changed our understanding of how molecules behave and interact. The capacity to examine molecular structures and dynamics with exceptional accuracy has facilitated novel insights across multiple fields, including chemistry, biology, and materials science. Microwave spectroscopy is an important tool for figuring out the unique properties of complex molecular species because it can show how molecules rotate and vibrate. The addition of advanced spectroscopic techniques has also made it more powerful as a tool, making it possible to study difficult systems that were previously out of reach. This nuanced understanding enables both scholarly investigation and practical applications across various domains, underscoring the essential role of microwave spectroscopy in contemporary science. Furthermore, the possibility of parity violation research, especially in chiral molecules, underscores the unexploited capabilities of microwave spectroscopy in tackling essential scientific inquiries. The investigation of chiral oxorhenium complexes, as indicated by recent research, demonstrates the prospective role of microwave spectroscopy in facilitating critical experimental validations concerning biohomochirality. The foundation established by extensive spectroscopic investigations, encompassing meticulous analysis of entities such as methyltrioxorhenium, illustrates the methodologies' adaptability and efficacy in revealing nuanced variations in molecular behavior. As scientists keep making enantiopure forms and improving experimental methods, microwave spectroscopy is going to be very important in answering long-standing scientific questions. In the end, this will help us understand molecular asymmetry and its effects better. Furthermore, the convergence of microwave spectroscopy with other sophisticated techniques, such as imaging spectroscopy, highlights its ability to amalgamate multidimensional data for a comprehensive understanding of molecular phenomena. Microwave spectroscopy clarifies the complex connections between energetic processes and molecular interactions, enhancing both fundamental science and practical applications, including environmental monitoring and pharmaceutical development. Likewise, research on high-energy electron behavior during solar flares demonstrates the adaptability of spectroscopic methods, providing essential understanding of electron dynamics and transport mechanisms. These examples all point to the fact that microwave spectroscopy will continue to have a big effect on scientific research, encouraging new ideas and driving progress in technology and applied sciences. This is an exciting area for future research.

Microwave spectroscopy is a major step forward in analytical methods. It can be used in many fields, such as chemistry and materials science. The main benefit of this method is that it gives researchers detailed information about molecular rotational transitions, which lets them study molecular structures and dynamics with great accuracy. Microwave spectroscopy lets you see very small changes in energy levels that happen when molecules interact with each other. This is especially important for looking at complicated mixtures, where traditional spectroscopic methods might not work. Moreover, the integration of statistical calibration techniques improves the dependability of microwave spectroscopic data, mitigating challenges related to nonlinear correlations in instrument responses, as evidenced by current investigations in analytical chemistry and radiometry (Boone *et al.,* 2014). The advancement of enantiomer-specific microwave spectroscopy underscores its importance in chiral molecule analysis, which is essential for pharmaceutical applications. These improvements make it easier to find and separate chiral compounds and also make it easier to measure enantiomeric excess accurately. Recent advancements have utilized a reaction chamber featuring a supersonic free jet source to enhance experimental conditions, thereby increasing the technique's efficiency. We used this setup to study 1-indanol using resonance-enhanced multiphoton ionization spectroscopy. This gave us information about the different conformers of this important chiral molecule. The findings from these studies greatly enhance our comprehension of chiral discrimination and molecular dynamics in microwave spectroscopy (Langenhan *et al.,* 2019). Furthermore, the continuous advancements in experimental configurations, exemplified by the novel slit valve design employing piezoelectric stack actuators, are augmenting the viability of high-resolution microwave spectroscopy experiments. The slit valve design not only improves the properties of the molecular beams, but it also makes it easier to control operational parameters, which improves the signal-to-noise ratio and spectral resolution. These kinds of new ideas will make it possible to do more detailed spectroscopic studies, especially with chiral molecules, because they let scientists change the conditions to get the best molecular interactions. As data continues to be collected from these advanced setups, the ramifications for microwave spectroscopy will likely surpass fundamental comprehension, resulting in advancements in drug development and materials science (Langenhan *et al.,* 2019).

REFERENCES

1. Akhmerov, A. R., Bakkers, E. P. A. M., Bruno, A., de Lange, G., et al. (2015). Realization of microwave quantum circuits using hybrid superconducting-semiconducting nanowire Josephson elements. https://core.ac.uk/download/388667344.pdf (accessed on 29th October 2025).
2. Alexander, D., Alexander, D., Alexander, D., Alexander, D., Alissandrakis, C., Arnoldy, R., Aschwanden, M., et al. (2013). Implications for electron acceleration and transport from non-thermal electron rates at looptop and foot point sources in solar flares. https://core.ac.uk/download/9652084.pdf (accessed on 29th October 2025).
3. Antonsen, I., Bell, S. A., Benyon, R., Boese, F., et al. (2013). A new challenge for meteorological measurements: The meteoMet project-Metrology for meteorology. https://core.ac.uk/download/pdf/70604032.pdf (accessed on 29th October 2025).
4. Astier, P., & Pain, R. (2012). Observational evidence of the accelerated expansion of the universe. https://core.ac.uk/download/580036003.pdf (accessed on 29th October 2025).
5. Astorino, M. D., Fastampa, R., Frezza, F., Maiolo, L., et al. (2018). Polarization-maintaining reflection-mode THz time-domain spectroscopy of a polyimide based ultra-thin narrow-band metamaterial absorber. https://core.ac.uk/download/153366124.pdf (accessed on 29th October 2025).
6. Bacsa, B., Baird, E. E., Bando, T., Baptiste, B., Chenoweth, D. M., Chenoweth, D. M., Chenoweth, D. M., et al. (2012). Microwave assisted synthesis of Py-Im polyamides. https://core.ac.uk/download/7432535.pdf (accessed on 29th October 2025).
7. Beckley, L. E. (2025). Research in Space Science and Technology. https://core.ac.uk/download/pdf/42877671.pdf (accessed on 29th October 2025).
8. Bellatreccia, F., Della Ventura, G., & Marcelli, A. (2013). SR-FTiR microscopy and FTIR imaging in the earth sciences. http://arxiv.org/abs/1312.5750 (accessed on 29th October 2025).
9. Bieber, A., Bücker, D., & Drescher, M. (2018). Light-induced dipolar spectroscopy - A quantitative comparison between LiDEER and LaserIMD. http://arxiv.org/abs/1806.04190 (accessed on 29th October 2025).
10. Blake, G. A., Bumgarner, R. E., Stockman, P. A., Suzuki, H., et al. (1992). Microwave and tunable far-infrared laser spectroscopy of the ammonia–water dimer. https://core.ac.uk/download/4870835.pdf (accessed on 29th October 2025).

11. Boone, E. L., & Rivers, D. L. (2014). Dynamic Bayesian nonlinear calibration. http://arxiv.org/abs/1411.3637 (accessed on 29th October 2025).

12. Bruhat, L. E., Cottet, A., Datta, S., Delbecq, M. R., et al. (2012). Photon mediated interaction between distant quantum dot circuits. http://arxiv.org/abs/1303.2639 (accessed on 29th October 2025).

13. Cadorim, H. R., Campo Menoyo, J. D., Carvalho, G. S., Duarte, F. A., et al. (2017). Development of a fast-screening method for the direct determination of chlorinated persistent organic pollutants in fish oil by high-resolution continuum source graphite furnace molecular absorption spectrometry. https://core.ac.uk/download/153371322.pdf (accessed on 29th October 2025).

14. Carbone, E., D'Isa, F., Fantz, U., Hecimovic, A., et al. (2020). Analysis of the C_2 ($d^3\Pi_g$–$a^3\Pi_u$) Swan bands as a thermometric probe in CO_2 microwave plasmas. https://core.ac.uk/download/323182989.pdf (accessed on 29th October 2025).

15. Cerny, P., Kabourek, V., Kania, P., Piksa, P., et al. (2013). Prague's Emission Fourier Transform Microwave Spectrometer–Design and preliminary results. https://core.ac.uk/download/30312093.pdf (accessed on 29th October 2025).

16. Cheinet, P., Gallagher, T. F., Lehec, H., Luc-Koenig, E., et al. (2018). Laser and microwave spectroscopy of even-parity Rydberg states of neutral ytterbium and Multichannel Quantum Defect Theory analysis. http://arxiv.org/abs/1805.02712 (accessed on 29th October 2025).

17. Cheong, L. Y., Jilani, M. T., Rehman, M. Z. U., Wen, H., et al. (2014). Equivalent circuit modeling of the dielectric loaded microwave biosensor. https://core.ac.uk/download/30312218.pdf (accessed on 29th October 2025).

18. Chu, S. T., Little, B. E., Mitchell, A., Morandotti, R., et al. (2018). Harnessing optical micro-combs for microwave photonics. http://arxiv.org/abs/1710.08611 (accessed on 29th October 2025).

19. Cole, J. H., Lisenfeld, J., & Müller, C. (2019). Towards understanding two-level-systems in amorphous solids–Insights from quantum circuits. http://arxiv.org/abs/1705.01108 (accessed on 29th October 2025).

20. Coy, S., Flannery, C., Foy, B., Hetzler, C., et al. (1990). Determination of molecular spectroscopic parameters and energy-transfer rates by double-resonance spectroscopy. https://core.ac.uk/download/pdf/10419419.pdf (accessed on 29th October 2025).

21. Duran, V., Tainta, S., & Torres-Company, V. (2015). Ultrafast electrooptic dual-comb interferometry. https://core.ac.uk/download/334420079.pdf (accessed on 29th October 2025).

22. Efstathiou, G., Ellis, J. R., Leibundgut, B., Lilly, S., et al. (2006). Report by the ESA-ESO Working Group on Fundamental Cosmology. https://core.ac.uk/download/44148497.pdf (accessed on 29th October 2025).

23. Enss, C., Fleischmann, A., Gastaldo, L., Kempf, S., et al. (2013). Multiplexed readout of MMC detector arrays using non-hysteretic rf-SQUIDs. http://arxiv.org/abs/1309.4929 (accessed on 29th October 2025).

24. Fanale, F. P., Kaula, W. M., Mccord, T. B., Trombka, J., et al. (2025). Report of the Terrestrial Bodies Science Working Group. Volume 9: Complementary Research and Development. https://core.ac.uk/download/pdf/42877057.pdf (accessed on 29th October 2025).

25. Fleischer, S., Hwang, H. Y., Lu, J., Nelson, K., et al. (2016). Nonlinear two-dimensional terahertz photon echo and rotational spectroscopy in the gas phase. https://core.ac.uk/download/83234499.pdf (accessed on 29th October 2025).

26. Fragner, A., Izmalkov, A., Izmalkov, A., Lupascu, A., Wallraff, A., Berkley, A. J., Kerman, A. J., et al. (2009). Large-amplitude driving of a superconducting artificial atom: Interferometry, cooling, and amplitude spectroscopy. https://core.ac.uk/download/36210949.pdf (accessed on 29th October 2025).

27. Gilfanov, M., Medvedev, P., Sazonov, S., Shtykovskiy, P., et al. (2014). Impact of thermal diffusion and other abundance anomalies on cosmological uses of galaxy clusters. http://arxiv.org/abs/1312.4170 (accessed on 29th October 2025).

28. Gottlieb, C. A., Lee, K. L. K., Martin-Drumel, M. A., McCarthy, M. C., et al. (2018). Vibrational satellites of C_2S, C_3S, and C_4S: Microwave spectral taxonomy as a stepping stone to the millimeter-wave band. http://arxiv.org/abs/1804.04126 (accessed on 29th October 2025).

29. Gurer-Orhan, H., Saso, L., & Suzen, S. (2017). Detection of reactive oxygen and nitrogen species by electron paramagnetic resonance (EPR) technique. https://core.ac.uk/download/154947312.pdf (accessed on 29th October 2025).

30. Karki, K., Marcus, A. H., Pullerits, T., Torbjornsson, M., et al. (2013). Digital cavities and their potential applications. http://arxiv.org/abs/1302.6044 (accessed on 29th October 2025).

31. Langenhan, J. (2019). Experimental Characterization of a New Piezoelectric Slit Valve and REMPI Spectroscopy of Cold, Chiral Indanol Molecules.
32. Lieberman, A., & Schipma, P. (2025). Air pollution monitoring instrumentation: A survey. https://core.ac.uk/download/pdf/85241513.pdf (accessed on 29th October 2025).
33. McDonnell, J. J., Orlowski, N., & Pratt, D. L. (2016). Intercomparison of soil pore water extraction methods for stable isotope analysis. https://core.ac.uk/download/131024311.pdf (accessed on 29th October 2025).
34. Miller, A., Young, B., Cabrera, B., & Golwala, S. (2009). Future developments in low temperature detectors for CMB and Submm astronomy. https://core.ac.uk/download/4886227.pdf (accessed on 29th October 2025).
35. Riehle, F. (2015). Towards a re-definition of the second based on optical atomic clocks. https://core.ac.uk/download/580036336.pdf (accessed on 29th October 2025).
36. Shelkovnikov, A., Amy-Klein, A., Amy-Klein, A., Arimondo, E., Asselin, P., Asselin, P., Auerbach, N., et al. (2010). High resolution spectroscopy of methyltrioxorhenium: Towards the observation of parity violation in chiral molecules. https://core.ac.uk/download/52319891.pdf (accessed on 29th October 2025).
37. Siegel, P. H. (2007). THz instruments for space. https://core.ac.uk/download/4876918.pdf (accessed on 29th October 2025).
38. Solano, E., Romero, G., García-Ripoll, J. J., Mandel, L., & Haroche, S. (2009). Microwave photon detector in circuit QED. https://core.ac.uk/download/36025507.pdf (accessed on 29th October 2025).
39. Zmuidzinas, J. (2007). Technology for submillimeter astronomy. https://core.ac.uk/download/4884493.pdf (accessed on 29th October 2025).

CHAPTER 4

ATOMIC SPECTROSCOPY

Contents

INTRODUCTION

Atomic spectroscopy is a very important analytical tool that helps us understand how electromagnetic radiation and matter interact. This method is based on how atoms absorb, emit, or scatter light. It is very useful for both qualitative and quantitative analysis of different substances. Scientists can learn important things about elemental composition, concentration, and even molecular structure by looking at the wavelengths that are emitted or absorbed during these interactions.

Atomic spectroscopy is a key tool in many fields, from environmental science to materials characterization, because it is so accurate and sensitive. It is an essential tool for both researchers and industry professionals. An investigation into atomic spectroscopy uncovers a complex array of methodologies, each designed to meet particular analytical requirements. Flame atomic absorption inductively coupled plasma optical emission spectroscopy, and laser-induced breakdown spectroscopy are some of the best techniques because they are useful and accurate. Each method has its own pros and cons, which are mostly determined by the physical and chemical properties of the substances being studied. For example, the choice of technique may depend on the sample matrix, the desired detection limits, and the elemental range.

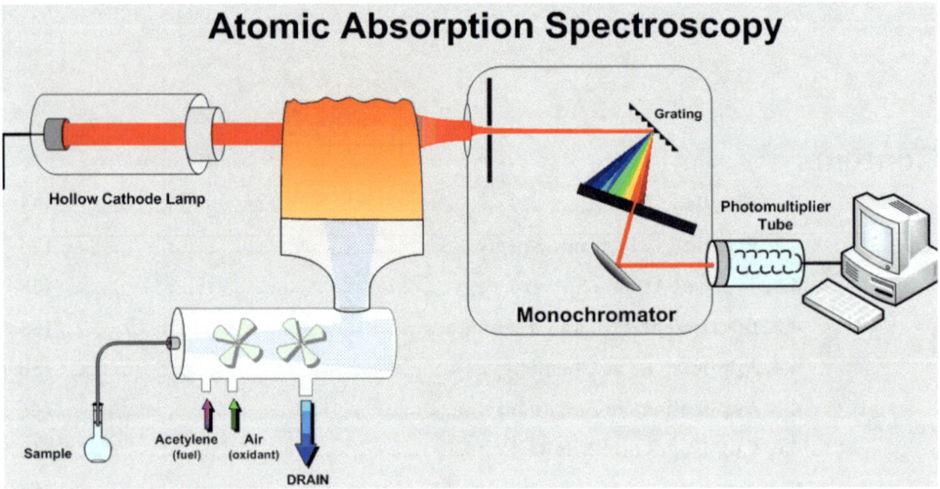

Figure 4.1. *Atomic spectroscopy.*

Source: https://3.bp.blogspot.com/-L7sl9qCskQc/U90FCeYAekI/AAAAAAAAATs/Nn9XU6Kn3Gk/s1600/2014-08-02_210120.png.

This shows how atomic spectroscopy can be used in many different ways. In general, learning about these methods gives researchers more options for how to set up their experiments and make sure that their analytical goals are in line with their needs. Atomic spectroscopy has effects that go beyond just academic research. It affects many real-world uses in fields like healthcare, environmental monitoring, and materials science. The capacity to identify and measure trace elements in intricate matrices is crucial for evaluating pollutant concentrations, guaranteeing food safety, and confirming pharmaceutical purity. This wide range of uses is becoming more important in today's world, where new technologies and regulations require more advanced analytical methods. Additionally, as new applications emerge in fields like nanotechnology and plasma-liquid interactions, researchers need to keep up with how atomic spectroscopy is changing. The continuous evolution and enhancement of methodologies underscore the dynamic character of this discipline, guaranteeing its position at the forefront of scientific research and application, thereby facilitating progress that tackles modern challenges across diverse sectors (Anderson *et al.,* 2014; Geyter *et al.,* 2019) (Figure 4.1).

Definition of Atomic Spectroscopy

Atomic spectroscopy is a strong way to analyze things that uses the interaction of electromagnetic radiation with matter to figure out what elements make up a substance. Atomic spectroscopy is based on the idea that atoms absorb and give off light at certain wavelengths when they get energy. Scientists can use this phenomenon to find and measure the amounts of elements in a sample by looking at the specific wavelengths of light that are absorbed or emitted. This method is used in many different fields, such as environmental science, chemistry, and materials science, to learn more about the elemental makeup of complicated samples. Atomic spectroscopy creates spectral lines for different energy levels in atoms. This gives it a unique fingerprint that can be used to identify elements with great accuracy. This makes it different from other spectroscopic methods that may not give the same level of detail in elemental analysis. Laser-Induced Breakdown Spectroscopy (LIBS) is one of the main methods used in atomic spectroscopy. It is unique because it can analyze multi-element compositions in real time with little sample preparation. In this method, a laser pulse is focused on a sample, which makes a plasma that is very hot and then gives off light that is unique to the elements in the sample. LIBS has some problems, though, like baseline interference from background radiation, even though it has some benefits, like being able to analyze things quickly and not needing reagents. This interference can make it hard to measure elements and needs careful correction methods to fix. Recent research has investigated novel methodologies for baseline correction, resulting in substantial enhancements in the precision of measurements acquired via LIBS. The ability to fine-tune the analysis enhances the reliability of results and solidifies LIBSs role in contemporary atomic spectroscopy applications (Boas *et al.,* 2018). Atomic spectroscopy also includes other methods, such as Atomic Absorption Spectroscopy (AAS) and Inductively Coupled Plasma Optical Emission Spectroscopy (ICP-OES). These methods give different but useful information about elemental analysis. AAS measures how much light free atoms in a gas state absorb. This lets you find trace metals and other elements in a wide range of samples. ICP-OES, on the other hand, uses an inductively coupled plasma to make atoms and ions excited, which then give off light at specific wavelengths. This method is especially useful for finding several elements at once, which is why it is often used in environmental monitoring and industrial settings. The interaction of different atomic spectroscopy techniques, such as new ways to analyze samples like baseline correction in LIBS, shows how dynamic the field is and how it keeps adding to our understanding of science (Lennard *et al.,* 2017).

4.1. PRINCIPLES OF ATOMIC SPECTROSCOPY

The basic ideas behind atomic spectroscopy are based on how electromagnetic radiation interacts with the electronic states of atoms. This method uses the specific wavelengths of light that atoms absorb or give off to analyze different elements in both qualitative and quantitative ways. When an atom goes from a higher energy state to a lower one, it gives off a photon with a wavelength that is unique to that element. On the other hand, absorption happens when light energy matches the energy difference between quantum states, which makes electrons move to higher energy levels. This duality is what makes emission and absorption spectroscopy possible. These two types of spectroscopies are very important for many uses, from monitoring the environment to diagnosing patients. Atomic spectroscopy is a powerful tool for figuring out what simple and complex samples are made of because it can accurately measure these interactions. Instrumentation is very important for atomic spectroscopy because it is based on the basic ideas of atomic transitions. The light source, the system for introducing samples, and the detection unit are all important parts. For example, in flame atomic absorption spectroscopy, a flame acts as the atomizer, turning samples into free atoms that can be analyzed by vaporizing and ionizing them. After that, the light source sends out certain wavelengths that are linked to the elements being studied, and the detectors measure how strong the light that is sent or received is. There are different types of spectrometers, like grating and prism spectrometers, which can break up light into its different wavelengths. This lets you accurately identify and measure elements. To make measurements more reliable in fields like chemistry and materials science, you need to understand the subtleties of these parts. This will improve analytical sensitivity and specificity. The incorporation of contemporary technologies has enhanced the range and accuracy of atomic spectroscopy. Inductively coupled plasma optical emission spectroscopy (ICP-OES) and laser-induced breakdown spectroscopy (LIBS) are two examples of how this type of analysis has changed over time. ICP-OES uses a high-temperature plasma to break down samples into atoms. This makes it possible to detect lower levels of elements and analyze multiple elements at once. LIBS, on the other hand, uses focused laser pulses to make plasma, which lets you quickly analyze solid samples in different settings.

These improvements show how atomic spectroscopy is becoming more and more important for analyzing complicated materials and doing forensic work. Interdisciplinary collaboration has also brought new ideas and methods to the field, which has helped scientists in all areas understand and protect the environment better (Baianu, 2011; Lennard *et al.*, 2017). So, the principles of atomic spectroscopy not only show how scientifically strong it is, but they also show how it is constantly changing to meet new analytical needs (Figure 4.2).

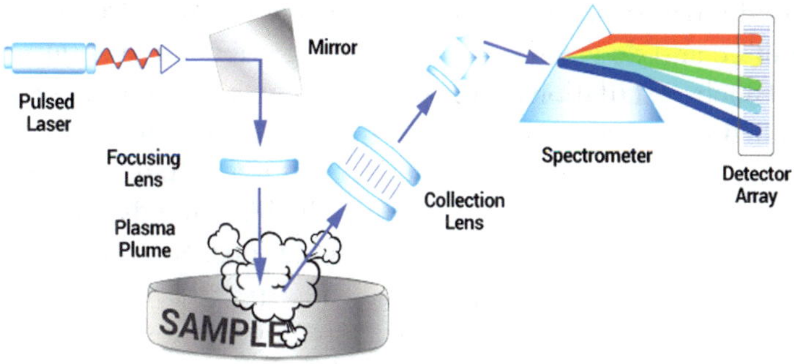

Figure 4.2. *Laser-induced breakdown spectroscopy (LIBS).*

Source: https://cdn.gophotonics.com/community/1_638223181524693414.jpg.

4.1.1. Interaction of Light with Matter

Quantum mechanics is what really controls how light interacts with matter. It tells us how photons, or light particles, interact with atoms and molecules. This interaction can cause different things to happen, such as light being absorbed, emitted, or scattered. When an atom absorbs light, its electrons may move to higher energy states. When an atom emits light, its electrons move back to lower energy states, releasing energy as photons. This process is very important in atomic spectroscopy because it lets you identify atomic and molecular structures by their unique spectral lines. Every element has its own energy levels, which create a unique atomic spectrum that can be used to identify and analyze the matter in question. Recent technological advancements have enhanced our comprehension of light-matter interactions, especially through the innovation of laser-cooled atoms integrated with nanophotonic structures. These systems provide an exceptional framework for investigating innovative modalities of light-matter interaction. Researchers can change the internal quantum states and center of mass motion of atoms to levels never seen before by putting them in controlled environments. Degenerate Raman cooling has become an important way to control the movement of atoms. Atoms are concentrated near the surface of optical fibers, which makes their interactions with guided light fields very precise. These techniques improve the accuracy of atomic spectroscopy and make it possible to study new phenomena like light-induced self-organization and heat transfer on a nano scale. This connects quantum physics to real-world uses in nanotechnology. The investigation of quantum dynamics in light-matter interactions has led to the development of novel concepts, including THz spectroscopy, which broadens the application of atomic spectroscopy into uncharted territories. In systems with strong or ultra-strong coupling, the interaction of ac and dc electric fields has a big effect on quantum transitions. Rabi and Bloch oscillations are two examples of mutually interacting oscillations that can create complex spectra with many narrow high-amplitude resonant lines. These spectral characteristics are essential for applications in nanoelectronics and nanophotonics, as they elucidate the dynamic behavior of quantum systems subjected to external perturbations. So, the many ways that light interacts with matter not only make atomic spectroscopy more interesting, but they also push technology forward by using quantum principles to improve how materials are handled and characterized.

4.1.2. Energy Levels and Electron Transitions

The quantization of energy levels in atoms is essential for comprehending atomic spectroscopy and the dynamics of electrons within an atom. The energy level of each electron tells it where it is in relation to the nucleus. When an electron moves between these energy levels, it can either take in or give off energy as photons. The energy difference between these levels is equal to certain wavelengths of light, which makes the spectral lines look different. The fact that energy levels are separate is very important for figuring out what elements are based on their unique spectral signatures. So, atomic spectroscopy is a very useful tool for both studying atomic structure and analyzing elements. There are many different types of energy transitions, and environmental factors can make these transitions less stable. For example, when excited atomic states interact with nearby atoms, they can change the pressure of the spectral lines that are seen. A thorough analysis of hydrogen's excited states shows that pressure has a big effect on transitions between higher excited states, especially in the 4P manifold. It is important to include hyperfine structure in this case; ignoring these interactions can cause mistakes in high-precision measurements. The advanced modeling of these interactions via quantum mechanics is crucial for precise spectroscopic analysis and offers insights into atomic and molecular physics, as evidenced by recent advancements in the field (Adhikari *et al.,* 2019). Atomic spectroscopy reveals the interactions between atomic and nuclear properties, as well as transitions between quantized energy levels. Atomic electrons are not only affected by the nucleus's electromagnetic forces, but they are also affected by its other properties, such as its spin and charge distribution. There is an interesting connection between atomic transition energies and nuclear properties. This connection lets researchers get information that would be hard to get through nuclear physics alone. High-precision atomic spectroscopy techniques can elucidate subtleties in nuclear structure and dynamics through phenomena such as hyperfine splitting and isotope shifts. The theoretical and experimental insights acquired in this domain persist in enhancing our comprehension of atomic and nuclear interactions, rendering this field a crucial area of study in contemporary physics (Adriana *et al.,* 2010) (Figure 4.3).

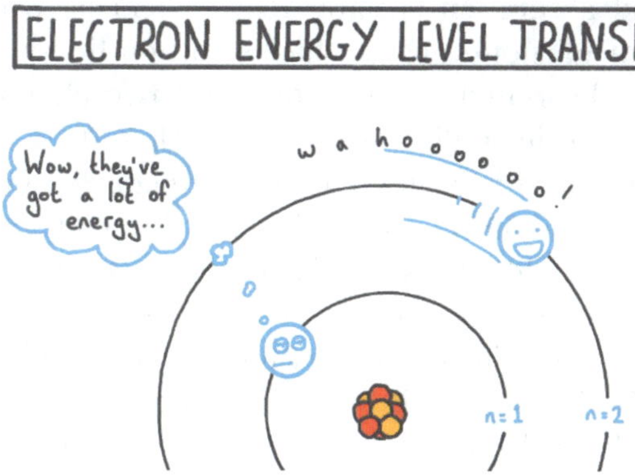

Figure 4.3. *Electron energy level transitions.*

Source: https://media.nagwa.com/269128640393/en/thumbnail_l.jpeg.

4.1.3. Emission and Absorption Spectra

The phenomena of emission and absorption spectra are essential for comprehending atomic structure and behavior within the context of atomic spectroscopy. Atoms absorb energy, which causes electrons to move from lower to higher energy levels. This creates unique absorption spectra. The unique electronic configuration of each element causes it to absorb certain wavelengths of light. This makes dark lines in the spectrum where light has been taken out of the continuum. This interaction shows that energy levels in atoms are quantized and gives us important information about the chemical makeup and physical properties of substances. Not only does a close look at these patterns help us figure out what elements are, but it also helps us learn more about how atoms interact with each other in different states of matter. Absorption spectra show the wavelengths that an atom absorbs, while emission spectra show the light that comes out when electrons go back to lower energy states. This change gives off energy in the form of photons, which makes bright lines on a dark background in the spectrum. By observing the emitted light, scientists can determine the energy levels within an atom and understand atomic structure more deeply. It is crucial to recognize that the attributes of both spectra are fundamentally interconnected by the principles of quantum mechanics, especially the selection rules that dictate transitions between states. For example, research on rubidium atoms in dense argon demonstrated the applicability of the Kennard-Stepanov relation, thereby validating the correlation between absorption and emission spectra in collisionally broadened atomic systems (Klaers *et al.,* 2014). Recent improvements in spectroscopy methods have also changed the way we get and study emission and absorption spectra. Researchers have been able to get Doppler-free spectra of atomic gases thanks to non-linear spectroscopy methods like saturated absorption spectroscopy and two-photon absorption spectroscopy. This is shown in studies of rubidium. These new ideas make it easier to measure atomic transitions and hyperfine structures more accurately. This is helpful for figuring out how atoms behave in different situations, such as when there are magnetic fields. These kinds of progress make learning about quantum mechanics and atomic spectra more interesting by showing how well theoretical predictions and experimental observations work together (Allafort *et al.,* 2009). Consequently, the ongoing investigation of emission and absorption spectra is essential for both theoretical inquiry and practical applications in atomic spectroscopy.

4.2. TYPES OF ATOMIC SPECTROSCOPY

There are many different methods used in atomic spectroscopy, and each has its own strengths and uses. One of these is atomic absorption spectroscopy (AAS), which measures how much light is absorbed by a vaporized sample to figure out how much of a certain element is present. In AAS, a light source sends out wavelengths that are unique to certain elements. This lets analysts figure out the concentrations by looking at how much light is absorbed. This method works especially well for finding trace elements in complicated mixtures, which is why it is so popular in food and environmental testing. Its main benefit is its sensitivity, which lets it look at lower concentrations than many other methods. In some cases, though, AAS can only look at a certain number of elements at once, which can be a problem in some types of comprehensive analysis. Atomic emission spectroscopy (AES) is a complementary method to atomic spectroscopy that looks at the light that excited atoms give off at certain wavelengths. This method can accurately measure a wide range of elements in a sample, usually with the help of inductively coupled plasma (ICP). The ICP-AES variation uses high temperatures to make a stable plasma that ionizes the atoms in the sample. The emissions create clear spectral lines that can be measured to find out how much of each element is present. This flexibility makes ICP-AES good for analyzing multiple elements at once, which is useful for samples that are complicated, like biological or geological materials. In fields like materials science and environmental monitoring, where precise compositional analysis is important for understanding basic processes and interactions, it is important to be able to measure elements over a wide range. Atomic fluorescence spectroscopy (AFS) is another important advancement in atomic spectroscopy. It uses the principle of fluorescence to find out how much of each element is in a sample. In AFS, a light source first excites atoms, and then they give off light as they go back to lower energy states. This phenomenon creates a very sensitive way to find certain elements. AFS is especially good for looking at trace metals because it is much more sensitive and specific than many other methods. Its creation has helped make progress in chemical analysis and environmental monitoring. As the technique progresses, its integration with other spectroscopic methods, such as Raman spectroscopy, may augment its analytical capabilities, facilitating a more thorough comprehension of molecular chemistry and environmental contaminants, as demonstrated by the implementation of collective molecular structure fingerprinting approaches in recent studies (Bryndzia *et al.,* 2016) (Figure 4.4).

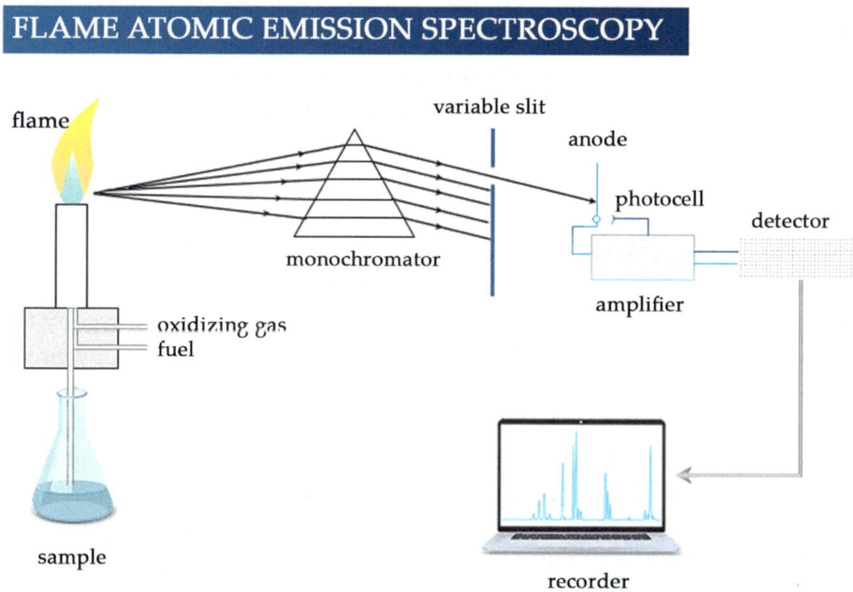

Figure 4.4. *Atomic emission spectroscopy (AES).*

Source: https://www.priyamstudycentre.com/wp-content/uploads/2021/11/flame-atomic-emission-spectroscopy.png.

4.2.1. Atomic Absorption Spectroscopy (AAS)

Atomic absorption spectroscopy (AAS) is an important part of analytical chemistry because it is a key method for measuring trace elements in different types of samples. The methodology is based on the idea that atoms of a certain element in a gas state can absorb light at certain wavelengths that come from that element. AAS is especially useful because it is very sensitive and specific, which makes it great for finding small amounts of metal ions in food, biological fluids, and environmental samples. For example, it has become an important tool for keeping an eye on heavy metal contamination, which is important for figuring out how safe the environment and people's health are. In places where industrialization has had an effect, the need for reliable detection methods has grown. This is because heavy metal effluents need to be carefully analyzed to avoid harming the environment and protecting public health. AAS is useful in more than just its usual applications; it can be used in many fields, such as food science, environmental monitoring, and clinical diagnostics. In food science, AAS is used to find trace metals that could be bad for people's health if they are present in large amounts.

In the same way, the method is used in environmental monitoring to check the levels of pollutants in soil and water samples. Using new sample preparation methods makes the method more useful because it guarantees accurate results while reducing the risk of contamination. For instance, improvements in sample digestion methods and the addition of automated systems have made the analysis process easier, which has increased throughput and lowered the chance of human error. The ongoing improvement of AAS also makes it possible to find lower detection limits, which is in line with stricter rules about metal levels in different products and places. AAS is still widely used, but new analytical methods are giving it a run for its money, which has led to ongoing debates about its future usefulness. Inductively Coupled Plasma Optical Emission Spectroscopy (ICP-OES) and Fourier Transform Near-Infrared (FT-NIR) spectroscopy are two techniques that have been suggested as possible replacements. They have some advantages, such as being able to process more samples at once and using fewer reagents. For example, FT-NIR has been promoted as a cleaner option to AAS for predicting calcium content in wine, as it can analyze several factors at once (Barroso *et al.,* 2018). This brings up important questions about when AAS is still the best choice, especially for certain niche uses where it is still the best at detecting certain metals. Ultimately, the ongoing evolution and adaptation of AAS are crucial to preserving its status as a fundamental component of atomic spectroscopy, especially in light of advancing technology and emerging analytical challenges in environmental safety and health sciences (Nanjunda *et al.,* 2005).

4.2.2. Atomic Emission Spectroscopy (AES)

Atomic emission spectroscopy (AES) is an important analytical method in atomic spectroscopy as a whole. It lets you look at the elemental composition of different samples in both qualitative and quantitative ways. The basic idea behind AES is that atoms are excited, which means they take in energy and then give off specific wavelengths of light as they go back to their ground state. This emission spectrum is like a fingerprint for each element, which helps scientists tell the difference between different elements and ions. AES's high sensitivity and specificity are important in many fields, from environmental monitoring to biochemistry, where accurate elemental analysis is essential. Comprehending the mechanisms and applications of AES furnishes essential knowledge for the progression of both scholarly and practical endeavors in chemistry and associated fields. The operational efficiency of AES is influenced by several factors, including the type of excitation source and the geometrical configuration of the spectrometer.

Flame, plasma, and electric discharge are all common sources of excitation. Each has its own benefits, depending on what the analysis needs. Inductively coupled plasma (ICP) is a popular choice for AES. It has higher temperatures and better ionization capabilities, which makes it more sensitive and able to analyze trace elements. Also, improvements in detector technology and data processing algorithms have made it easier to improve spectral resolution and analysis speed. As practitioners investigate the diverse options for implementing AES, it is essential to consider not only the instrumentation but also the expected sample matrix and elements of interest, thereby ensuring optimal analytical performance outcomes. AES is more than just a way to find things; it can also tell you about the physical and chemical properties of materials. AES can be used to study things like energy transitions and how external magnetic fields affect how atoms behave. Studying atomic transitions helps us understand some of the basic ideas of quantum mechanics, such as the rules that govern electron transitions. Researchers often use the different spectral lines that come from these transitions to learn more about the atomic environment when they study atomic structures using complementary methods.

These capabilities demonstrate the significance of AES not only in elemental analysis but also in the progression of theoretical frameworks in atomic physics. Studies that look at engineered nanoparticles using complementary characterization techniques (Anderson *et al.*, 2014) show that new technologies related to AES are also opening up new uses in cutting-edge fields like nanotechnology. Continued innovation in AES methodologies promises to enrich our understanding of atomic interactions and enhance analytical precision in various scientific domains.

4.2.3. Inductively Coupled Plasma Mass Spectrometry (ICP-MS)

Inductively coupled plasma mass spectrometry (ICP-MS) is a key method in atomic spectroscopy because it is the most sensitive and accurate way to analyze trace elements. This method uses a high-temperature plasma to ionize samples, which makes it possible to find and measure metals and other elements at very low levels, often in the parts per trillion range. After that, the ionized particles are sent to a mass spectrometer, which sorts them and counts them based on their mass-to-charge ratio. This process lets researchers look at complex matrices, which is very important for a wide range of uses, from biomedical research to environmental monitoring. ICP-MS is very useful for finding contamination and figuring out the elemental makeup of samples. This is especially important now that people are more worried about heavy metal pollution and its effects on health.

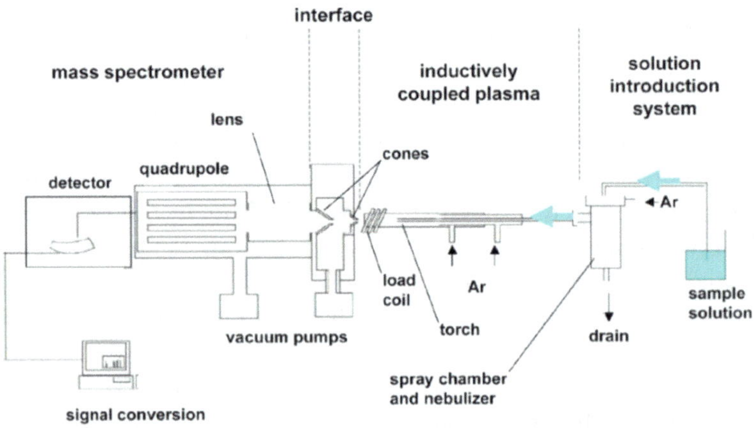

Figure 4.5. *Schematic of an inductively coupled plasma mass spectrometer (ICP-MS).*

Source: https://www.researchgate.net/profile/Richard-Gilstrap-2/publication/44226526/figure/fig5/AS:606775002796032@1521677842621/Schematic-of-an-Inductively-Coupled-Plasma-Mass-Spectrometer-ICP-MS.png.

The operational efficiency of ICP-MS is attributable to both its exceptional sensitivity and its ability to deliver rapid analysis. ICP-MS can look at more than one element at a time, which makes it much faster than traditional atomic spectrometry methods for full assessments. This feature is especially useful for keeping an eye on trace metal levels in environmental samples, where high levels of dangerous elements can have a big effect on public health. The ability of the technique to analyze different types of samples, like blood serum and plasma, also shows how important it is in clinical settings. As emphasized by the necessity to monitor trace elements due to metabolic disturbances, ICP-MS is an essential instrument for comprehending the equilibrium of vital elements in biological systems, thereby promoting improved health outcomes and the management of potential toxicities (Nanjunda *et al.,* 2005). Moreover, ongoing improvements in ICP-MS technology are making it even more useful for studying engineered nanoparticles (ENPs), which are becoming more common in many fields. To accurately describe ENPs, you need to know exactly what elements they are made of, how big they are, and what shape they are in. ICP-MS and other analytical methods work well together to give you this information. By combining ICP-MS with microscopy and separation methods, scientists can get a full picture of the properties of nanoparticles, which is important for figuring out how they might affect the environment and be toxic to living things. ICP-MS's ability to help with these kinds of complex characterization projects makes it an essential tool for the progress of nanotechnology and environmental science. It gives us information that is important for both public health policy and regulatory frameworks (Anderson *et al.,* 2014) (Figure 4.5).

4.3. INSTRUMENTATION AND TECHNIQUES

Instrumentation and techniques in atomic spectroscopy are essential for the efficient analysis and identification of elemental compositions in diverse samples. The atomic absorption spectrometer (AAS) is one of the most important tools in this field. It works by using the principle of free atoms absorbing light. In this method, a sample is burned in a flame or graphite furnace, and the light that comes out is measured at certain wavelengths that match the elements of interest. This method is very sensitive and selective, which makes it perfect for analyzing trace metals. But there can be problems, like spectral interferences and the need for careful calibration to get accurate results. AAS is a basic technique that makes it possible to use more advanced spectroscopic methods that improve elemental analysis in many areas, especially when studying biological systems and environmental samples (Nanjunda *et al.,* 2005). AAS is useful, but other atomic spectroscopy methods greatly improve analytical capabilities. Inductively Coupled Plasma Optical Emission Spectroscopy (ICP-OES) and Inductively Coupled Plasma Mass Spectrometry (ICP-MS) are well-known for their higher sensitivity and wider dynamic range. ICP-OES uses a plasma at a high temperature to ionize samples, which lets it find many elements at once. The spectral lines that come from this give us a lot of information about the sample's makeup. On the other hand, ICP-MS uses both mass spectrometry and plasma ionization to look at trace and ultra-trace elements with an accuracy that is unmatched. These methods not only raise the limits of detection but also speed up the process of getting samples, which is important in fields like geology, biology, and material science. The efficacy of these technologies highlights the necessity for continuous advancement and incorporation of novel methodologies to enhance analytical performance (Lennard *et al.,* 2017). The progress of atomic spectroscopy instruments is always changing the ways that elemental analysis is done. Recent advancements encompass the integration of laser ablation with ICP-MS, enhancing spatial resolution and facilitating the analysis of heterogeneous samples at a microscale. Also, the use of advanced calibration methods and multi-dimensional spectral mapping makes it possible to tell the difference between overlapping spectral lines, which is a common problem in complex matrices. These improvements help us understand how elements are distributed in both art conservation and environmental monitoring, where accurate analysis is very important. Studies of painted textiles show how scientists and conservators can work together to use modern spectroscopic techniques to learn a lot about materials, how they were made, and how they break down over time. Thus, the ongoing enhancement of instruments and methodologies is essential for the advancement of atomic spectroscopy and its utilization in various scientific disciplines (Lennard *et al.,* 2017; Nanjunda *et al.,* 2005) (Figure 4.6).

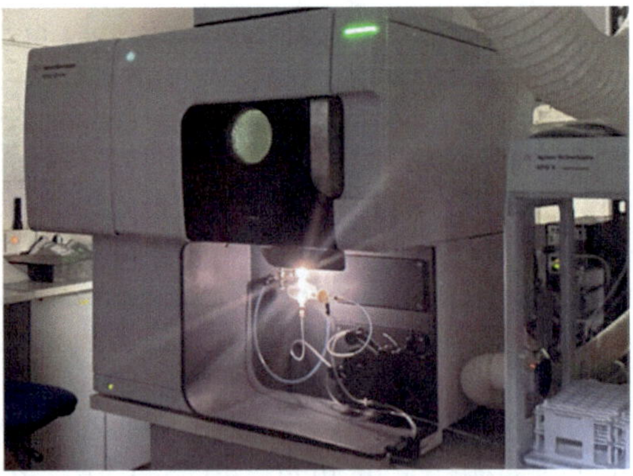

Figure 4.6. *Inductively coupled plasma optical emission spectroscopy.*

Source: https://www.impact-solutions.co.uk/wp-content/uploads/2020/03/ICP-OES-3-1.jpg.

4.3.1. Components of Atomic Spectroscopic Instruments

The basic parts of atomic spectroscopic instruments are very important for making it possible to accurately analyze the composition of elements and the structure of molecules. The light source is the most important part of an atomic spectrometer because it makes the electromagnetic radiation that causes electrons to move between atoms. Hollow cathode lamps are often used for atomic emission spectroscopy, while more advanced laser systems are used for techniques like laser-induced breakdown spectroscopy. Choosing the right light source is very important because it affects the resolution and sensitivity of the measurements. The sample introduction system works with the light source to move atoms or molecules into the path of the radiation, making sure that the interaction is as good as it can be. This joining of parts is very important for getting accurate spectral data, which is the basis for later analytical processes that use this data.

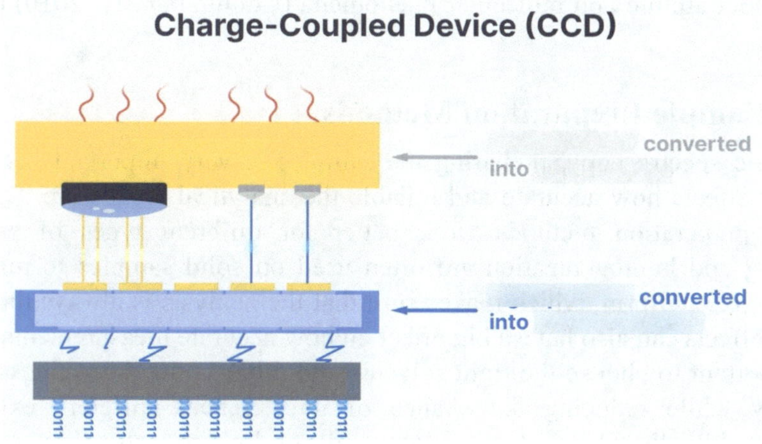

Figure 4.7. *Charge coupled device (CCD).*

Source: https://o.quizlet.com/HWodyaNdu3-J7MYpea8mXg_b.png.

When the radiation interacts with the atoms and causes them to become excited, the detection phase is very important for how spectroscopic instruments work as a whole. For further study, detectors turn the light that is emitted, transmitted, or absorbed into electrical signals. Photomultiplier tubes, charge-coupled devices (CCDs), and photodiodes are some of the most common types of detectors used in atomic spectroscopy. Each type of detector has its pros and cons. CCDs are often preferred because they are sensitive and can record a wide range of wavelengths at the same time. The ability of detectors to tell the difference between weak signals and noise is greatly affected by their conversion efficiency and response time. Modern atomic spectroscopic systems are also using digital signal processing technologies more and more to improve the data they collect. This makes the spectral readings from the instrument clearer and more reliable. The creation of advanced spectroscopic parts helps to improve instruments that are designed to solve difficult analytical problems. The LUVOIR Surveyor mission, which was brought up in the context of possible astronomical discoveries, is an example of how high-resolution spectroscopy can be used in space (Fleming *et al.,* 2016). In this context, the design of multiplexed spectrographs is a big step forward because it lets you look at more than one spectral line at a time. These new methods not only make observations more efficient, but they also open up new areas of scientific research that can be done with atomic spectroscopy. Researchers are pushing the limits of technology, and new solutions to old problems keep coming up. This shows how important instruments, and their parts are for helping us learn more about atomic and molecular phenomena (Contursi *et al.,* 2010) (Figure 4.7).

4.3.2. Sample Preparation Methods

In atomic spectroscopy, preparing the samples is very important because it directly affects how accurate and reliable the analytical results are. Different sample preparation methods work better for different types of samples. Grinding and homogenization are often used on solid samples to make the particle size uniform, which makes sure that the analysis is always the same. Matrix effects can also have a big effect on how accurate measurements are, so it's important to choose the right solvents and diluents that keep the sample's integrity while reducing the chance of interference. Different extraction methods, like liquid-liquid extraction and solid-phase extraction, can also be used to separate analytes from complicated matrices. This makes atomic spectroscopy more accurate for elemental analysis. In atomic spectroscopy, the way samples are introduced is closely related to how they are prepared, so you need to know a lot about both. Common ways to introduce samples include nebulization, which turns liquid samples into fine droplets, and solid sampling methods, which may use laser ablation or electrothermal vaporization. Each method has its own benefits, so you need to choose the one that works best for the samples' physical and chemical properties. For example, Rangarajan et al. (2014) talk about the problems with traditional sample preparation methods and stress the need for better protocols that make the analysis process more efficient while lowering the chance of contamination or loss of analytes during preparation. Along with standard preparation methods, new developments in the field are leading to new ways that make things faster and more accurate. New methods, like microwave-assisted acid digestion, make it easier to break down complex matrices quickly, which makes it easier to get to the analytes of interest. In addition, using automation in sample preparation can greatly reduce human error and make results more consistent. These kinds of improvements align well with studies such as those by Flynn et al. (2018), which examine the nuances of sample integrity in various contexts. As atomic spectroscopy continues to develop, using more advanced sample preparation techniques not only improves analytical performance, but also opens up new possibilities for atomic spectroscopy to be used in many different areas, such as environmental monitoring and biomedical research.

4.3.3. Calibration and Standardization Procedures

In atomic spectroscopy, proper calibration and standardization are very important because they make sure that quantitative measurements are correct and trustworthy. Calibration means making a connection between the concentration and the instrument's response. This connection can be affected by different types of instrument errors and environmental factors. To set up this relationship, you usually need a set of standard solutions with known concentrations. This lets you make a calibration curve. A well-defined calibration curve is important for accurate analysis because it takes into account how instrument sensitivity might change over time. Also, standardization procedures need to be used on a regular basis to reduce the effects of drift and systemic errors, which will make spectroscopic measurements more accurate (Ebert *et al.,* 2005). It is important to use high-quality calibration standards in atomic spectroscopy to keep the calibration processes accurate. It is very important to choose the right standards because they need to be very similar to the sample matrix being studied. Analytical results can be very different if there are impurities or changes in the chemical composition. Fluorescence intensity standards and spectral fluorescence standards are also very important for making sure that spectroscopic data from different instruments and labs can be compared. These standards are used to check performance and make sure that quantitative analysis is correct. They meet the needs of a wide range of atomic spectroscopy applications. Using such strict standardization methods makes research results more reliable and makes people trust the data (Ebert *et al.,* 2005). The reliability of atomic spectroscopy measurements is further enhanced by the statistical validation of calibration models. High correlation coefficients are often signs of a good calibration, but they don't always mean that there is a strong linear relationship between the variables. In fact, the linearity that is seen can sometimes hide nonlinear features that are really there. To make sure that models are fully validated, more statistical tests, like the Lack-of-fit test or the Mandel's fitting test, are needed. For instance, cadmium calibration curves in atomic absorption spectroscopy have demonstrated elevated correlation coefficients in conjunction with a thorough assessment of their linearity. Researchers can greatly improve the reliability and strength of their spectroscopic analyses by using these advanced statistical methods to avoid possible problems with data interpretation. In this way, careful calibration and standardization procedures make atomic spectroscopy possible, which gives us deep insights into the composition and concentrations of elements (Figure 4.8).

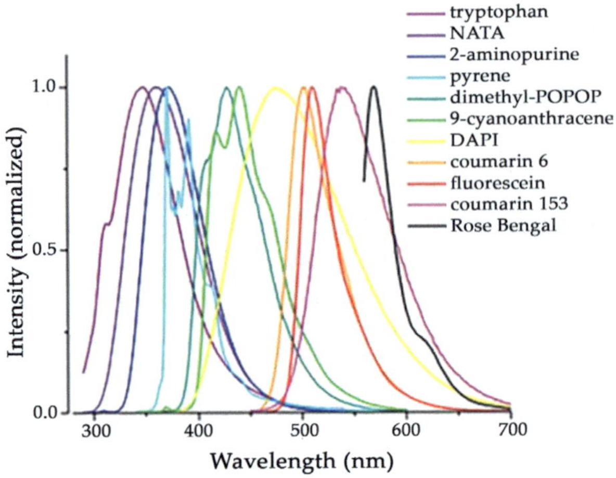

Figure 4.8. *Fluorescence spectroscopy.*

Source: https://static.horiba.com/fileadmin/Horiba/_processed_/7/f/csm_2_fluorescence_emission_spectra_145f0b1d58.jpg.

4.4. APPLICATIONS IN CHEMISTRY

Atomic spectroscopy is useful in many areas of chemistry, but it is especially useful for analyzing materials. It is very useful for environmental monitoring, clinical diagnostics, and quality control in manufacturing processes because it can identify and measure elements. For example, atomic spectroscopy helps environmental chemists find heavy metals and pollutants in soil and water samples. This information helps them make sure that rules are followed and that public health is protected. Chemists often use methods like atomic absorption spectroscopy (AAS) and inductively coupled plasma optical emission spectroscopy (ICP-OES) to find trace elements and figure out how they might affect ecosystems. Moreover, the adaptability of these techniques highlights their significance in enhancing our comprehension of both inorganic and organic compounds, illustrating the essential function of atomic spectroscopy in modern chemical analysis. Atomic spectroscopy is important in materials science as well as environmental science. It is especially useful for characterizing engineered nanoparticles (ENPs). These nanoparticles have special properties that change how they interact with light and matter. To fully realize their potential in nanotechnology and medicine, we need to use precise characterization methods. Atomic spectroscopy gives us important information about the elemental makeup and distribution of these nanoparticles, which is important for figuring out how they will affect and be safe in different uses. The improvements in atomic techniques make it easier to get morphological and structural information that helps design ENPs for specific uses in drug delivery and catalysis. The incorporation of atomic spectroscopy into materials science not only deepens the comprehension of nanoparticle dynamics but also expedites the creation of novel materials tailored for specific functions in advancing technologies, as indicated by Anderson et al. (2014). Additionally, the convergence of atomic spectroscopy with two-dimensional (2D) materials has created novel opportunities in analytical chemistry, especially by improving methods such as surface-enhanced Raman spectroscopy (SERS). The addition of 2D materials like graphene and transition metal dichalcogenides has a big impact on many spectroscopic uses because they have unique electronic properties and large surface areas. Studies indicate that these materials can markedly enhance the sensitivity of Raman signals, facilitate the detection of low-concentration analytes and yield extensive insights into complex samples. The synergistic effects seen when combining 2D materials with traditional spectroscopic methods show how they could change chemistry today. This opens the door to new sensor designs and molecular characterization, as shown by Xia et al. (2017) (Figure 4.9).

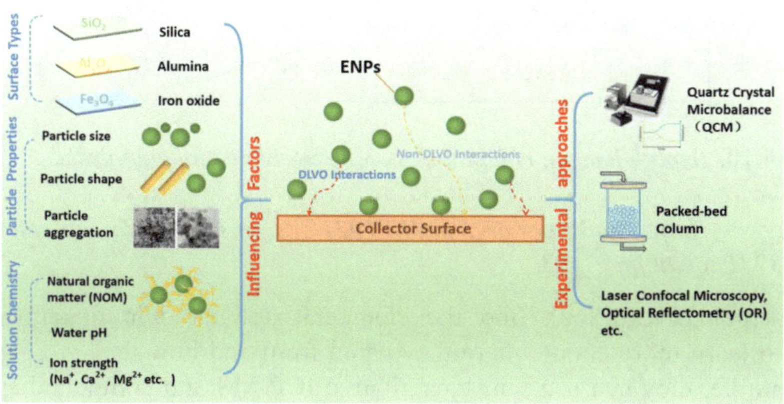

Figure 4.9. *Deposition of engineered nanoparticles (ENPs).*

Source: https://media.springernature.com/original/springer-static/image/art:10.1007%2Fs11356-018-3225-2/MediaObjects/11356_2018_3225_Figa_HTML.png.

4.4.1. Elemental Analysis in Environmental Samples

Evaluating elemental composition in environmental samples is essential for comprehending ecological interactions and evaluating anthropogenic effects. Atomic spectroscopy, especially methods like inductively coupled plasma mass spectrometry (ICP-MS) and inductively coupled plasma-optical emission spectroscopy (ICP-OES), is a great way to look at a lot of different elements in different types of materials. These techniques allow for the identification of both major and trace elements, thereby aiding the analysis of pollutants, nutrients, and essential minerals in environmental matrices such as soil, water, and biological specimens. The capacity to produce multivariate elemental fingerprints enhances the comprehension of environmental conditions and potential alterations induced by human activities, thus serving a crucial function in environmental monitoring and assessment protocols (Husted *et al.,* 2012). Atomic spectroscopy is useful for more than just standard environmental monitoring. It is also being used more and more in new ways, like to verify the authenticity of food products.

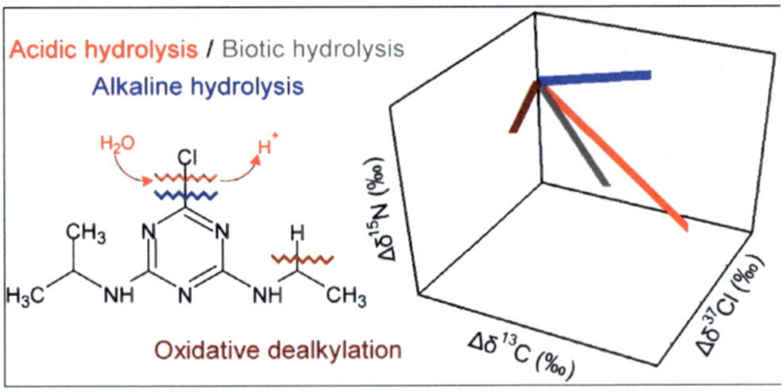

Figure 4.10. *Triple-element, compound-specific stable isotope analysis.*

Source: https://pubs.acs.org/cms/10.1021/acs.est.1c03981/asset/images/acs.est.1c03981.social.jpeg_v03.

Researchers can look into the elemental and isotopic fingerprints of plants to learn more about where they came from and how they were grown. For example, they can use semi-quantitative ICP-MS and compound-specific isotope analysis (CSIA). Atomic spectroscopy is an important tool in both environmental science and agricultural studies because it can find a wide range of elements, including essential nutrients and possibly harmful heavy metals. These analyses help us evaluate the quality and safety of food, and they also help us learn more about how nutrients move through the environment and how healthy it is (Husted *et al.,* 2012). Also, new atomic spectroscopy techniques have made it possible to study engineered nanoparticles (ENPs) and how they interact with the environment in new ways. It is very important to characterize ENPs because their unique properties can affect how they act, how toxic they are, and how available they are to living things in ecosystems. There are many ways to characterize something, such as microscopy and separation methods that use hyphenated detection systems. These methods let you look at all of the physical and chemical properties, such as size, shape, and elemental makeup. By combining atomic spectroscopy methods with these analytical techniques, researchers can better understand how ENPs behave in the environment. This greatly helps us understand how pollution works and how different species interact with each other (Figure 4.10).

4.4.2. Pharmaceutical Applications

To make the drug discovery process better, the pharmaceutical industry has been using more and more new technologies, like atomic spectroscopy. This method makes it possible to accurately measure the concentrations of elements, which is important for figuring out what drug compounds are and how they work together. One important use is in screening protein-protein interactions, which is very important for finding possible drug targets. Recent advancements indicate that atomic force microscopy (AFM) force spectroscopy has been utilized to enhance the signal-to-noise ratio in these screenings, facilitating the detection of binding events, including those between focal adhesion kinases and protein kinase B. The improvements in probe and substrate functionalization, along with a new statistical-based data processing method, have made atomic spectroscopy look like a useful tool for automating drug screening. This shows how important atomic techniques are for speeding up drug discovery, especially for cancer drugs (Basson *et al.,* 2019). Additionally, the incorporation of atomic spectroscopy into biosensor technology has created new possibilities for the real-time monitoring and detection of pharmaceutical compounds. The creation of an all-electronic biosensor for opioids, which uses a chemically bonded receptor protein on a graphene field-effect transistor, is an example of this trend. This biosensor not only shows high sensitivity and specificity for opioid antagonists like naltrexone, but it also uses atomic interactions to detect ligand binding. The engineered receptor, which has been redesigned to be more soluble and stable, can change the way pharmaceutical agents are monitored in real time. This biosensor could change the way drugs are used to treat pain and addiction by using atomic-level information to make opioid medications more effective and safer (Crook *et al.,* 2014). Atomic spectroscopy is very important for making sure that drugs are safe and of good quality, in addition to helping with drug discovery and monitoring. Pharmaceutical companies can use spectroscopic methods to thoroughly check raw materials and finished products for elemental impurities and make sure they meet regulatory standards. For example, heavy metals and other elements can be very dangerous in pharmaceutical products, causing health problems. Companies can use atomic spectroscopy to get an exact number for these contaminants, which lowers the risks to customers. Atomic techniques also make it possible to do quick analyses, which is very important for keeping strict quality control processes in drug development. So, atomic spectroscopy not only helps find new drugs, but it also makes the pharmaceutical supply chain safer and more reliable, making sure that patient safety is always the most important thing (Figure 4.11).

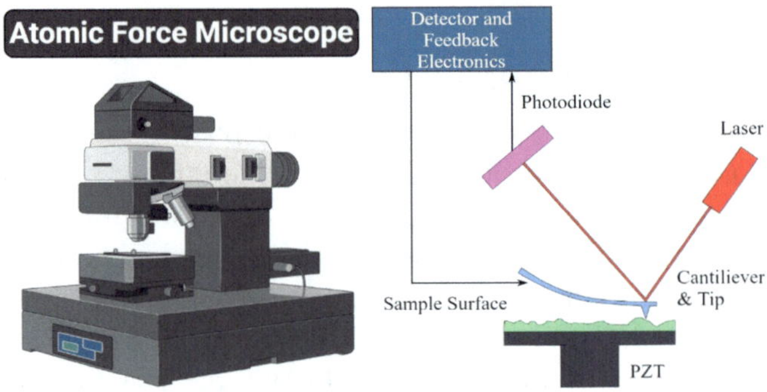

Figure 4.11. *Atomic force microscope.*

Source: https://microbenotes.com/wp-content/uploads/2020/03/Atomic-Force-Microscope-AFM.jpeg.

4.4.3. Industrial Quality Control

Quality control in industry is very important for making sure that products made in different fields are safe and reliable. This is especially true in places where accuracy is very important. Quality control methods are mainly meant to find any differences from the set standards during the production process. This lowers the risk of producing low-quality goods that could lead to bigger problems in the system. When it comes to atomic spectroscopy, strong quality control procedures are needed to make sure that the analytical methods used give accurate and precise measurements. These measurements are important for a wide range of uses, from environmental monitoring to biomedical analysis. Industries can improve their operational efficiency and stay in line with government rules by using strict quality checks like calibrations and standardization procedures. This protects both the health of consumers and the environment. Moreover, the adoption of sophisticated analytical methodologies in industrial quality control, such as atomic spectroscopy, has transformed the methods by which manufacturers assess material composition and quality. Techniques like scanning electron microscopy combined with energy dispersive spectroscopy (SEM-EDS) are now essential for getting a clear picture of the particles that come from industrial emissions. This is especially important when looking at how emissions affect public health, since many studies have found links between airborne particulate matter and bad health outcomes (Wilkinson *et al.,* 2013). Effective particle characterization not only helps us understand toxicity, but it also improves risk assessment models. This lets manufacturers change their processes to reduce harmful emissions without sacrificing product quality. The use of atomic spectroscopy in industrial quality control processes not only raises health and safety standards, but it also helps companies reach operational excellence. Techniques like UV Raman spectroscopy are special because they can accurately identify the properties and flaws of materials, which is important for keeping the quality of different products at the level you want. For instance, the ability to identify nanomaterials by changes in Raman spectral features gives manufacturers useful information for judging quality (Alexander *et al.,* 2009). As a result, industries are under more pressure from regulators and customers to use sustainable practices. Atomic spectroscopy not only meets quality standards, but it also encourages new ways of thinking about materials science. This combination of quality control and advanced analytical methods pushes industries to use their resources more efficiently and take better care of the environment.

4.5. APPLICATIONS IN ASTRONOMY

Astronomy heavily depends on the principles of atomic spectroscopy to figure out what celestial bodies are made of, how they behave, and how they move. Astronomers can find out a lot of important things about stars, like their temperature, density, and pressure, by looking at the light that elements in stars give off or take in. Understanding spectral lines, which are unique signatures of the elements, is very important for understanding how stars evolve and how galaxies form. Recent improvements in radiation transport models, especially for late-type stars, have greatly improved our understanding of how these processes work. Research indicates that the analysis of observed spectra and photometric magnitudes is fundamentally grounded in theoretical models of radiation transport, which delineate the correlation between stellar atmospheric conditions and their emitted light (Chiavassa *et al.,* 2014). Atomic spectroscopy is therefore essential for confirming and improving these models, which will eventually lead to progress in astrophysics. Atomic spectroscopy can be used to study exoplanets and their atmospheres, not just stars. Astronomers can see the starlight that goes through a planet's atmosphere using methods like transit spectroscopy. This lets them figure out what chemicals make up the planet. This method is very helpful for finding bio-signatures or other signs that a place is livable. Atomic spectroscopy has become much more sensitive and accurate, making it possible to find even very small amounts of substances in the atmospheres of distant exoplanets. Furthermore, advanced devices like optical lattice clocks, which employ atomic transitions for accurate measurement, illustrate the relationship between atomic spectroscopy and technological advancement in astronomy (Ablewski *et al.,* 2015). The combination of atomic spectroscopy with modern observational methods makes it possible to make important discoveries in the search for life outside of our solar system. Atomic spectroscopy is also very important for studying cosmic events like supernovae and stellar nucleosynthesis. When a supernova goes off, it sends out materials that were made in the cores of stars. This adds a lot of different elements to the space between stars. Astronomers can use spectroscopy to find these elements in observational data. This gives them a better understanding of what happens when stars die and how galaxies become more chemically rich. For example, spectral analysis of supernova remnants shows that there are heavy elements, which are important for figuring out where the elements in our universe came from. These studies show how important atomic spectroscopy is for linking theoretical astrophysics with real-world observations. As the field continues to change, the ongoing improvement of spectral analysis methods will surely give us a better understanding of how our universe works. This will show once again how important atomic spectroscopy is in modern astronomy (Figure 4.12).

Figure 4.12. *Applications in astronomy.*

Source: https://d12oja0ew7x0i8.cloudfront.net/images/news/ImageForNews_59547_16577217095598985.jpg.

4.5.1. Analysis of Stellar Compositions

Analyzing the compositions of stars is an important part of atomic spectroscopy because it helps us understand the chemical elements that make up stars and how they form. Astronomers can learn more about how stars evolve and how elements are made by looking at the light that stars give off or take in. In this situation, new methods like 3D hydrodynamical model atmospheres have become very useful. These methods consider a range of stellar parameters, such as temperature and density inhomogeneities, which have a big impact on the predicted line strengths of spectral lines. Studies show that the differences between 1D hydrostatic models and 3D simulations can cause big changes in the amounts of elements that are found, especially in places with very little metal. In this way, atomic spectroscopy is not only a way to observe things, but also a way to improve theoretical models of stellar atmospheres (Prieto *et al.,* 2007). The analysis of spectral lines via atomic spectroscopy is fundamentally based on the concepts of local thermodynamic equilibrium (LTE) and the dynamics of ions and molecules in stellar atmospheres. For instance, research indicates that in red giant stars with diverse metallicities, molecular species like CH, NH, and OH display varying intensities based on the utilization of either a 1D or 3D model for analysis.

This difference shows how the complicated nature of stellar atmospheres can change abundance calculations, which affects our understanding of the ratios of elements like carbon, nitrogen, and oxygen. As a result, using 3D models is very important for making accurate interpretations, especially in stars with low metallicity where the LTE can be very different. These discoveries highlight the importance of improving our observational methods and comprehension as we endeavor to chart the complex terrain of stellar chemistry (Prieto *et al.,* 2007). The function of volatiles in the formation of stars and planets adds another layer to the study of stellar compositions. Volatiles are essential constituents of the condensable mass in planet-forming environments, predominantly comprising hydrogenated compounds derived from abundant elements such as carbon, nitrogen, and oxygen.

To understand how planetary atmospheres form and what they are made of it is important to know how these volatiles move around, where they are found, and how they interact with each other in protoplanetary disks. Scientists have been able to find and measure these molecules thanks to better observational methods. This has revealed complicated chemical processes that set the stage for planet formation. As the connection between stellar material and volatiles in disk environments becomes clearer, it shows how atomic spectroscopy and cosmochemistry are linked and how they work together to explain the birth and evolution of our solar system and beyond.

4.5.2. Study of Cosmic Phenomena

Atomic spectroscopy is a key part of studying cosmic phenomena because it helps us understand the makeup and behavior of celestial bodies. Scientists can learn about the elements present, their physical states, and the conditions under which these substances exist in space by looking at the light that atoms in different astronomical objects give off or absorb. For instance, atomic spectroscopy has been very useful for figuring out what the atmospheres of faraway planets and moons are made of. This has let scientists guess what geological and maybe even biological processes are happening there. Gamma-ray and X-ray fluorescence, along with atomic spectroscopy, have become essential instruments in planetary geology, showcasing their versatility across various celestial bodies, as evidenced by comparative analyses conducted on the Moon, Mars, and other locations (Blasius *et al.,* 2025).

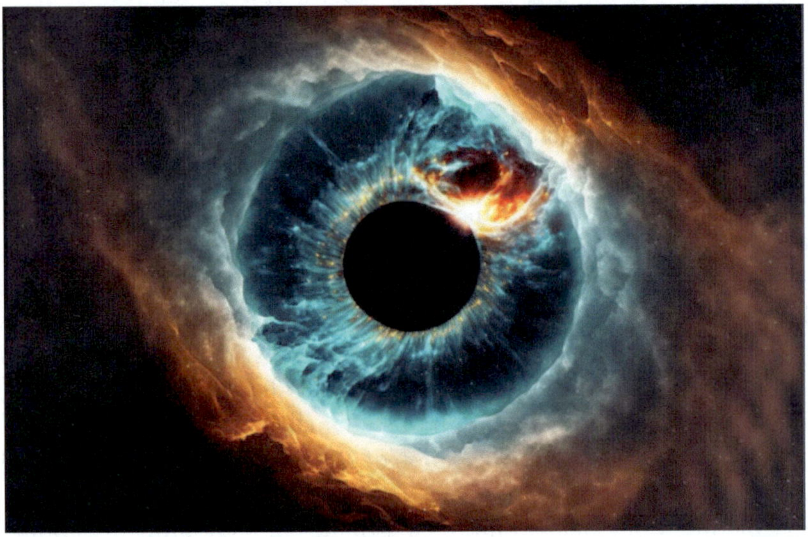

Figure 4.13. *Cosmic celestial phenomenon.*

Source: https://img.freepik.com/premium-photo/cosmic-celestial-phenomenon-that-is-black-hole-singularity_124507-31894.jpg.

Atomic spectroscopy is also very important for understanding the universe's big picture, especially when it comes to looking at strong interactions in nuclear physics. The recent European initiatives, as detailed in the Forward Look into Nuclear Physics, underscore the importance of elucidating the origin, evolution, and phases of strongly interacting matter, primarily constituted of atomic components (Bracco *et al.,* 2010). This investigation encompasses cosmic phenomena, including supernovae, neutron stars, and black holes, where atomic interactions determine the behaviors that regulate their lifecycles and emissions. As a result, improvements in spectroscopy not only help us understand how atoms interact with each other better, but they also help us make theoretical models that predict what will happen when these cosmic events happen. Additionally, atomic spectroscopy has implications for cosmic evolution and history, shedding light on processes that have influenced the universe over billions of years. Scientists can use different atomic spectroscopic methods to go back to the first nucleosynthesis after the Big Bang and study how elements formed and spread out across the universe. This knowledge is essential for understanding how galaxies and star systems form and how the elements needed for planets to form are spread out. By continuing to put money into nuclear physics and advanced spectroscopic techniques, researchers can answer these basic questions, which will help us understand cosmic phenomena and our place in this huge universe (Bracco *et al.,* 2010) (Figure 4.13).

4.5.3. Understanding the Universe's Evolution

The universe's evolution is closely tied to the basic constants of nature that control how matter and energy behave. Recent studies underscore the influence of dark matter fields on the cosmological evolution of these constants, indicating that a coherently oscillating scalar or pseudoscalar dark matter field can effectuate temporal changes. Atomic spectroscopy has been used to study this phenomenon. It makes it possible to measure atomic interactions and changes in fundamental constants very accurately. Researchers have established rigorous constraints on quadratic interactions between dark matter and particles, including photons and electrons, by employing atomic dysprosium spectroscopy and analyzing primordial helium abundance data. This groundbreaking research elucidates the evolution of the universe by demonstrating the impact of dark matter on the structure of space-time and elemental composition, thereby deepening our understanding of cosmological processes and their fundamental physics (Flambaum *et al.,* 2015). Along with dark matter, the stars in galaxy clusters have a big impact on how the universe's chemistry changes over time. The elemental abundances in the intracluster medium (ICM) show how stars have changed over time in these gravitational wells. Nevertheless, a paradox emerges when existing models undervalue the elemental enrichment in the ICM, resulting in a discrepancy that necessitates additional scrutiny into the dynamics of stellar formation. Recent research indicates that stars originating from a specific initial mass function may not correspond with observed abundance patterns, requiring modifications in star formation efficiency and supernova production efficiency. By examining these discrepancies, astronomers can deepen their comprehension of stellar nucleosynthesis and metal production, which are crucial to the evolution of galaxies and the broader universe. To understand the variety of star populations and the processes that shape cosmic evolution, it is important to deal with these kinds of paradoxes (Loewenstein *et al.,* 2013). Atomic spectroscopy becomes an essential instrument in the pursuit of comprehending the universe's evolution. Atomic spectroscopy helps astronomers learn about the makeup and behavior of stars and interstellar matter by letting them look at the electromagnetic emissions and absorptions from celestial bodies. Spectroscopy helps scientists find small changes in how atoms behave, which helps them learn more about dark matter and how the universe's chemical makeup changes over time. Researchers can investigate fundamental constants and correlate these behaviors with the overarching narratives of the universe's formation and evolution through precise measurements. As the field progresses, the convergence of atomic spectroscopy and astrophysics will persist in uncovering profound insights into the interactions of matter and energy, ultimately influencing the cosmos as we comprehend it (Flambaum *et al.,* 2015; Loewenstein *et al.,* 2013).

4.6. CHALLENGES AND LIMITATIONS

Atomic spectroscopy has many benefits for elemental analysis, but it also has some problems and limits. One major worry is the instrument's sensitivity and resolution, which can have a big impact on the quality of the results. Different spectroscopic methods, like atomic absorption and emission spectrometry, can give different results, which makes it hard to get the same results every time, especially when looking at complicated samples. These differences are caused by things like sample matrix effects, which can make it harder to find target elements. Also, detection limits can make it harder to analyze trace elements, especially when the concentration is close to the instrument's detection threshold. For example, even though technology has made sensitivity better, some elements are still hard to work with because they have low ionization potentials or tend to form stable complexes that are hard to analyze with standard atomic spectroscopy methods (ANONYMOUS, 2010). Calibration and method validation are other problems that come up when using atomic spectroscopy. To make sure that measurements are accurate and precise across a wide range of samples, strict calibration protocols are needed. This can be boring and take a lot of time. You need to use different standard solutions to make each calibration curve. This is especially important when working with samples that are not all the same or have different matrices. Also, laboratories have to deal with more operational problems because they have to do regular maintenance and updates to calibration standards, which can lead to both downtime and higher costs. Depending on the atomic spectroscopy method used, there may be more problems if overlapping spectra can't be found, especially when trying to find more than one element in a single sample. So, calibrating is still an important but difficult part of using atomic spectroscopy well (Lennard *et al.,* 2017). Lastly, atomic spectroscopy has a big problem with data interpretation because it's so complicated. Even though the instruments can give you detailed spectral data, it takes a lot of skill to analyze these results. To get reliable results, you need to be able to handle things like changes in the signal-to-noise ratio and the possibility of spectral interference. It is even harder to do multi-element analyses when elements interact with each other, both physically and chemically. Consequently, expert interpretation is contingent upon a profound comprehension of both physical chemistry and the underlying spectral phenomena. Consequently, the requirement for proficient scientists to accurately interpret the data underscores a critical limitation in atom spectroscopy, emphasizing the necessity for ongoing advancements in both technology and training to improve analytical results (Lennard *et al.,* 2017).

4.6.1. Sensitivity and Detection Limits

In atomic spectroscopy, sensitivity and detection limits are two very important factors that determine how well an analytical method works. Sensitivity is the ability of a method to tell the difference between small changes in concentration. Detection limits are the lowest concentration of an analyte that can be reliably found. The pursuit of enhanced sensitivity has resulted in the creation of sophisticated instrumental methodologies, such as resonance ionization spectroscopy (RIS). This method has shown promise for improving detection because it selectively ionizes certain elements, which cuts down on interference from matrix effects. As this process continues, lowering detection limits makes atomic spectroscopy more useful in many areas, from environmental monitoring to biomedical analysis. This helps us learn more about how different materials behave chemically. The combination of different analytical methods has made the field even more revolutionary by making it more sensitive and allowing for more accurate detection limits. For example, the combination of secondary neutral mass spectrometry (SNMS) and resonance ionization spectroscopy (RIS) shows how using more than one method together can improve analytical power. Researchers have been able to do simultaneous multielement analyses with better detection limits by using the best parts of both methods. A significant instance pertains to the detection of trace elemental impurities in solids, where meticulous design of laser configurations facilitated the measurement of multiple elements within a singular experimental framework. These kinds of improvements not only make analyses more useful, but they also make it easier to study complicated samples like meteorites in depth, which gives us information about their elemental makeup and how they formed (Burnett *et al.,* 1995). To get the most out of atomic spectroscopy applications, you need to know how to find the right balance between detection limits and analysis throughput. Researchers are having a hard time keeping high sensitivity while also making sure that data is collected quickly as the need for quick and accurate analyses grows, especially in places with few resources. Research indicates that attaining reduced detection limits frequently necessitates a compromise in analytical speed, as demonstrated in numerous experimental configurations. To get around these problems, it is important to keep improving methods and come up with new ideas. For example, improvements in tuning resonance frequencies with computer-controlled systems have made the multielement analysis process much easier, allowing for quick data collection without lowering the quality of the results. Consequently, the examination of sensitivity and detection limits continues to be a vital focus in atomic spectroscopy, shaping the advancement and utilization of next-generation analytical methodologies (Aoyama *et al.,* 2012) (Figure 4.14).

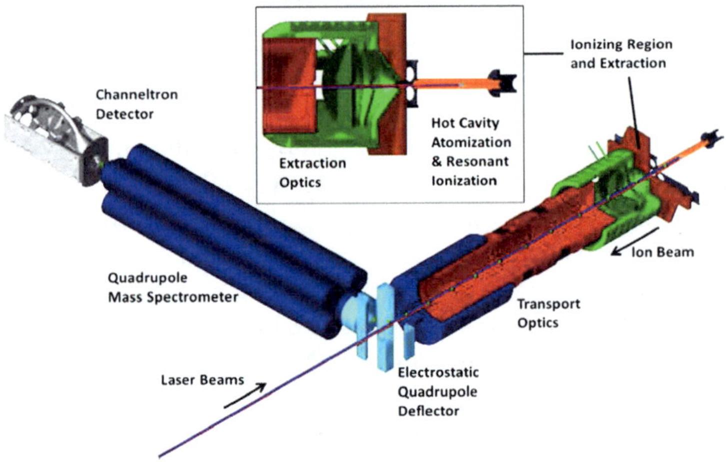

Figure 4.14. *Resonance ionization spectroscopy (RIS).*

Source: https://www.readkong.com/static/c3/6d/c36d3e45e26b-279cf5ddd19a99f3c684/resonance-ionization-spectroscopy-ris-on-ytterbium-cern-7281624-2.jpg.

4.6.2. Interference from Other Elements

In atomic spectroscopy, interference from other elements can have a big effect on how accurate spectral measurements are. Because different types of atoms can emit or absorb light at similar wavelengths, spectral lines can overlap, making the analysis harder. This spectral congestion can make it hard to find specific elements in a sample, which can lead to wrong conclusions about the data. For instance, when there are a lot of different elements at different concentrations, the resulting spectrum can become confusing, making it harder to get accurate readings. The difficulty of separating overlapping spectral lines highlights the need for better instrumental techniques that can pick out closely spaced transitions. This will improve the accuracy of measurements in a variety of fields, such as environmental monitoring and material analysis. Quantum interference effects make atomic spectra more complicated by changing the shapes and positions of the lines. Recent studies, including those investigating lithium isotopes, indicate that these interference phenomena can produce non-Lorentzian line profiles, influenced by factors such as excitation polarization. If these effects are not taken into account, they can cause significant changes in the inferred line strengths and line center positions, shifting them by as much as 1 MHz. This change, which is caused by quantum mechanical interactions, shows how important it is to have strict correction methods in spectroscopy. Better accuracy in finding optical transition frequencies can lead to better measurements of the nuclear charge radius. This shows how interference can have a big impact on spectral analysis and atomic physics as a whole (Brewer *et al.,* 2013). Furthermore, the interference observed in higher-order transitions, such as quadrupole excitations in X-ray absorption spectroscopy, illustrates the difficulties arising from interactions among diverse atomic states. These transitions, which are often hidden by more obvious dipole allowed transitions, hold important information about the electronic structure of materials. If a crystal lattice doesn't have inversion symmetry, orbital states can mix on site, which causes quantum mechanical interferences that make it even harder to understand spectra. To get useful information from spectroscopic data, you need to understand how these interconnected dynamics work. This is especially true in systems that are going through phase transitions, like the metal-insulator transition seen in vanadium-chromium compounds. Such analyses underscore the importance of comprehensive investigative methodologies in elucidating the complex network of interference to enhance our understanding of material properties (Toschi *et al.,* 2011).

4.6.3. Cost and Accessibility of Equipment

The cost and availability of atomic spectroscopy equipment are two of the most important things that determine how widely it is used in schools and research labs. Inductively coupled plasma mass spectrometry (ICP-MS) and atomic absorption spectrophotometers (AAS) are two examples of high-performance spectrometers that can be very expensive. Institutions that want to buy such high-tech tools have trouble with money, which can make it hard for them to do cutting-edge research or teach in new ways. Because of this difference, students from schools with limited budgets may not be able to get the resources they need, which will only make the gap in scientific literacy and technological progress between different groups of people bigger. It is very important to solve these money problems so that atomic spectroscopy techniques can be used more widely.

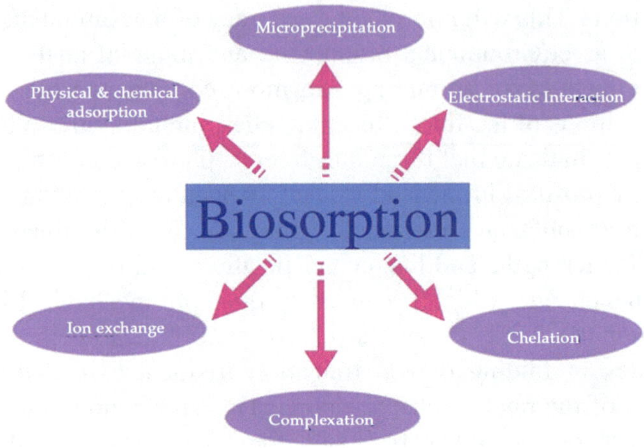

Figure 4.15. *Different mechanisms involved in biosorption phenomenon.*

Source: https://www.researchgate.net/profile/Ajay-Kumar-266/publication/336218116/figure/fig1/AS:809944093835264@1570117128692/Different-mechanisms-involved-in-biosorption-phenomenon-Asgher-2012.ppm.

This will encourage both inclusivity and new ideas in scientific research. Atomic spectroscopy equipment is not only expensive to buy, but it also costs money to keep up and run. Institutions have to pay more because these tools are so specialized that they need trained staff to use and maintain them. Also, consumable materials like gases or reagents that are needed for the system to work well can add to costs that make it hard for institutions to stay within their budgets. Programs that help institutions share resources, like collaboration programs or centralized facilities, may be other ways to lower these costs. Institutions can improve accessibility and create a learning environment that encourages active participation in scientific exploration through atomic spectroscopy techniques by making equipment use more collaborative. Investigating affordable alternatives and novel techniques can greatly enhance the accessibility of atomic spectroscopy. For example, research has examined biosorption techniques for the extraction of heavy metals, employing agro-industrial waste materials like the epidermis of Agave atrovirens. These studies show not only how to deal with environmental pollutants in a useful way, but they also show how important it is to have cheap and easy-to-use technologies for practical uses. The results show that cheaper tools can work well with traditional atomic spectroscopy methods. This encourages a culture of innovation that leads to more research and development. The creation of these affordable alternatives significantly enhances the potential for broader utilization of atomic spectroscopy in laboratories and fieldwork, facilitating progress in diverse scientific fields while fostering sustainability and cost-effective practices (Barriada *et al.,* 2019; ANONYMOUS, 2010) (Figure 4.15).

SUMMARY

The study of atomic spectroscopy has shed light on the complex processes that govern the interaction between light and matter, especially in elemental analysis. Scientists can now qualitatively and quantitatively find and study different elements by looking at their characteristic spectral lines. Atomic spectroscopy is not just useful for basic research; it is also very useful for monitoring the environment and characterizing materials. For instance, in research on the movement and toxicity of arsenic in subsurface environments, knowing spectral signatures makes it possible to accurately track and measure pollutants. Due to improvements in methods, atomic spectroscopy is very important for finding pollutants and figuring out how to clean them up and protect the environment. This shows how useful this type of analysis can be in real life (Perez *et al.,* 2020). Furthermore, improvements in atomic spectroscopy methods, like the creation of more sensitive detection methods, have greatly improved researchers' ability to study complicated systems. The use of modern instruments has made it possible for researchers to do more detailed analyses with more accuracy and precision. For example, explaining how green rust phases and arsenic species interact shows how powerful atomic spectroscopy is in environmental geochemistry. By integrating diverse analytical methodologies, researchers can achieve a thorough comprehension of metallic mobility and the stability of mineral phases in contaminated settings. This comprehensive perspective is essential for formulating efficacious environmental policies and conservation strategies, highlighting the significance of atomic spectroscopy in modern scientific research (Lennard *et al.,* 2017). Lastly, atomic spectroscopy has effects on other fields, combining analytical chemistry with materials science and environmental science, for example. The joint work of scientists and conservators shows how atomic spectroscopy can help protect cultural heritage, especially when it comes to studying painted textiles. Researchers can figure out the materials and methods used in old works of art by using different types of spectroscopies. This helps us learn more about how they break down and how to protect them. As atomic spectroscopy progresses, it will undoubtedly yield significant insights across various fields, ultimately enhancing both scholarly understanding and practical implementations in multiple sectors. Consequently, the importance of atomic spectroscopy is reaffirmed not only in its scientific applicability but also in its extensive societal implications (Lennard *et al.,* 2017; Perez *et al.,* 2020).

In atomic spectroscopy, it is very important to know how light and matter interact in order to figure out the structures of atoms and molecules. The technique uses different methods, like absorption and emission spectroscopy, to get important information about the energy levels of atoms. Researchers can find different spectral lines in the spectrum that is made when atoms absorb or emit light. These lines correspond to specific transitions between energy states. This spectral fingerprinting is very useful in many areas of science, such as chemistry and materials science, where scientists study the exact makeup and behavior of substances. The capacity of spectroscopic techniques to elucidate complex details regarding electronic configurations and molecular dynamics has established a foundation for progress in quantum mechanics and the emerging domain of materials research. Recent improvements in spectroscopic methods have greatly improved our understanding of complicated systems, especially when it comes to high-temperature superconductors. The work to improve scanning tunneling microscopy/spectroscopy (STM/STS) has been very important in solving problems that have come up with these special materials in the past. These methods not only enable the analysis of intricate characteristics at the atomic level but also aid in the clarification of phenomena such as the pseudogap and vortex core dynamics in superconductors. Studies cited in the literature indicate that mastery of these techniques has produced consistent outcomes, thereby enhancing comprehension of the unique characteristics displayed by high-temperature superconductors. This research has implications that go beyond pure science; they are becoming more important for creating new materials and technologies for electronics and energy storage. The development of optical lattice clocks is another example of how atomic spectroscopy is always trying to be more precise and accurate. These clocks achieve unprecedented frequency stability by simultaneously interrogating multiple atoms. Enhanced atomic interactions help with problems like density-dependent frequency shifts, making it possible for both precision and accuracy to exist in many-particle systems. Researchers have reported progress that suggests that new methods, like using strong atomic interactions to reduce collisional shifts, can greatly expand the limits of spectroscopy. This dual improvement not only makes measurements more reliable, but it also moves forward our theoretical understanding of how atoms interact. This shows how atomic spectroscopy is still evolving and important in modern scientific research.

REFERENCES

1. Ablewski, P., Bartoszek-Bober, D., Bielska, K., Bober, A., et al. (2015). Strontium optical lattice clocks for practical realization of the meter and secondary representation of the second. http://arxiv.org/abs/1506.04680 (accessed on 29th October 2025).
2. Abrikosov, A. A., Abrikosov, A. A., Alff, L., Anderson, P., et al. (2006). Scanning tunneling spectroscopy of high-temperature superconductors. http://arxiv.org/abs/cond-mat/0610672 (accessed on 29th October 2025).
3. Adhikari, C. M., Dawes, R., Jentschura, U. D., Kolachevsky, N., et al. (2019). Pressure shifts in high-precision hydrogen spectroscopy: I. Long-range atom-atom and atom-molecule interactions. http://arxiv.org/abs/1904.03025 (accessed on 29th October 2025).
4. Allafort, A., Hingant, B., Bordé, C., Cagnac, B., Cahuzac, P., Grynberg, G., Grynberg, G., et al. (2009). Non-linear spectroscopy of rubidium: An undergraduate experiment. http://arxiv.org/abs/0902.3115 (accessed on 29th October 2025).
5. Allende Prieto, C., Allende Prieto, C., Allende Prieto, C., Asplund, M., Asplund, M., Asplund, M., Asplund, M., et al. (2007). 3D hydrodynamical simulations of surface convection in red giant stars. Impact on spectral line formation and abundance analysis. http://arxiv.org/abs/astro-ph/0703652 (accessed on 29th October 2025).
6. Anderson de la LLana, S., Anderson, I. S., Capitán Vallvey, L. F., Cecchini, R., et al. (2014). A general perspective of the characterization and quantification of nanoparticles: Imaging, spectroscopic, and separation techniques. https://core.ac.uk/download/186617942.pdf (accessed on 29th October 2025).
7. Aoyama, T., Berengut, J. C., Berengut, J. C., Flambaum, V. V., Pettitt, E. C., Orzel, C., DeMille, D., Foot, C. J., et al. (2012). Searching for new physics through AMO precision measurements. http://arxiv.org/abs/1208.4506 (accessed on 29th October 2025).
8. Baianu, I. C. (2011). Food Physical Chemistry and Biophysical Chemistry. https://core.ac.uk/download/pdf/290434.pdf (accessed on 29th October 2025).
9. Balandin, A. A., Baranov, A. V., Calizo, I., Cardona, M., Liu, G., Bejenari, I., Calizo, I., et al. (2009). Ultraviolet Raman spectroscopy of single and multi-layer graphene. http://arxiv.org/abs/0903.1922 (accessed on 29th October 2025).
10. Barriada, J. L., Chanona-Pérez, J., Hernández-Botello, M. T., López-Cortez,

L., et al. (2019). Determination of biosorption mechanism in biomass of agave, using spectroscopic and microscopic techniques for the purification of contaminated water. https://core.ac.uk/download/323288149.pdf (accessed on 29th October 2025).

11. Barroso, J. M., Ferreira, H., Gaspar, L., Rato, J., et al. (2018). Predicting calcium in grape must and base wine by FT-NIR spectroscopy. https://core.ac.uk/download/187759521.pdf (accessed on 29th October 2025).
12. Basson, M. D., Chen, L., Dong, L., Hou, X., et al. (2019). Optimization of protein-protein interaction measurements for drug discovery using AFM force spectroscopy. https://core.ac.uk/download/305123513.pdf (accessed on 29th October 2025).
13. Beernaert, H., Croux, C., Elskens, M., Van Loco, J., et al. (2025). Linearity of calibration curves: Use and misuse of the correlation coefficient. https://core.ac.uk/download/pdf/6468999.pdf (accessed on 29th October 2025).
14. Bergin, E. A., Brittain, S., Marty, B., Mousis, O., et al. (2014). Volatiles in protoplanetary disks. http://arxiv.org/abs/1401.2423 (accessed on 29th October 2025).
15. Blasius, K. R., Cutts, J. A. J., Davis, D. R., Pang, K., et al. (2025). Geology orbiter comparison study. https://core.ac.uk/download/pdf/42877066.pdf (accessed on 29th October 2025).
16. Boas, P. R. V., Franco, M. A. M., & Milori, D. M. B. P. (2018). Comparison of algorithms for baseline correction of LIBS spectra for quantifying total carbon in Brazilian soils. http://arxiv.org/abs/1805.03695 (accessed on 29th October 2025).
17. Bracco, A., Chomaz, P., Gaardhøje, J. J., Heenen, P. H., et al. (2010). Perspectives of nuclear physics in Europe: NuPECC Long Range Plan 2010. https://core.ac.uk/download/19502483.pdf (accessed on 29th October 2025).
18. Brewer, S. M., Brown, R. C., Gillaspy, J. D., Porto, J. V., et al. (2013). Quantum interference and light polarization effects in unresolvable atomic lines: Application to a precise measurement of the 6,7 Li D2 lines. http://arxiv.org/abs/1212.2220 (accessed on 29th October 2025).
19. Bryndzia, L. T., Ferralis, N., Grossman, J. C., Liu, Y., et al. (2016). Genome-inspired molecular identification in organic matter via Raman spectroscopy. http://arxiv.org/abs/1602.02083 (accessed on 29th October 2025).
20. Burnett, D. S., Calaway, W. F., Gruen, D. M., Pellin, M. J., et al. (1995). Simultaneous dual-element analyses of refractory metals in naturally occurring matrices using resonance ionization of sputtered atoms. https://core.ac.uk/download/4873526.pdf (accessed on 29th October 2025).

21. Chiavassa, A., Chiavassa, A., Vögler, A., Beeck, B., Freytag, B., Freytag, B., Gustafsson, B., et al. (2014). Analysis of stellar spectra with 3D and NLTE models. http://arxiv.org/abs/1403.3089 (accessed on 29th October 2025).
22. Contursi, A., Coutens, A., Cros, A., Tielens, A. G. G. M., Lorenzani, A., Godard, B., Mookerjea, B., et al. (2010). Interstellar OH^+, H_2O^+ and H_3O^+ along the sight-line to G10.6-0.4. http://arxiv.org/abs/1005.5653 (accessed on 29th October 2025).
23. Crook, A., Han, G. H., Hong, S. J., Johnson, A. T. C., et al. (2014). Scalable production of highly-sensitive nano sensors based on graphene functionalized with a designed G protein-coupled receptor. https://core.ac.uk/download/76394235.pdf (accessed on 29th October 2025).
24. Dareau, A., Meng, Y., Rauschenbeutel, A., Schneeweiss, P., et al. (2017). Near-ground-state cooling of atoms optically trapped 300 nm away from a hot surface. http://arxiv.org/abs/1712.05749 (accessed on 29th October 2025).
25. De Geyter, N., Morent, R., Nikiforov, A., Rezaei, F., et al. (2019). Applications of plasma-liquid systems: A review. https://core.ac.uk/download/288817520.pdf (accessed on 29th October 2025).
26. Ebert, P., Engel, A., Hoffmann, K., Macdonald, D., et al. (2005). Journal of Fluorescence. https://core.ac.uk/download/210638942.pdf (accessed on 29th October 2025).
27. Flambaum, V. V., & Stadnik, Y. V. (2015). Can dark matter induce cosmological evolution of the fundamental constants of nature? http://arxiv.org/abs/1503.08540 (accessed on 29th October 2025).
28. Fleming, B., France, K., & Hoadley, K. (2016). CHISL: The Combined High-resolution and Imaging Spectrograph for the LUVOIR Surveyor. http://arxiv.org/abs/1605.05178 (accessed on 29th October 2025).
29. Flynn, C. (2018). Sensitivity of black phosphorus to O_2 and atmosphere measured by macroscopic four-point probe. https://core.ac.uk/download/215521003.pdf (accessed on 29th October 2025).
30. Husted, S., & Laursen, K. H. (2012). Fingerprint analysis of biological samples using ICP-MS and IR-MS. https://core.ac.uk/download/16386274.pdf (accessed on 29th October 2025).
31. Klaers, J., Moroshkin, P., Sass, A., Weitz, M., et al. (2014). Kennard-Stepanov relation connecting absorption and emission spectra in an atomic gas. http://arxiv.org/abs/1401.0731 (accessed on 29th October 2025).

32. Lennard, F., Smith, M., & Thompson, K. (2017). A literature review of analytical techniques for materials characterization of painted textiles–Part 2: Spectroscopic and chromatographic analytical instrumentation. https://core.ac.uk/download/84591645.pdf (accessed on 29th October 2025).

33. Levie, I., & Slepyan, G. (2017). The new concept of nano-device spectroscopy based on Rabi-Bloch oscillations for THz-frequency range. http://arxiv.org/abs/1706.10014 (accessed on 29th October 2025).

34. Loewenstein, M. (2013). Perspectives on intracluster enrichment and the stellar initial mass function in elliptical galaxies. https://core.ac.uk/download/pdf/42736740.pdf (accessed on 29th October 2025).

35. Pálffy, A., Borchert, G. L., Bouchiat, M. A., Bransden, B., Demtröder, W., Goldanskii, V., Otten, E. W., et al. (2010). Nuclear effects in atomic transitions. http://arxiv.org/abs/1106.3218 (accessed on 29th October 2025).

36. Perez, J. P. H. (2020). Green rust formation and reactivity with arsenic species. https://core.ac.uk/download/288114074.pdf (accessed on 29th October 2025).

37. Rangarajan, A. (2014). Graphene etch mask for silicon. https://core.ac.uk/download/29175004.pdf (accessed on 29th October 2025).

38. Rao Nanjunda, A. (2005). Trace element estimation–Methods & clinical context. https://core.ac.uk/download/pdf/107563.pdf (accessed on 29th October 2025).

39. Rey, A. M., Baym, G., Boyd, M., Campbell, G., Gibble, K., Haller, E., Ye, J., et al. (2010). Suppression of collisional shifts in a strongly interacting lattice clock. http://arxiv.org/abs/1007.0059 (accessed on 29th October 2025).

40. Toschi, A., Rodolakis, F., de Groot, F., Sangiovanni, G., Rueff, J. P., Stöhr, J., Held, K., et al. (2011). Atomic and itinerant effects at the transition metal x-ray absorption K-pre-edge exemplified in the case of V_2O_3. http://arxiv.org/abs/1111.1544 (accessed on 29th October 2025).

41. Wilkinson, K. (2013). Particulate airborne impurities. https://core.ac.uk/download/11698736.pdf (accessed on 29th October 2025).

42. Xia, M. (2017). A review on applications of two-dimensional materials in surface enhanced Raman spectroscopy. https://core.ac.uk/download/190749177.pdf (accessed on 29th October 2025).

CHAPTER 5

INFRARED (IR) SPECTROSCOPY

Contents

INTRODUCTION

Infrared (IR) spectroscopy is an important tool for chemical analysis, especially for figuring out what functional groups and molecular structures are. This method gives information about the molecular vibrations that happen at certain frequencies by measuring how much infrared light a sample absorbs. IR spectroscopy is used in many fields, including organic chemistry, material science, and even biomedical research. One of the best things about this method is that it doesn't damage samples, which means that their chemical makeup stays the same. This makes it much more useful in many fields. Also, improvements in instruments have made IR spectroscopy easier to use and more effective, which is why it is now used in so many labs and research centers. Infrared spectroscopy has many benefits, including the fact that it doesn't damage the sample and that it can analyze a lot of samples at once. Conventional chemical analysis techniques can be laborious and occasionally expensive, posing difficulties for expedited evaluations across multiple domains.

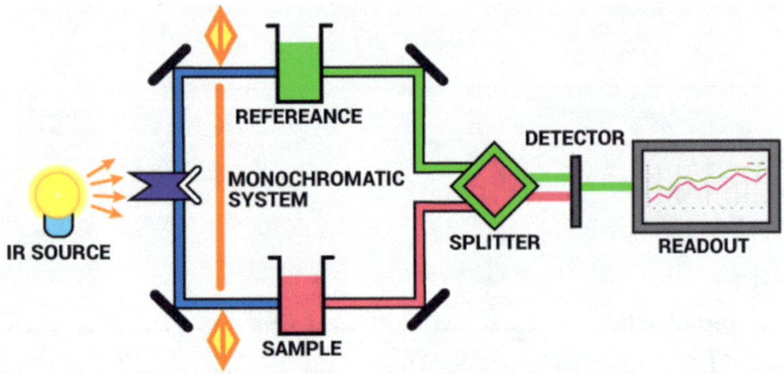

Figure 5.1. *Infrared spectroscopy.*

Source: https://cdn1.byjus.com/wp-content/uploads/2018/12/infrared-spectroscopy-2.jpg.

On the other hand, infrared spectroscopy, especially near-infrared (NIR) methods, has become a cheaper option that gives quick and accurate data for analyzing complicated mixtures. As noted in the literature, NIR technology has become a popular way to do large-scale chemical analysis, especially in the food industry (*see* Baianu *et al.,* 2011; Buning and Diller, 2000; Wilcox and Cavins, 1995; Williams and Norris, 1987). Infrared spectroscopy has become a key method in both qualitative and quantitative analysis because it can take measurements quickly and give accurate analytical results. Even though IR spectroscopy is very reliable, there are still some problems, especially when it comes to analyzing samples that aren't all the same, since light scattering can make spectral measurements less accurate. The inherent physical variations—like differences in particle size and sample surface—can cause effects that make it harder to understand what they mean. Consequently, it is imperative to distinguish the spectral contributions resulting from variations in chemical compositions from those originating from physical characteristics. Recent developments suggest methods to assess and alleviate these multiplicative effects, thereby improving the precision of quantitative analysis of heterogeneous mixtures. Research indicates that the use of advanced modeling techniques can greatly enhance the reliability of spectral data in these scenarios, thereby broadening the utility of infrared spectroscopy in the analysis of complex samples (Chen *et al.,* 2011). This ongoing research highlights the dynamic nature of the field, establishing infrared spectroscopy as an essential instrument in contemporary analytical chemistry (Figure 5.1).

Definition and Basic Principles

Infrared (IR) spectroscopy is a powerful way to analyze materials by looking at their molecular vibrations and rotational transitions. The basic idea behind IR spectroscopy is how infrared radiation interacts with matter. When a molecule takes in IR radiation, its vibrational energy levels change, which causes it to move between quantized vibrational states. The absorption spectrum of each functional group in a molecule is like a molecular fingerprint. Because of this specificity, IR spectroscopy is a very useful tool in many fields, such as chemistry, biochemistry, and environmental science. It lets researchers accurately figure out molecular structures and functional groups. The usefulness of IR spectroscopy is greatly increased by the fact that it can look at a wide range of samples, such as solids, liquids, and gases. The method usually involves measuring how much infrared light is absorbed or transmitted over a certain range of wavelengths, usually between 4000 cm^{-1} and 400 cm^{-1}. Fourier Transform Infrared (FTIR) spectroscopy, which is a more advanced type, can get fast, high-resolution spectra by taking a full spectrum in one measurement. This method is not only quick, but it also works well for studying complicated mixtures and solutions. Moreover, advanced techniques like attenuated total reflectance (ATR) enhance the versatility of IR spectroscopy by enabling researchers to analyze samples with minimal preparation, thereby facilitating novel approaches to material analysis, as evidenced by hyperbolic polariton modes in hexagonal boron nitride (Caldwell *et al.,* 2018). In IR spectroscopy, it is very important to know about vibrational modes and how molecules interact with each other. For example, diatomic molecules have certain vibrational modes that can be studied using both classical and quantum mechanics. Researchers can look into a smooth transition between vibrational and rotational spectra by using fractional calculus in these studies. This works especially well for diatomic molecules. These advanced methods help us learn more about molecular dynamics and the electronic structure of materials that we study with IR spectroscopy (Dirac *et al.,* 2012). Ultimately, the combination of traditional and new analytical methods shows how important IR spectroscopy is for both basic science and real-world uses in material science and chemical analysis.

Historical Development of IR Spectroscopy

Infrared (IR) spectroscopy has its roots in the early 19th century, when scientists started looking into the electromagnetic spectrum and how it interacts with matter. Sir William Herschel discovered infrared radiation in 1800 while doing experiments with sunlight. This led to the discovery of what we now know as the infrared region of the spectrum. After this first discovery, the field grew with important work from people like Auguste de la Rive and John William Strutt, also known as Lord Rayleigh, who helped people understand spectral techniques better.

By the end of the 19th century, improvements in optics and detection techniques made it possible to use IR light in chemical analysis. This was a major turning point in analytical chemistry. This historical basis laid the groundwork for the development of modern IR spectroscopy, including techniques such as Fourier Transform Infrared (FTIR) spectroscopy. The 20th century saw the rise of new technologies that changed infrared spectroscopy forever. This made it possible for it to be used in many different areas of science. The Fourier Transform algorithm, which came out in the 1970s, was a big step forward. It made it possible to quickly collect and analyze data, which made molecular characterization more efficient.

The emergence of advanced detectors and minicomputers further reinforced this transformation, enabling the practical implementation of IR spectroscopy in laboratories. A significant application of this technique is in the analysis of historical artifacts, such as barkcloths, where Fourier transform infrared spectroscopy with attenuated total reflection (ATR-FTIR) has been indispensable. This method not only helps to tell the difference between types of fabric, but it also helps art historians and conservators learn about their origins and how to care for them. This shows how technology and historical analysis are becoming more connected in IR spectroscopy (Holmes-Smith *et al.,* 2019). As its uses grew, IR spectroscopy became a key part of many scientific fields, connecting chemistry, biology, and material science in a meaningful way.

The evolution of IR spectroscopy, while significant, highlights the challenges encountered in the research-technology community, especially regarding its classification as a peripheral science. The dynamics of this community, marked by a lack of clear disciplinary and professional ties, demonstrate how certain specialized techniques, such as IR spectroscopy, can be overlooked despite their technological excellence (Johnston *et al.,* 2001). To understand the complex world of IR spectroscopy today, where it is still changing and adapting to new analytical challenges and uses, you need to know more about the historical and social contexts in which it developed.

Importance in Scientific Research and Industry

Infrared (IR) spectroscopy is an essential analytical method used in many fields of science and industry. Its importance cannot be overstated. In research environments, IR spectroscopy aids in the identification and quantification of chemical substances, supplying researchers with critical data necessary for the progression of knowledge. This analytical technique allows scientists to examine molecular structures, bonding environments, and chemical interactions without damaging them, which makes it essential in chemistry, biology, and materials science. Researchers can use IR spectroscopy to study complicated mixtures, which has led to advances in fields like pharmaceuticals and environmental monitoring. This method is useful in many different situations, which is why it is still an important part of modern scientific research. In industry, IR spectroscopy is often used in quality control and assurance processes to help make sure that products are always of the same high quality. For example, food and beverage companies use IR spectroscopy to check the composition of ingredients, find contaminants, and make sure their products meet safety standards.

The petrochemical industry also uses this method to check the quality and purity of fuels and other petroleum products. IR spectroscopy cuts down on downtime and makes operations more efficient by giving quick and accurate analysis. Because it doesn't hurt anything, it can also be used to monitor things in real time, which makes it an even more important tool for keeping products safe during the manufacturing process. These uses show that IR spectroscopy not only helps businesses be more productive, but it also protects the health and safety of consumers. Also, improvements in IR spectroscopy technologies keep making it more useful in science and business. Near-infrared (NIR) spectroscopy is an example of this evolution because it has improved capabilities in areas like seed-water interaction studies, where it can accurately classify seed viability and predict moisture content. Studies employing NIR spectroscopy have shown that multivariate calibration models can attain up to 99% accuracy in distinguishing viable seeds from non-viable ones, highlighting their potential for agricultural applications.

The capacity to employ particular NIR wavelength bands for precise predictions facilitates the incorporation of IR technologies into portable and economical sensor devices. These kinds of changes should make it less necessary to use big, complicated tools, which will make advanced analytical methods more accessible to both researchers and businesses (Johnston *et al.*, 2001; Lestander *et al.*, 2003). The continuous advancement of IR spectroscopy is set to augment its functionality, solidifying its status as a fundamental element of scientific inquiry and industrial implementation.

5.1. THE ELECTROMAGNETIC SPECTRUM

The electromagnetic spectrum includes a wide range of wavelengths, from radio waves on one end to gamma rays on the other. The infrared (IR) spectrum is one of these areas, and it is very important for infrared spectroscopy. The IR spectrum is usually split into three parts: near-infrared, mid-infrared, and far-infrared. Each part corresponds to a different way that molecules vibrate. In spectroscopic applications, it is important to know about these divisions because they are related to the energies of molecular transitions. This is why IR spectroscopy uses the specific absorption features in these areas to identify substances. This lets scientists learn about molecular structure and interactions. This collection of information about how molecules behave in the IR spectrum improves analytical chemistry and materials science, where it is very important to be able to tell exactly what chemical components are present. Direct uses of infrared spectroscopy show how useful it is for studying the electromagnetic spectrum. The method works by absorbing IR radiation, which makes molecules vibrate and rotate.

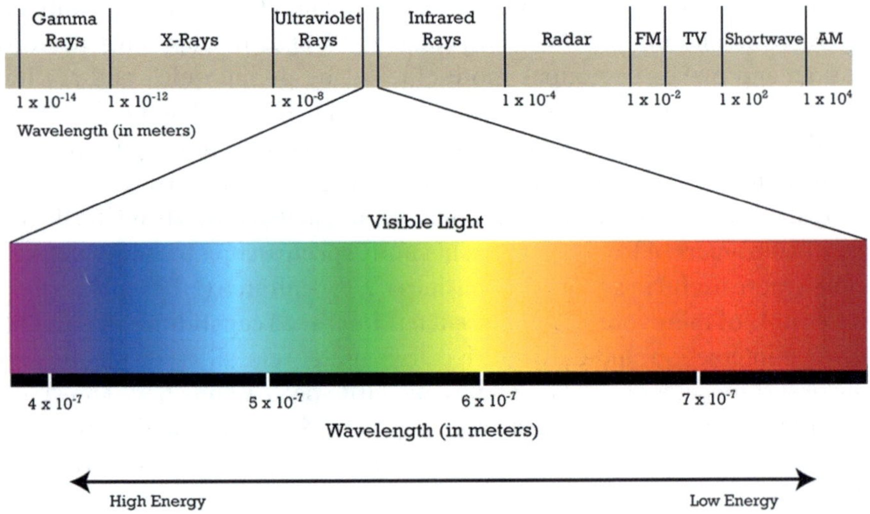

Figure 5.2. *Infrared radiation in the electromagnetic spectrum.*

Source: https://cdn1.byjus.com/wp-content/uploads/2023/03/Infrared-Radiation-in-the-Electromagnetic-Spectrum-Updated.png.

This creates unique spectral fingerprints for different chemicals. Researchers can look at changes that happen in the near-infrared, visible, and near-ultraviolet parts of the spectrum using Fourier transform spectroscopy, which makes measurements more accurate. Using advanced tools like the Michelson interferometric spectrometer makes it possible to study atmospheric gases that absorb light in these ranges. This technology shows how well the IR spectrum can be used for atmospheric studies, showing how spectroscopy can help with environmental monitoring and understanding how molecules interact (Gerlach *et al.,* 2025). Recent progress in materials science, particularly in graphene plasmonics, underscores the prospective applications of infrared spectroscopy within the electromagnetic spectrum. Graphene's unique properties let it interact with mid-infrared radiation, which opens up new possibilities for research and technology. Localized plasmonic modes in graphene can be used in a number of devices, including modulators and mid-infrared photodetectors. This pushes the field of spectroscopic analysis forward. It is important to note that making better spectroscopic tools with graphene can make vibrational spectroscopy applications more sensitive and accurate. As IR spectroscopy becomes more important, its use with new materials shows how analytical techniques are becoming more diverse and how they can help us learn more about science in many areas (Avouris *et al.,* 2014) (Figure 5.2).

5.1.1. Overview of the Electromagnetic Spectrum

The electromagnetic spectrum has a lot of different wavelengths, from gamma rays, which have the shortest wavelengths, to radio waves, which have the longest. This continuum is split into different areas based on how electromagnetic radiation interacts with matter and how it is used. The most important thing about these areas is infrared (IR) radiation, which has wavelengths between about 700 nanometers and 1 millimeter. The IR part of the spectrum is very important in areas like spectroscopy and thermography because it is linked to molecular vibrations and changes that happen in many chemical compounds. Researchers can learn about the different types of radiation by studying the EM spectrum. This helps improve technologies like IR spectroscopy, which is used to analyze chemicals and characterize materials. The principles of IR spectroscopy are based on how infrared radiation interacts with matter, so it is very important to understand how energy is transferred and absorbed. In this method, IR radiation makes molecular bonds vibrate, which makes them absorb certain wavelengths of light. This absorption creates unique spectral fingerprints that can be used to find and measure things. For example, in seed studies, near infrared (NIR) calibration models have proven very useful for predicting seed moisture content using multivariate analysis methods. This shows that IR spectroscopy can be used for more than just identifying chemical structures. The electromagnetic spectrum is therefore crucial for understanding how IR spectroscopy can be used in many different scientific fields (Lestander *et al.,* 2003). Also, new techniques like thermoplasmonics show how the IR region of the spectrum can be used in new ways. Researchers can raise the temperature from a distance by using focused infrared laser beams and plasmonic nanoparticles to heat specific areas. This method is being looked into using infrared nanospectroscopy, which lets you look at and analyze thermal images in detail at the nanoscale. These findings highlight the importance of infrared radiation in both material analysis and microscopic thermal management applications. As a result, the ongoing study of the electromagnetic spectrum improves our understanding of IR spectroscopy and adds new features to it, which helps progress in many fields of science and technology (Alabastri *et al.,* 2018).

5.1.2. Position of Infrared Radiation Within the Spectrum

The electromagnetic spectrum has a lot of different wavelengths and frequencies. Infrared (IR) radiation is in a certain part of the spectrum that connects the visible light and microwave parts. Infrared radiation is very important in both atmospheric science and spectroscopy. It is located between about 700 nanometers (nm) and 1 millimeter (mm). This area is split into three parts: near-infrared, mid-infrared, and far-infrared. Each part has its own physical properties and uses. The near-infrared band, which goes from 750 nm to 1400 nm, is very important in many fields, like agriculture and remote sensing, because water absorbs it and it can be used to check the health of plants. To use infrared spectroscopy as a powerful analytical tool, you need to know exactly where IR fits into the electromagnetic spectrum. The unique position of infrared radiation in the electromagnetic spectrum makes it useful for many scientific purposes, especially in spectroscopy. When IR radiation hits matter, it causes molecules to vibrate and rotate, which creates specific absorption patterns that show how the molecules are arranged. Analysts can identify specific functional groups in molecules by measuring these absorption bands. This helps with chemical identification and analysis. This property makes infrared spectroscopy very useful in a number of fields, such as pharmaceuticals, environmental science, and materials analysis. As researchers come up with more advanced methods, the ability to read IR spectra keeps getting better, which makes it easier to find gas emissions or pollutants. These kinds of progress show how important it is to put infrared radiation in the right place within the electromagnetic spectrum and what that means for scientific research. Many experimental protocols, including those in space-related projects, use IR for observational purposes, which further highlights how important it is to place infrared radiation correctly in the spectrum. Infrared is important for unmanned planetary missions because it is used by instruments that are designed to work in this range to collect data about celestial bodies and their atmospheres (Jones *et al.,* 2025). Scientists know that certain wavelengths of IR can go through gas atmospheres, which gives us more information about the makeup of planets and the features of their surfaces. Workshops on exobiology have also shown how useful IR techniques can be for experiments in Earth orbit and beyond, which has led to more research into their uses (Brownlee *et al.,* 2025). Consequently, comprehending the role of infrared radiation is not merely theoretical; it is essential to groundbreaking progress in both exploratory and applied scientific domains (Figure 5.3).

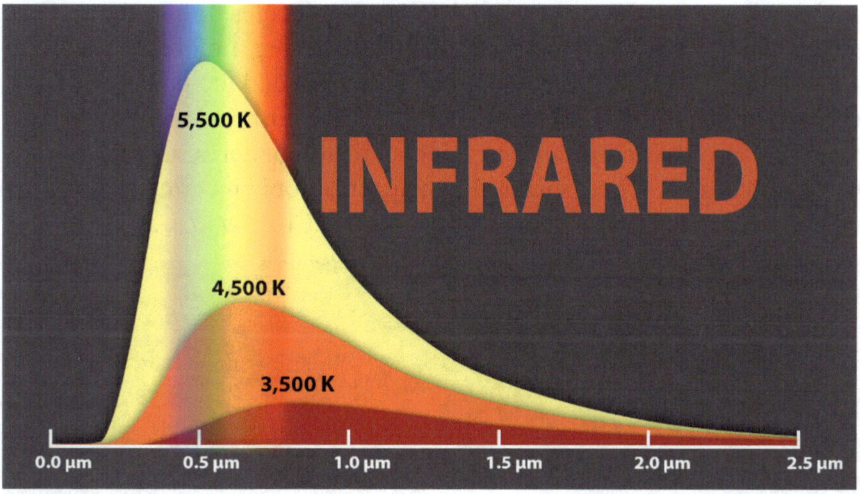

Figure 5.3. *Infrared light spectrum.*

Source: https://coolcosmos.ipac.caltech.edu/system/avm_image_sqls/binaries/193/original/Black_Body.jpg?1375209813.

5.1.3. Comparison of IR Radiation with Other Types of Radiation

Infrared spectroscopy provides a unique perspective in contrast to other forms of radiation, including ultraviolet (UV) and microwave radiation. Infrared radiation is found in a certain part of the electromagnetic spectrum, between microwaves and visible light. This arrangement gives IR radiation its own unique vibrational properties, which are especially important for how molecules interact with each other. Infrared radiation, on the other hand, is mostly responsible for exciting vibrational modes in molecules, while UV radiation is mostly responsible for electronic transitions. This makes it possible to look at functional groups and find molecular structures using spectral fingerprints. Consequently, the infrared region offers essential insights that are crucial in various applications, from chemical analysis to materials science, allowing researchers to detect finer structural details that may be concealed under UV or other high-energy radiation. When you compare infrared radiation to microwave radiation, you can see both similarities and important differences that are important for their different spectroscopic uses. Molecular vibrations are related to both types of radiation, but the ways they interact with each other are very different. Microwave radiation usually causes rotational transitions, which help figure out the shapes and details of molecules. Infrared radiation, on the other hand, looks at vibrational transitions, giving a more detailed picture of how molecules interact, including bond strength and the presence of certain functional groups. Mid-infrared spectroscopy has attracted significant interest due to its capacity to capture complex biochemical signatures, as demonstrated in biomedical research, where mid-infrared spectroscopic imaging is investigated as a diagnostic modality for cancer detection (McCrow *et al.,* 2011).

Figure 5.4. *Ultraluminous infrared galaxy.*

Source: https://external-content.duckduckgo.com/iu/?u=https%3 A%2F%2 Fpre-view. redd.it% 2Fic-4553-ultraluminous-infrared-galaxy-by-nircam-cycle-1-v0-2k6 hg9tpa3ca1.png%3Fauto%3Dwebp%26s%3D885e7ddc6933d4b6d49500adf70bb560cce27ea7&f=1&nofb=1&ipt=415d1054456eaa22b29331f7bfe9eb7e7083972a667c91eac6d71bbdac46244d.

This difference in how IR and microwave radiation behave shows how many different ways there are to characterize molecules across the electromagnetic spectrum. Additionally, improvements in infrared spectroscopy technology, like those made possible by infrared space observatories, have opened up new ways to compare it to optical techniques. For example, using mid-infrared spectroscopy to sort celestial bodies shows that IRs can do more than just work on Earth. Studies on ultraluminous infrared galaxies (ULIRGs) indicate that classifications derived from mid-infrared spectra can provide profound insights into galactic evolution and the composition of obscuring materials, in contrast to conventional optical techniques (Aussel *et al.,* 1999). This shows not only how useful IR radiation is for breaking down complicated systems, but also how useful it is when used with other types of spectroscopies to get a complete picture of molecular and cosmic phenomena. This is why the interaction between infrared radiation and other types of radiation keeps helping us understand chemical and astrophysical processes better (Figure 5.4).

5.2. TYPES OF INFRARED (IR) SPECTROSCOPY

Infrared (IR) spectroscopy includes a number of different methods; each designed for a certain use in chemical analysis and material characterization. Fourier Transform Infrared Spectroscopy (FTIR) is a well-known technique that uses a Michelson interferometer to measure how much infrared radiation a sample absorbs. FTIR is especially useful because it can quickly collect a wide range of spectral data, which gives samples detailed chemical fingerprints. This method is widely used in many fields of science, from finding functional groups in organic compounds to studying complicated biological materials. Moreover, improvements in FTIR have made it possible to make portable devices, which makes it more useful for fieldwork and on-site analyses. Another important type of IR spectroscopy is Near Infrared (NIR) spectroscopy, which is becoming more popular because it is fast and doesn't damage things. NIR spectroscopy works by measuring how much light is absorbed. The vibrations of molecules in the near-infrared range (usually 800–2500 nm) give information about the chemical makeup of a sample. It is widely used in the food and agriculture industries, where quick and accurate evaluations of ingredients are very important. NIR is a very useful tool for quality control in food production because it can quickly give quantitative results. This holds true for a wide range of products, such as meats, grains, and dairy (*see* Buning and Diller, 2000; Wilcox and Cavins, 1995; Williams and Norris, 1987 – for reviews of the development of NIR techniques prior to 1998, when diode arrays were introduced; *see also* Baianu *et al.,* 2011). Another important type of IR is mid-infrared spectroscopy, which is known for being very sensitive to molecular vibrations and overtones. This method has been increasingly utilized in biomedical domains, especially for the early detection of diseases like cancer. Recent research indicates that mid-infrared spectroscopic imaging can reveal essential morphological and biochemical data, thereby improving diagnostic precision (McCrow *et al.,* 2011). The addition of high-brightness laser sources, such as quantum cascade lasers (QCLs), has made the signal-to-noise ratios even better, which is very important in clinical settings. As mid-infrared technology continues to improve, its use in medicine is likely to close the gap between research and real-world use, which will lead to better patient outcomes through better diagnostic tools (Figure 5.5).

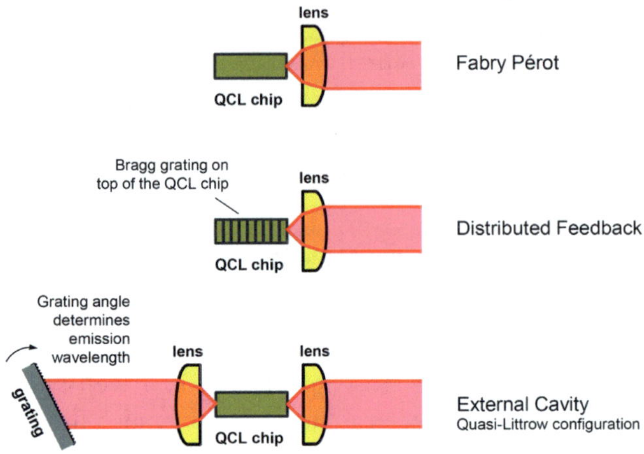

Figure 5.5. *Quantum cascade lasers (QCLs).*

Source: https://pubs.rsc.org/services/images/RSCpubs.ePlatform.Service.FreeContent.ImageService.svc/ImageService/Articleimage/2017/CS/c7cs00403f/c7cs00403f-f2_hi-res.gif.

5.2.1. Fourier Transform Infrared (FTIR) Spectroscopy

Fourier transform infrared (FTIR) spectroscopy is a key part of infrared (IR) spectroscopy. It is great because it can quickly and accurately give detailed information about molecules. This method uses the ideas behind Fourier Transform to turn time-domain signals into frequency-domain data. This makes it possible to do a lot of research on molecular vibrations and interactions. The capacity to acquire complete spectra in a singular measurement augments its applicability across diverse domains, including chemistry, materials science, and biomedicine. Recent developments, including FTIR-attenuated total reflectance (FTIR-ATR) and FTIR imaging, have greatly expanded the uses of FTIR, making it a useful tool for both qualitative and quantitative analysis in the lab (Armăşelu *et al.,* 2017). These innovations also make it easier to do research at the micro and nanoscale levels, which lets scientists look into materials and phenomena that were previously out of reach. FTIR spectroscopy is well-known for its use in analyzing biological and biomedical samples because it can be used for so many different things. Because the technique is sensitive to functional groups, it is very useful for finding biomolecules like proteins, lipids, and nucleic acids. FTIR is an important tool for medical research and diagnosis because it helps us understand how diseases work by giving us information about molecular conformations, interactions, and dynamics. The emergence of imaging techniques utilizing FTIR, including FTIR imaging spectroscopy, facilitates spatially resolved chemical information, which is essential for histopathological analyses (Bellatreccia *et al.,* 2013). FTIR has not only changed the way spectroscopy works, but it has also become an important tool for biological research. The ongoing advancement of FTIR methodologies, particularly their integration with synchrotron radiation, significantly improves the techniques' spatial and temporal resolution capabilities. These improvements have made FTIR useful for more mineralogical and environmental studies, giving scientists the ability to look at the molecular structure of geological materials with more accuracy than ever before (Bellatreccia *et al.,* 2013). Researchers can study very small amounts of samples when they use synchrotron sources and FTIR techniques together. This is important in fields that need detailed evaluations of complex matrices. So, the development of FTIR spectroscopy shows not only how technology has improved, but also how important this method is for solving modern scientific problems in many fields.

5.2.2. Attenuated Total Reflectance (ATR) Spectroscopy

Attenuated total reflectance (ATR) spectroscopy has become an important method in infrared (IR) spectroscopy because it makes it easier to analyze a wide range of samples. ATR spectroscopy makes it easy to directly measure solid, liquid, and even powdered samples by using an internal reflection element. This is different from traditional methods that often require complicated sample preparations. The principle of ATR is based on total internal reflection, which happens when an IR beam hits a crystal with a high refractive index at a certain angle. This creates an evanescent wave that goes into the sample that is touching the crystal surface. This one-of-a-kind method makes it possible to get high-quality spectral data with very little change to the sample. This makes ATR especially useful in fields like materials science and biological analysis. ATR spectroscopy is even more useful because it can be used on a wide range of materials, such as complex biological samples and nanostructured materials. For instance, in a recent study, scientists used ATR micro-spectroscopy to look into hyperbolic polariton modes in hexagonal boron nitride. This progress shows that ATRs can look at materials made in different ways without having to do a lot of nanostructuring or complicated sample handling. Also, ATRs are sensitive to surface and near-surface phenomena, which makes it possible to study thin films and layered structures that are becoming more important in nanophotonics and materials engineering. As a result, ATR spectroscopy not only increases the number of samples that can be used, but it also expands the range of scientific research that can be done in the field of IR spectroscopy (Caldwell *et al.,* 2018). ATR spectroscopy is useful for more than just practical purposes; it is also useful for multivariate data analysis, which has become more important in the field of vibrational spectroscopy. Researchers can now measure specific parts of complex mixtures by combining IR spectral data with complicated analytical methods like partial least squares regression. For example, a study showed that FT-MIR, NIR, and Raman spectroscopy could be used to predict the amount of trans fatty acids in dairy products. This showed how ATR could be used in food science to improve these methods. The ability to analyze and distinguish between naturally occurring and industrially induced substances further emphasizes the crucial role ATR plays in the advancement of analytical chemistry. The combination of ATR spectroscopy and data analysis makes it even more important in the field of IR spectroscopy as a whole, opening up new possibilities for use in many fields (Beattie *et al.,* 2015) (Figure 5.6).

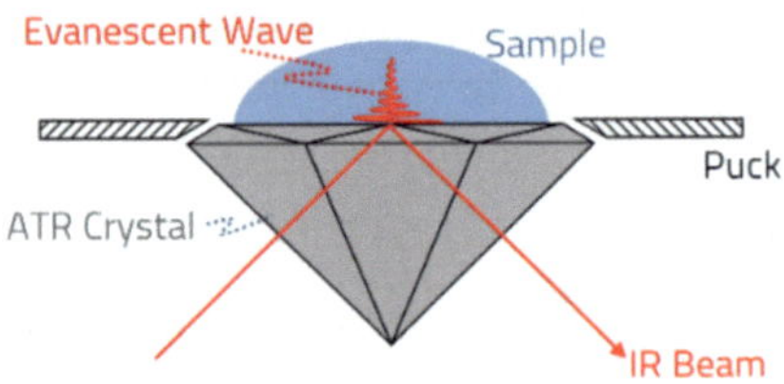

Figure 5.6. *Attenuated total reflectance ATR-FTIR spectroscopy principles.*

Source: https://external-content.duckduckgo.com/iu/?u=https%3A%2F%2Fspecac.com%2Fwp-content%2Fuploads%2F2022%2F04%2FTN21-01F1.jpg&f=1&nofb=1&ipt=75520254ee31059c6e6625d98e55c0c85a935302037b2d1964ab4423bd804635.

5.2.3. Near-Infrared (NIR) Spectroscopy

Near-infrared (NIR) spectroscopy is a powerful analytical method that is well-known for how well it can tell you about the molecular makeup of different substances. NIR spectroscopy uses the overtone and combination vibrations of molecular bonds to work in the wavelength range of about 750 to 2500 nanometers. It is especially good at analyzing organic compounds because biomolecules have C–H, N–H, and O–H functionalities that vibrate in certain ways. The method doesn't damage the sample, so it can be analyzed in real time without losing its integrity. This makes it especially useful for pharmaceuticals and agriculture. Also, improvements in NIR instruments have made them more sensitive and higher resolution, which allows researchers to find very small amounts of analytes. This makes them even more useful in a wide range of fields. NIR spectroscopy is useful in many different industries, especially for quality control and process monitoring. Food and drink, cosmetics, and petrochemicals are some of the industries that use NIR to quickly check the quality of their products and make sure they meet strict rules. NIR spectroscopy is very important for reducing waste and improving production processes because it makes it easy to quickly check the moisture content, fat levels, and chemical composition of a sample. Moreover, the combination of NIR spectroscopy and chemometric methods greatly improves the analysis, making it possible to get useful information from complicated data sets. As mentioned, the technology's ability to work without damaging the sample being tested fits with the industry's growing need for sustainable practices because it allows for continuous monitoring (Mitra *et al.,* 2018). As sensor technologies continue to improve, NIR spectroscopy is likely to improve scientific research beyond its current uses on Earth. For example, its ability to be used in remote sensing shows promise for lunar exploration, where NIR imaging spectrometers could tell what minerals are on the Moon's surface by looking at sunlight that reflects off of them. Researchers can get important geological information by using advanced data analysis tools, which helps us understand environments outside of Earth (Pieters *et al.,* 2025). These applications highlight the significant ramifications of NIR spectroscopy, positioning it at the confluence of important scientific research and technological progress. As research in this field advances, NIR spectroscopy is expected to significantly influence future analytical methodologies across various scientific disciplines.

5.3. INSTRUMENTATION AND TECHNIQUES

Infrared (IR) spectroscopy has made a lot of progress in terms of both instruments and methods, which has made it more useful in many areas of science. One important development is the rise of synchrotron radiation sources, which make it possible to do high-resolution IR spectroscopic studies. Synchrotron radiation makes it easier to study very small samples, which are often only a few micrometers in size. This feature broadens the range of research to include analytes that would be difficult to examine using traditional IR techniques, thereby offering researchers enhanced spatial and temporal resolution. These advancements highlight the increasing adaptability of infrared methodologies, establishing it as an essential instrument in contemporary analytical chemistry and associated fields, as elaborated in the extensive examination of infrared theory and instrumentation presented by Bellatreccia *et al.,* (2013). Another important area of IR spectroscopy is the use of portable instruments, which are becoming more and more important in studies of archaeology and cultural heritage. This mobile technology makes it possible to use spectroscopic methods on site, which protects valuable artifacts while allowing for real-time analysis. Researchers use nondestructive methods to lower the risk of damaging cultural heritage objects, which can be very fragile. Fourier transform infrared spectroscopy, and other similar techniques have been used a lot in these situations because they can characterize things without damaging them. As a result, mobile IR spectroscopy not only helps preserve historical artifacts, but it also helps us learn more about them by giving us important information about their materials and how they were made, as described by Donais *et al.,* (2016). Moreover, advancements in imaging techniques have enhanced the analytical potential of infrared spectroscopy, enabling time-resolved studies crucial for comprehending intricate geo-mineralogical processes. Imaging techniques let you see how the parts of a sample are spread out in space. This information is very important for both mineralogy and environmental science. These kinds of improvements let scientists' study how different mineral components interact with environmental factors, which opens up new areas for research in both basic and applied fields. New methods have the potential to work with other types of spectroscopies, which would make infrared methods even better at analyzing things. The continuous advancement of instrumentation and methodologies in IR spectroscopy not only expands the range of possible analyses but also creates a solid foundation for interdisciplinary research and applications within the scientific community (Bellatreccia *et al.,* 2013) (Figure 5.7).

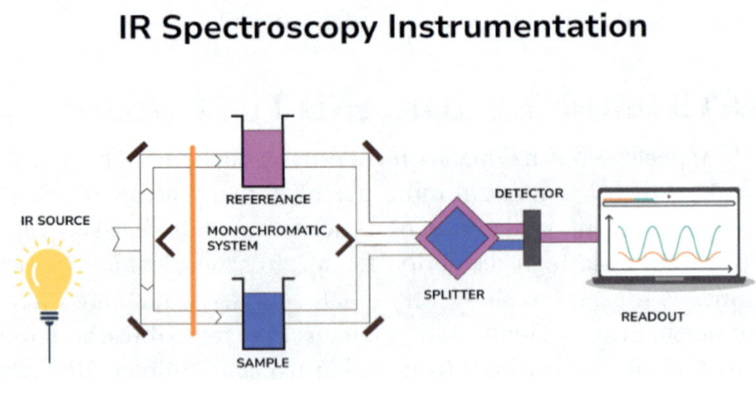

Figure 5.7. *IR-spectroscopy instrumentation.*

Source: https://media.geeksforgeeks.org/wp-content/uploads/20240506183015/IR-Spectroscopy-Instrumentation.png.

5.3.1. Components of an IR Spectrometer

The basic parts of an infrared (IR) spectrometer are very important for how well it works and how good the spectral data it gets is. The IR source is what makes the spectrometer work. It sends out infrared radiation over a wide range. Incandescent lamps, globars, and laser systems are all common sources. Each has its own benefits depending on the application. For example, quantum cascade lasers (QCLs), which are high-brightness, broadly tunable laser sources, have shown promise in improving signal-to-noise ratios and spatial resolution when analyzing biological tissues. This ability is very important in biomedical fields, like cancer diagnosis, where precise spectral data is needed to tell the difference between healthy and diseased tissues. Research has shown that advanced spectroscopic techniques can improve imaging capabilities (McCrow *et al.,* 2011). The second important part is the sample holder, which needs to make sure that the samples are in the path of the IR beam while causing as little interference as possible. Different holders are used for different types of samples, such as solids, liquids, and gases. The sample holder's job goes beyond just holding the sample; it must also provide stable and repeatable conditions for spectral measurement, making it easier to compare results from different analyses. The spectrometer has an optical system after the sample holder. This system may include mirrors, beamsplitters, and lenses that help focus and direct the IR light onto the detector. The efficiency and setup of this optical system are very important for the quality of the spectra it produces, which affects data reliability and analysis (Baianu *et al.,* 2011). Finally, the detector is the last important part of an IR spectrometer; it turns incoming IR radiation into electrical signals for data processing. Thermocouples, bolometers, and photoconductive detectors are all types of detectors that are commonly used. Each one has its own set of features, like how quickly it responds and how sensitive it is to different IR wavelengths, which makes them better for some uses than others. The choice of detector has a big impact on how well the spectrometer can make accurate and high-quality spectral data. This is important for a wide range of uses, such as food analysis, pharmaceuticals, and environmental monitoring. To make IR spectroscopy work better in a variety of analytical settings, it is important to understand these parts and how they work together (Baianu *et al.,* 2011; McCrow *et al.,* 2011).

5.3.2. Sample Preparation Methods

The details of sample preparation methods are very important for the accuracy of infrared (IR) spectroscopy, as they affect the quality of the analytical data collected from different materials. Because samples can be anything from solid minerals to painted textiles, it's important to use the right preparation methods to get the best results. In particular, when the amount of material is small, like when meteorites separate minerals, strict sample preparation steps must be followed to make sure that the sample characteristics are accurately represented. These specialized techniques can include grinding, pelleting, and the careful application of mounting media. All of these things make it easier for the IR beam to interact with the sample, which improves the quality and resolution of the spectrum (Grady *et al.,* 2004). Researchers can make IR spectroscopy results much more reliable by improving the ways samples are prepared. This can lead to important conclusions in studies in many different fields of science. When it comes to painted textiles, using IR spectroscopy requires special ways to prepare samples that take into account how fragile these historical objects are. For meaningful spectroscopic analysis, it is important to get the materials out of the fabric without damaging it. Recent developments have indicated the incorporation of non-invasive methodologies that facilitate sampling without compromising the integrity of the textile, thus safeguarding its historical significance. Techniques like mapping with Fourier transform infrared spectroscopy improve the spatial resolution of spectral data, which lets us make more detailed guesses about the textile's composition (Lennard *et al.,* 2017). These advancements highlight the interplay between scientific research and conservation efforts, resulting in enhanced methodologies that improve analytical outcomes and support the safeguarding of cultural heritage. The development of sample preparation techniques in IR spectroscopy exemplifies a wider trend toward interdisciplinary collaboration in materials analysis. Combining vibrational spectroscopy with other methods, like Raman spectroscopy and mass spectrometry, has been helpful for getting a full picture of the composition. This integrated approach not only underscores the significance of selecting suitable sample preparation techniques but also demonstrates how the limitations of one method can be alleviated by utilizing the strengths of complementary methods (Lennard *et al.,* 2017). For example, using surface-enhanced Raman spectroscopy with IR techniques can help us learn more about materials that were hard to analyze before. Collaborative frameworks like these give researchers more analytical tools and make it easier for them to make big strides in understanding complex materials. This will ultimately improve IR spectroscopy and its uses.

5.3.3. Data Acquisition and Analysis

The way that data is collected in infrared (IR) spectroscopy is very important because it has a direct effect on the quality of the analytical results. Usually, the acquisition means getting spectral data, which means being very careful with things like wavelength and intensity. Modern IR spectroscopic techniques use cutting-edge digital technology to make sure that data is collected correctly. This lets them create high-resolution spectra that show how molecules interact with each other in great detail. Different sample states can change the quality of the signal, so spectral acquisition has to take into account things like temperature, humidity, and sample thickness. Because Near Infrared (NIR) spectroscopy and other non-invasive methods can be used in many different fields, it is important to have reliable data collection methods that can always give the same results, especially in complicated settings. This is a big advantage over traditional methods (Baianu *et al.,* 2011). After the data has been collected, the analysis phase is when the captured spectra are interpreted to get useful information about the sample. To figure out what the spectral data means, scientists often use complex algorithms and mathematical models, such as multivariate analysis.

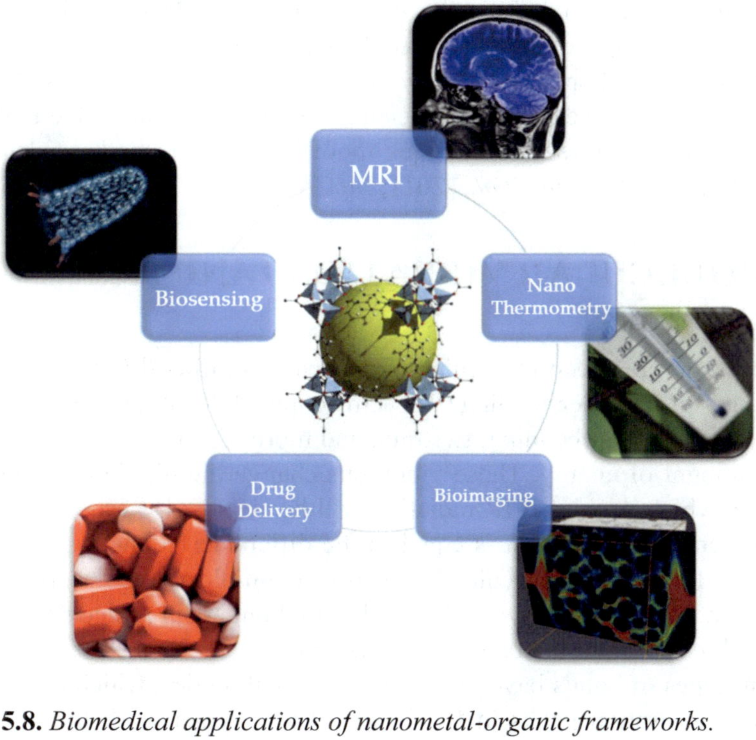

Figure 5.8. *Biomedical applications of nanometal-organic frameworks.*

Source: https://www.researchgate.net/publication/344062162/figure/fig1/AS:1143 1281419281963@1746192974119/Overview-of-biomedical-Applications-of-nano-metal-organic-frameworks-nanoMOFs.tif.

This lets them find and measure the chemical parts. The analysis also makes it easier to tell the difference between spectral patterns that match certain molecular structures. This is very useful in many fields, from material science to biomedicine. For instance, mid-infrared spectroscopy can help find out about biochemical states that could help find diseases early or keep an eye on treatment. Nonetheless, challenges like noise reduction and baseline correction are crucial to guarantee the precision of the analysis, as ongoing improvements in software tools enhance these functions, thereby increasing the dependability of the results obtained (McCrow *et al.,* 2011). In addition to traditional methods of processing data, new ways of analyzing data are also important for making infrared spectroscopy more useful. Machine learning and artificial intelligence are being used more and more in the data processing workflow. This makes it easier to analyze large amounts of data quickly. These technologies can find patterns and connections that normal analytical methods might not be able to see, which can improve the accuracy of predictions in many different areas. For instance, they can help make diagnostic protocols better by using IR spectroscopy to help explain the molecular causes of diseases. Researchers can make the most of IR spectroscopy by improving how they collect and analyze data. This makes it useful in more scientific fields and helps solve real-world problems, which makes the technique more important in both academic and industrial settings (Baianu *et al.,* 2011; McCrow *et al.,* 2011) (Figure 5.8).

5.4. MOLECULAR VIBRATIONS AND IR ABSORPTION

The connection between molecular vibrations and infrared (IR) absorption is very important for understanding IR spectroscopy. Molecules vibrate in different ways, like stretching, bending, twisting, and more. Each way corresponds to a certain amount of energy. These vibrations change the dipole moment of the molecule, which decides if IR radiation will be absorbed. Resonance happens when the energy of IR light is equal to the difference in energy between the vibrational levels of a molecule. This causes the molecule to absorb the light. In this way, IR spectroscopy can be used to find out about molecular structure, functional groups, and interactions in a wide range of chemical environments. Different types of bonds have different vibrational modes, which can show up as unique absorption peaks in IR spectra. These peaks work like molecular fingerprints that chemists can use to identify substances and study their structural properties. To make sense of the absorbance data from IR spectroscopy, you need to know how these molecular vibrations work. Recent improvements in methods like time-resolved vibrational spectroscopy have made it much easier to study molecular vibrations and their related IR absorption. These new methods let scientists' study fast-moving processes at the molecular level and get vibrational spectra with never-before-seen time resolution. For example, ultrafast laser spectroscopy has helped us learn more about biochemical and physical processes by giving us important information about how molecules move and how they are structured during reactions. Conferences such as the one organized by *Ludwig-Maximilians-Universität* and *Technische Universität München* show that people in this field remain committed to developing new ideas. They highlight the work done in time-resolved studies of molecular vibrations. The merging of traditional IR techniques with new laser methods shows how important it is to keep looking for new ways to use IR spectroscopy to learn more about molecular dynamics and interactions (Laubereau *et al.*, 2007). Furthermore, the interaction between localized heating effects in nanoscale systems and vibrational absorption has created opportunities for improved molecular characterization. Surface-enhanced infrared absorption (SEIRA) is an example of this dynamic because it makes the vibrational signals of molecules that are close to plasmonic nanoparticles stronger. Recent studies have shown that the localized thermal changes caused by mid-infrared laser interactions can greatly increase molecular vibrations, making it easier for certain vibrational modes to absorb energy. The significant rise in temperature at nanoscale hotspots validates the possibility that minor fluctuations in vibrational energy levels can result in detectable alterations in IR absorption spectra (Alabastri *et al.*, 2018). This progress opens up a new area of material science for both basic research and practical use. It makes it possible to use highly sensitive detection methods and adds to the impact of IR spectroscopy in modern chemistry (Figure 5.9).

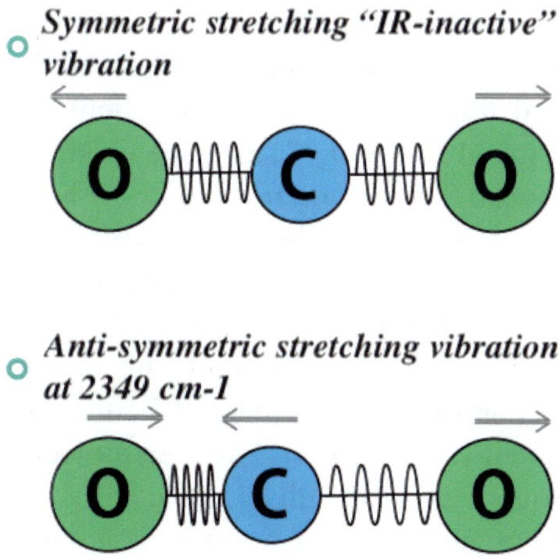

Figure 5.9. *Molecular vibrations.*

Source: https://www.jasco-global.com/wp-content/uploads/2019/02/IR1-2.png.

5.4.1. Types of Molecular Vibrations (Stretching and Bending)

To understand infrared (IR) spectroscopy, especially the two main types—stretching and bending vibrations—it's important to study molecular vibrations. There are two types of stretching vibrations: symmetric and asymmetric. These are changes in bond lengths. When two atoms that are next to each other move in the same direction while staying the same distance apart, this is called symmetric stretching. Asymmetric stretching, on the other hand, has these atoms moving in opposite directions, which changes the bond length in an uneven way. This distinction is important because it changes the frequency of the IR absorption bands that can be detected, which gives us information about the molecular structure and environment. The analysis of *planchéite* illustrates that the study of specific stretching vibrations allows chemists to infer essential characteristics related to SiO_3 units, highlighting the importance of these vibrational modes in substance identification via IR spectroscopy (Frost *et al.,* 2012). Bending vibrations, on the other hand, change the angle between bonds instead of the length of the bonds. There are two types of bending: in-plane and out-of-plane. Scissoring, or in-plane bending, happens when two atoms move closer together and then farther apart, changing the angle between two bonds in the same plane. Out-of-plane bending, which is also called "wagging" or "twisting," is when atoms move in a way that changes their positions outside of the plane defined by the bonds next to them. Both kinds of bending vibrations give us important information about how molecules are shaped and how they interact with each other. For example, research on organoclays has employed IR spectroscopy to examine bending modes in conjunction with thermal stability (Frost *et al.*, 2007). These vibrational modes are essential for both identification and for comprehending stability and reactivity in diverse chemical environments. To get the most out of IR spectroscopy, you need to know how to understand both stretching and bending vibrations. Finding these vibrations makes it possible to do detailed qualitative and quantitative analysis of molecular compounds, which shows structural and functional group characteristics. Vibrational frequencies are like fingerprints for different substances because they are linked to bond strengths and molecular shape. Chemists can also learn about changes in the environment or interactions between molecules by looking at the shifts in resonance frequency. This ability is especially useful in complicated systems, like those with organoclays or certain hydrated mineral structures like *planchéite*, where changes in molecular vibrations reveal important information about how these materials behave and stay stable. The investigation of stretching and bending vibrations not only aids in identification but also deepens the comprehension of the fundamental physicochemical properties of molecules in the context of IR spectroscopy (Frost *et al.,* 2012).

5.4.2. Factors Affecting IR Absorption

Different molecular properties, such as bond type, molecular symmetry, and functional groups, have a big effect on how well infrared (IR) absorption works. For example, different kinds of chemical bonds, like C-H, O-H, and N-H, have different vibrational frequencies. This causes them to absorb light in different ways in the IR spectrum. Also, molecular symmetry is very important; molecules that are not symmetric tend to have more absorptive bands than molecules that are symmetric. The idea behind this is that for a molecular vibration to absorb IR radiation, it has to cause a dipole moment change. As a result, molecules with polar functional groups absorb IR light better than those with nonpolar bonds. This shows how important molecular structure is in determining how molecules vibrate. The environment in which the molecules are found is another important factor that affects IR absorption.

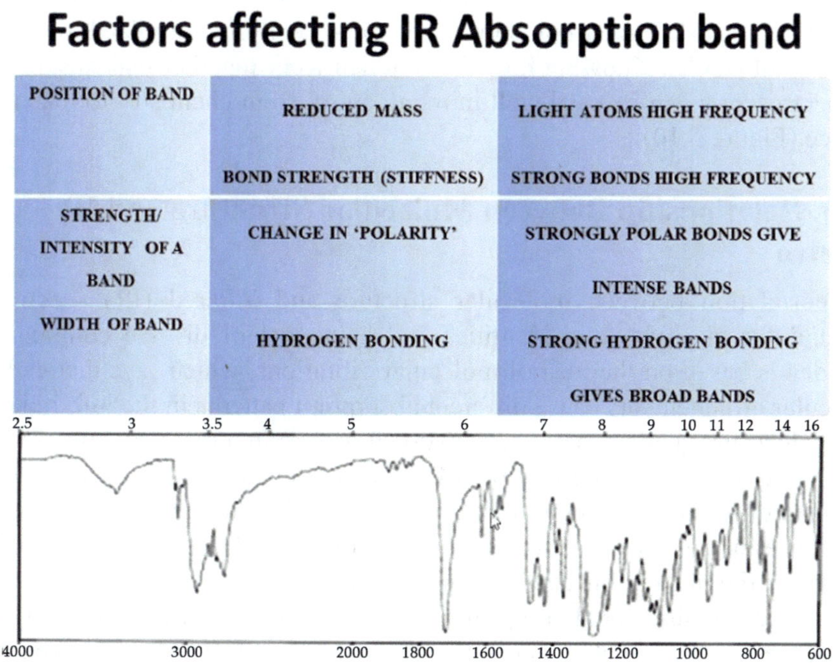

Figure 5.10. *Factors affecting IR absorption frequency.*

Source: https://i.ytimg.com/vi/SFumYTr0OA0/maxresdefault.jpg.

The solvent effect is when solvents change the frequency and intensity of a certain absorption band. Solvent polarity can change the positions of absorption peaks, especially for polar functional groups because of solvation effects. This change is very important for quantitative IR analysis because differences in how solvents interact with each other can cause errors in spectral data. A methodical examination of various solvents and their interactions with the analyte can elucidate the molecular environment, thereby enhancing the precision of IR spectroscopy as a formidable analytical instrument. Temperature changes can also change IR absorption because they can change how molecules move, which in turn changes how the molecules vibrate. Finally, things like concentration and path length outside the system are very important in figuring out how much IR is absorbed. Beers Law says that the absorbance is directly related to the concentration of the absorbing species and the distance that the IR light travels. So, higher amounts of an analyte usually lead to more absorption, which makes spectral readings clearer. But it's important to remember that samples that are too concentrated can cause saturation effects, where the absorption levels off and makes the response less linear. Also, the thickness of the sample can change the recorded spectrum, so it needs to be carefully optimized to get the most useful results. Knowing how these things work together makes sure that IR spectroscopy can be used well in many areas, from chemistry to materials science (Figure 5.10).

5.4.3. Relationship Between Molecular Structure and IR Spectra

The correlation between molecular structure and infrared (IR) spectra is essential for clarifying the identities and behaviors of diverse compounds. The idea is based on the idea of molecular vibrations, which says that certain molecular arrangements cause different absorption patterns in the infrared part of the electromagnetic spectrum. Functional groups in a molecule vibrate in a certain way because of the way their bonds are arranged and how strong they are. This difference in vibrational frequency is what makes IR spectroscopy such a useful tool for chemists to use to identify and describe substances. By looking at the absorption bands in an IR spectrum, you can learn important things about the functional groups that are present, how they are arranged, and even some information about the molecular symmetry. This helps us learn more about the compound's identity and function. Advanced techniques that improve spatial resolution can greatly improve the accuracy of IR spectroscopy in showing molecular structure. For example, a high spatial resolution IR imaging method that uses atomic force microscopy lets scientists look closely at organic matter in meteorite samples. This method shows how IR spectroscopy can be used in new ways, showing how functional groups are spread out unevenly at the nanometer scale. These improvements in spatial resolution are important for looking at organic materials and for helping us understand how molecular interactions affect spectral characteristics. The techniques have demonstrated a remarkable correlation with established associations between organic matter and minerals, highlighting the crucial connection between molecular structure and the resultant IR spectral data (Baden *et al.,* 2019). Recent studies have shown that different molecular conformations have a big effect on the vibrational spectra, which shows how molecular architecture and IR spectra are connected. For instance, mechanosensitive membrane channels show different IR responses depending on how they are gated. We used linear dichroism and sum-frequency generation spectroscopies to show how the structure of these channels changes as they open. This shows how IR methods can be used to pick up on complex molecular dynamics. Linear absorption might miss small structural differences, but advanced spectroscopic methods show important information about changes in shape and function. As a result, these insights into how molecules behave show how different structural features directly affect IR spectral patterns. This makes IR spectroscopy an important tool for both qualitative and quantitative molecular analysis (Birkner *et al.,* 2013).

5.5.APPLICATIONSOFINFRAREDSPECTROSCOPY

Infrared (IR) spectroscopy is used for much more than just chemical analysis; it has a big effect on many areas of science. For example, in environmental science, IR spectroscopy has become an important tool for measuring soil properties. Researchers can use visible-near infrared (vis–NIR) spectroscopy to look at important soil properties like the amount of organic matter, the types of minerals, and the levels of nutrients. Scientists can get quantitative data from spectral readings by using multivariate calibration methods. This helps them learn more about soil properties and how they affect ecosystems. As Mouazen et al. (2010) explain, IR spectroscopy's ability to do strong calibration and predictive modeling makes it possible to evaluate soil health and composition without damaging it. This is important for sustainable farming and environmental monitoring. In the field of material science, infrared spectroscopy also has many uses that make it easier to describe new materials. This method is very good at finding molecular structures, functional groups, and chemical bonds, which is why it is so important for developing new materials and making sure they are of high quality. Researchers can learn about the thermal stability, mechanical properties, and chemical resistances of polymers, composites, and nanomaterials by looking at their IR spectra. Furthermore, progress in photonic science, as noted by Aggarwal et al. (2016), has created new opportunities for IR spectroscopy in technological applications. Combining fiber optics and adaptive optics with IR instruments makes measurements more accurate and makes spectroscopy more useful in new technologies like smart materials and sensors. Finally, infrared spectroscopy plays a very important role in biological and medical research, with great potential for use in diagnostics and therapeutic monitoring. The capacity to swiftly analyze biological specimens for distinct molecular signatures can enhance the early identification of diseases, including cancer. Researchers can create targeted therapeutic strategies by using IR spectroscopy to find unusual biomolecular interactions or changes. Additionally, the non-invasive nature of the techniques is beneficial for *in vivo* studies, facilitating real-time monitoring of drug delivery and metabolism. The transformative potential of IR spectroscopy in biomedical applications underscores its versatility, solidifying its essential role in contemporary research across various disciplines (Figure 5.11).

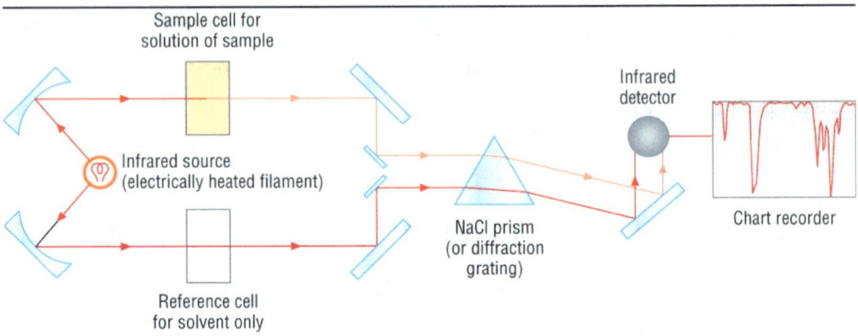

Figure 5.11. *Infrared (IR) spectroscopy theory.*

Source: http://www.chemhume.co.uk/ASCHEM/Unit%202/Ch10%20Alkenes/infrared_spectrometer.jpg.

5.5.1. Use in Organic and Inorganic Chemistry

Infrared (IR) spectroscopy is an important analytical method in both organic and inorganic chemistry. It lets you find and describe different molecular structures. In organic chemistry, IR spectroscopy mainly looks at the vibrational transitions of functional groups. This helps chemists figure out which bonds are present, such as O-H, C-H, and N-H, which are important for understanding how molecules behave. For example, researchers can use Fourier Transform Infrared Spectroscopy (FTIR) to get information about molecular vibrations and their spectral signatures. This makes it easier to identify complex organic compounds. This method is used a lot in fields like pharmaceuticals and petrochemicals. It gives important information that helps with synthetic strategies and quality control in organic synthesis, as shown by the results of literature reviews that looked at the analytical capabilities of these techniques (Lennard *et al.,* 2017). In the field of inorganic chemistry, IR spectroscopy reveals the complex interactions between metals and ligands and coordination complexes. Transition metal compounds, especially metal sulfides, have unique vibrational properties that can be easily studied with IR techniques. The unique IR spectra produced by these compounds provide essential information about their structure and reactivity, contributing to progress in material science and catalysis concerning inorganic compounds. Additionally, the combination of IR spectroscopy with other techniques, like Raman spectroscopy, makes it possible to study inorganic materials in depth, which helps us understand their complicated surface chemistry. This combined approach is necessary to get the best functional properties out of inorganic compounds. This is especially important for bioimaging, where the stability and behavior of metal sulfides as colloidal suspensions are very important (Dengo *et al.,* 2017). Recent improvements in analytical methods, such as mapping capabilities that let you take measurements in different places, have also made IR spectroscopy more useful for studying both organic and inorganic compounds. These new tools let scientists do in-depth studies of painted textiles, which often use both organic dyes and inorganic pigments. Using FTIR mapping and other methods makes it easier to study degradation processes, which is important for conservation and material characterization. The partnership between chemists and conservators shows how important interdisciplinary research is. For example, IR spectroscopy not only helps us understand what materials are made of, but it also improves the methods used to preserve cultural heritage artifacts. This connection highlights the importance of IR spectroscopy as a key instrument in integrating organic and inorganic chemistry in practical applications (Lennard *et al.,* 2017).

5.5.2. Applications in Pharmaceuticals and Biochemistry

Infrared (IR) spectroscopy is a game-changer in the fields of pharmaceuticals and biochemistry, especially when it comes to drug development and molecular characterization. IR spectroscopy shows what functional groups are in a compound by looking at the unique vibrational modes of molecular bonds. This ability is very important for figuring out what new pharmaceutical compounds are, since it can confirm the structure and purity of active ingredients. Also, IR spectroscopy has been used to watch chemical reactions happen in real time, which has helped us learn more about the kinetic and mechanistic parts of drug synthesis. The fact that the methods don't damage the samples makes them especially useful for studying sensitive biological samples, since they can collect useful information without harming the samples. In biochemistry, IR spectroscopy is a very useful tool for looking at how proteins are put together and how they interact with each other. By looking at the fingerprint region of IR spectra, scientists can figure out what functional conformations are and how they might change when a ligand binds. This application encompasses the evaluation of protein folding, dynamics, and post-translational modifications essential for comprehending biological processes. For example, a study on organotin(IV) compounds used ATR-FTIR spectroscopy to identify changes in the molecules of cell lines. This shows how IR can give important biophysical information that works well with traditional biochemical methods. The capacity to observe cellular reactions to pharmaceuticals via IR spectroscopy augments our comprehension of drug efficacy and mechanisms of action. Moreover, the incorporation of IR spectroscopy in the examination of metal complexes signifies a pivotal advancement in pharmaceutical research. Recent studies have shown how metal ions and organic ligands interact with each other in complex ways, which could lead to new ways to treat infections. One example is the synthesis of mixed-ligand metal complexes, where IR spectroscopy showed how ligands such as *sulphadoxine* and *pyrimethamine* bond with copper and iron ions. The resulting complexes exhibited superior antibacterial properties relative to their precursor compounds, underscoring the promise of metal-based technologies in pharmaceutical formulation. These results show that IR spectroscopy is not only useful for identifying compounds, but it also helps us understand their biopharmaceutical properties and safety profiles. This shows how important it is in modern pharmaceutical sciences (Figure 5.12).

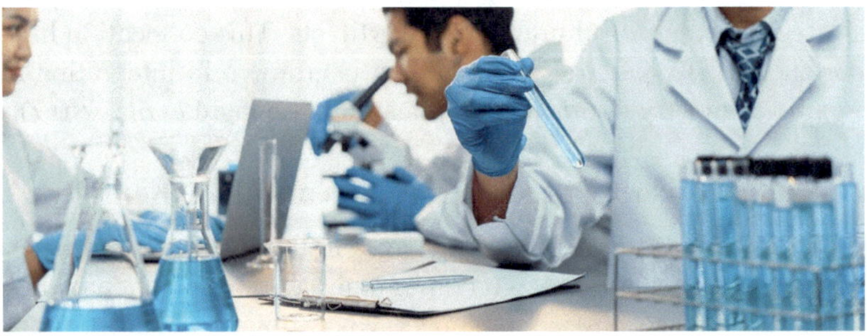

Figure 5.12. *Pharmaceuticals and biochemistry.*

Source: https://img.freepik.com/premium-photo/group-dedicated-scientist-conduct-chemical-experiments-research-medical-lab-as-groundbreaking-developing-vaccine-drug-antibiotic-pharmaceutical-biochemistry-laboratory-neoteric_31965-243313.jpg?w=2000.

5.5.3. Role in Environmental Monitoring and Analysis

Infrared (IR) spectroscopy is very important for monitoring and analyzing the environment because it lets us find and measure different compounds without hurting them. The growing worries about climate change and pollutants in the environment make it necessary to have analytical methods that work well and are trustworthy. IR spectroscopy, especially non-dispersive infrared absorption spectroscopy (NDIR), is a leading technology that works well even in tough situations. This feature makes it more useful for real-time monitoring of greenhouse gases and other pollutants, giving us important information that we need to understand and deal with environmental problems. For example, improvements in NDIR technologies have made it possible to make sensors that can accurately find emissions from industrial processes and exhaust systems. This has led to better compliance with regulations and more proactive environmental stewardship (Mouazen *et al.,* 2010). Additionally, IR spectroscopy can be used to look at the properties of soil, which is important for judging how land is used and how crops are grown. Visible-near infrared (vis-NIR) spectroscopy is especially useful for figuring out what minerals and organic matter are in soil and how available nutrients are. This part of environmental analysis is important for long-term land management and keeping an eye on soil health over time. As research advances, the incorporation of IR spectroscopy into soil analysis facilitates the creation of robust multivariate calibrations that enhance the prediction of essential soil properties. These predictive abilities help people in agriculture make smart choices about how to amend and manage soil, which helps protect soil and boost agricultural productivity (Mouazen *et al.,* 2010). IR spectroscopy helps monitor the quality of water in both land and water environments by giving detailed information about water quality. IR spectroscopy can find contaminants in water bodies, like heavy metals and organic pollutants, by measuring the vibrational transitions in molecular bonds. This analysis is useful for figuring out how well water treatment processes work and making sure that drinking water is safe. Using the latest IR technology can greatly improve the speed and accuracy of environmental assessments, making it possible to set up full monitoring programs in both cities and the countryside. As the need for more environmentally friendly methods grows, the combined use of IR spectroscopy in different areas of the environment makes it a key technology for future ecological research and policymaking (Bauke *et al.,* 2018).

5.6. LIMITATIONS AND CHALLENGES

There are some problems and limits with using infrared (IR) spectroscopy in analytical settings. One major problem is that sample matrices are naturally complex, which can make spectral analysis harder. In materials characterization, especially of painted textiles, overlapping absorption bands can make it hard to find important parts. Authors have observed that significant discoveries frequently result from meticulous collaborations between conservators and scientists, highlighting the essential role of contextual knowledge in the interpretation of infrared data. This challenge underscores the necessity of refining techniques such as Fourier transform infrared spectroscopy (FTIR) and leveraging advancements in mapping methodologies to improve material analysis, considering specific sample characteristics, as articulated in literature reviews on analytical techniques for textile characterizations (Lennard *et al.*, 2017).

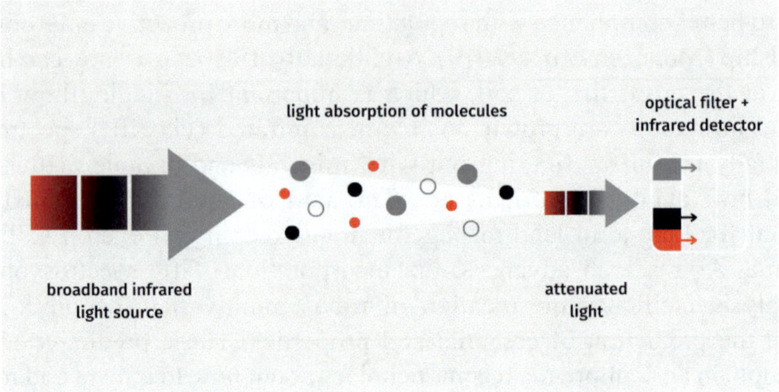

Figure 5.13. *Non-dispersive infrared spectroscopy (NDIR).*

Source: https://www.trace-gas.com/wp-content/uploads/2022/08/NDIR-EN.png.

The sensitivity of IR spectroscopy to sample preparation and environmental conditions presents considerable challenges. Sample degradation during preparation can change the chemical makeup of the sample, which can lead to wrong interpretations of the spectrum. The need for precise measurements in difficult situations has led to the creation of new methods. Non-dispersive infrared absorption spectroscopy (NDIR), for instance, seeks to mitigate the constraints imposed by environmental variables, facilitating non-invasive detection of gases and pertinent compounds (Bauke *et al.,* 2018). These new ideas show that people are still trying to solve the problems that traditional IR spectroscopy faces by coming up with solutions that work in different situations and make the results more reliable. Finally, the full potential of IR spectroscopy can be limited by the availability and use of the right tools. Not everyone can use advanced instruments that can do high-resolution analysis, and they often require expensive investments. This financial barrier can make it hard for underfunded institutions or smaller labs to do research, which slows down the progress of knowledge in the field. Also, differences between instruments can make it harder to compare data from different studies, which can make it harder to reproduce results. Therefore, creating standardized protocols and making sure that advanced methods are available may help get around some of these problems, making IR spectroscopy more widely used and accepted in different research fields while also dealing with its problems (Figure 5.13).

5.6.1. Limitations of IR Spectroscopy in Certain Analyses

Infrared (IR) spectroscopy is a useful analytical tool, but it has some drawbacks that can make it less useful for some purposes. One big problem is that it is sensitive to water, which often messes up the spectra of organic compounds. When samples have moisture in them, the water vapor can hide important absorption bands, making it harder to find and measure the target molecules. This problem is especially bad in biochemistry and environmental analysis, where samples often have a lot of water in them. Because of this, researchers may have to use drying or other methods to prepare samples, which could make the analysis take longer and be more complicated. These limitations require a thorough comprehension of the sample matrix and meticulous evaluation of the conditions in which IR spectroscopy is utilized to produce dependable and interpretable results. The structural complexity of some compounds can also make it much harder for IR spectroscopy to do its job. This method mostly picks up on molecular vibrations that are linked to functional groups. This means that compounds with similar functional groups might have absorption peaks that overlap. Because of this, it can be hard to tell the difference between closely related chemical species, especially in complex mixtures that are often found in real-world situations like analyzing painted textiles. The examination of analytical methodologies, particularly regarding painted textiles, underscores these intricacies, as vibrational spectroscopy may yield inconclusive outcomes, thereby requiring supplementary techniques for precise differentiation. The collaboration between conservators and scientists in interpreting findings is essential, as it deepens the understanding of material compositions and their historical context, highlighting the importance of integrating various analytical methodologies for thorough analysis (Lennard *et al.,* 2017). Besides these problems, IR spectroscopy has built-in limits on the kinds of materials it can analyze well. The method usually works best with polar molecules that have strong dipole moments, which makes the spectra clear and strong. Non-polar compounds, like hydrocarbons, might not show any IR activity or only show it weakly, which makes it hard to analyze them. This limitation constrains the range of substances that can be precisely characterized by IR spectroscopy, especially within the food and agricultural sectors where a variety of organic materials require evaluation. As noted in the literature, alternative methods like Near Infrared (NIR) spectroscopy provide a beneficial option for non-invasive and swift analysis, especially in studies of food composition. However, IR spectroscopy is still very important because it has unique benefits for identifying and characterizing molecules (Baianu *et al.,* 2011). Consequently, although IR spectroscopy is essential in numerous analytical contexts, acknowledging its limitations is vital for choosing suitable methodologies in various scientific investigations.

5.6.2. Challenges in Interpreting Complex Spectra

Interpreting complex spectra in infrared (IR) spectroscopy is very hard, especially when it comes to telling apart overlapping spectral features. Different materials can make complex spectra that can hide important information about the sample's composition. This complexity frequently stems from the numerous vibrational modes linked to various chemical bonds; consequently, professionals must have a profound comprehension of molecular interactions to accurately interpret the data. Additionally, environmental variables such as temperature and pressure can affect spectral results, making the analysis even more difficult. Because of this, successful interpretation requires not only knowledge of spectral libraries but also an understanding of the physical laws that control how molecules behave in different situations. This can make it hard to figure out what materials are in a sample. The presence of overlapping spectral peaks and the difficulty of telling the difference between compositional and non-compositional factors make it much harder to understand IR spectra. The primary concern is that some spectral features can be caused by non-chemical factors like grain size, phase angle, and temperature, which can hide the chemical information that researchers are looking for. Recent progress in the field has made it easier to measure these outside factors more accurately, which has led to new ways to lessen their impact on spectral readings. Nonetheless, the enduring characteristics of these non-compositional influences, particularly in instances involving heterogeneous samples, highlight a continual challenge for spectroscopists. This challenge is similar to what has been found in reflectance spectroscopy, where systematic differences can make it hard to get accurate spectral information, so interpretive methods need to be constantly improved (McCord *et al.,* 2025). Although considerable advancements have been achieved in spectral calibration and interpretative methodologies, specific spectral types, especially those displaying minimal features, persist in posing formidable challenges. For instance, figuring out what minerals and physical properties are in spectrally featureless asteroids is very important. Improvements in laboratory spectral calibrations have helped us understand how different things show up in spectral data, which has made it easier to find asteroids in the main asteroid belt (Burbine *et al.,* 2015). Analysts frequently contend with an absence of distinct spectral markers, which can impede efforts to ascertain mineral chemistries and abundances. These challenges exemplify a significant research domain in IR spectroscopy and underscore the necessity of creating innovative interpretive tools capable of accurately clarifying spectral data where conventional techniques are inadequate (Figure 5.14).

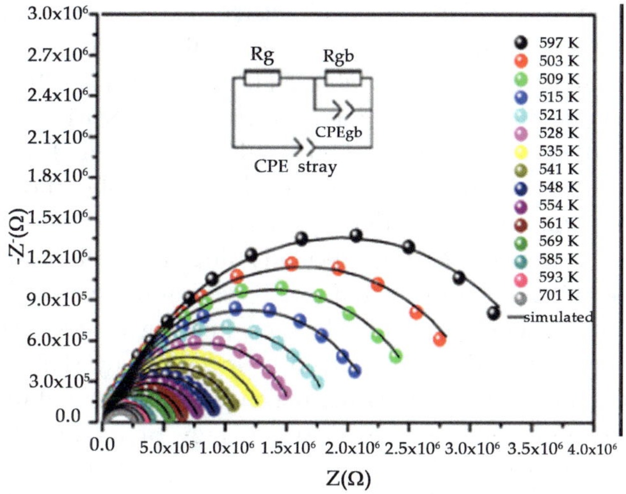

Figure 5.14. *Complex spectra at different temperatures.*

Source: https://www.researchgate.net/profile/Marwa-Enneffati/publication/320204321/figure/fig12/AS:941610980356115@1601508960327/Complex-spectra-at-different-temperatures.gif.

5.6.3. Advances to overcome limitations

There have been big improvements in infrared (IR) spectroscopy that have helped it get around some of the problems that come with older analytical methods. In the past, IR spectroscopy was limited in terms of sensitivity and resolution, which made it hard to use for analyzing complex materials like painted textiles. Recent advancements, especially in Fourier Transform Infrared Spectroscopy (FTIR), have alleviated these limitations by improving spectral resolution and facilitating the accurate characterization of molecular vibrations. The incorporation of mapping techniques into FTIR enables spatially resolved spectral analysis, thereby enhancing comprehension of the distribution of various components within a sample, which improves interpretative accuracy in the examination of historical artifacts (Lennard *et al.,* 2017). These improved abilities not only make it easier to find chemical compounds, but they also make it easier to study the composition of materials, which leads to better conservation strategies. Also, the use of cutting-edge technologies has changed how IR spectroscopy is used in ways that go beyond what was previously possible. The advent of surface-enhanced resonance Raman spectroscopy, when used with IR methods, is a combined approach that greatly lowers detection limits. This combination of techniques makes it easier to analyze very small samples that standard IR methods often miss. Additionally, improvements in sample preparation and the creation of portable FTIR systems make IR spectroscopy more useful in the field, allowing for *in situ* analysis of cultural heritage artifacts (Lennard *et al.,* 2017). The increased ease of use and sensitivity make it possible to get results that were previously impossible, allowing for real-time monitoring and assessment during conservation processes. The future of IR spectroscopy looks very bright, especially since research is still working to fix its problems. One area of research that is getting more attention is making spectroscopic techniques more flexible for exploring planets. This is shown by the improvements in Fourier Transform Spectroscopy (FTS) for studying celestial bodies. As scientists study the makeup of planets and other heavenly bodies, the structural integrity and data accuracy of modern FTS can help them make important new discoveries. Spectroscopists and researchers from different fields working together all the time will help improve these methods even more, making it easier to learn more about materials on Earth and in space. These kinds of collaborative efforts not only make IR spectroscopy more useful in a wider range of scientific fields, but they also make it more important in a number of important analytical fields.

SUMMARY

The improvements in infrared (IR) spectroscopy have made analytical methods much better in many areas. This method, which is known for its quick and non-destructive analysis, has been very useful in fields like pharmaceuticals, environmental monitoring, and material sciences. The incorporation of near-infrared diffuse reflectance spectroscopy (NIR-DRS) facilitates the identification and quantification of complex compounds, as demonstrated by the successful measurement of *Shengmaxinside I* content in honey-fried processing. The model created had a high determination coefficient ($R^2 = 0.9878$), which shows that the methods are reliable and effective for quality control (Cheng *et al.,* 2018). In the end, the accuracy and speed of IR spectroscopy tools are very useful for both research and business. The effects of IR spectroscopy go beyond science; they also influence the arts and conservation, where the ability to identify materials is important for preserving history. Fourier transform infrared spectroscopy with attenuated total reflection (ATR-FTIR) and principal component analysis (PCA) has made it possible to distinguish between old barkcloths. This has given art historians and conservators important new information. The study emphasized the difficulties in developing a non-destructive and reliable analytical method; however, the identification of distinct groups based on molecular vibrations illustrates the potential of FTIR to improve our comprehension of textile conservation. This new method not only helps to verify the authenticity of artifacts, but it also helps to develop conservation strategies that are specific to the materials used, which improves the field of cultural heritage preservation (Holmes-Smith *et al.,* 2019). To sum up, infrared spectroscopy is a useful tool for many fields, from chemical analysis to studies of cultural heritage. The ongoing improvement of methods and the use of more advanced statistical techniques have made IR spectroscopy much better at finding and measuring different substances. IR spectroscopy is very useful for checking the quality of pharmaceutical compounds or studying old textiles because it is quick, accurate, and non-invasive. As research and technology continue to improve, IR spectroscopy will likely have more uses, which will lead to better methods in both science and art. This analytical method has a big effect on many fields, helping people understand complicated materials better and adding to the intellectual legacy of those fields.

Infrared (IR) spectroscopy has become a revolutionary analytical method that is well-known for its ability to find molecular structures and functional groups. The interaction of infrared light with matter is what makes it work. This causes certain wavelengths to be absorbed, which are related to molecular vibrations. This property allows researchers to learn important things about the makeup of a substance. In recent years, improvements in instrumentation, especially the development of high-brightness, tunable laser-based sources like quantum cascade lasers (QCLs), have made spectroscopic analyses much more sensitive and precise. These new ideas solve problems that come up with traditional methods, like low signal-to-noise ratios at high spatial resolutions. This makes IR spectroscopy more useful in fields like biomedical diagnostics and material science. The use of IR spectroscopy in the biomedical field is especially interesting, especially when it comes to diagnosing cancer. Mid-infrared spectroscopic imaging has demonstrated its efficacy in providing both morphological and biochemical information essential for the identification of pathological conditions. Researchers want to use very sensitive IR systems for healthcare by adding advanced data processing algorithms and improving the contrast of spectroscopic images. A key study shows that QCL systems are better than other systems at giving accurate numerical absorbance values, which shows that they are good at getting detailed information about biological samples. This move to use IR spectroscopy in medical diagnostics shows not only the technique's potential but also the need for ongoing innovation and improvement in how instruments are made and how data is interpreted (McCrow *et al.,* 2011).

IR spectroscopy is not only useful in medicine, but it is also very important for studying mineralogy and geochemistry. Recent literature underscores the incorporation of synchrotron radiation in infrared studies, thereby enhancing the analysis of minute samples frequently essential in geological research. This method improves both spatial and temporal resolution, making it possible to study geo-mineralogical processes and environmental changes in depth. Also, the advancement of advanced imaging methods makes it possible to study the complex hydrous parts of minerals. The continuous advancement of these methodologies indicates an increasing acknowledgment of the significance of IR spectroscopy across various scientific fields, confirming its role not only as a fundamental research instrument but also as a crucial element in enhancing our comprehension of intricate geological and environmental systems (Bellatreccia *et al.,* 2013).

REFERENCES

1. Adekoya, J. A., Ajanaku, K. O., Ajani, O. O., Ehi-Eromosele, C. O., et al. (2012). Cu(II) and Fe(III) complexes of sulphadoxine mixed with pyrimethamine: Synthesis, characterization, antimicrobial and toxicology study. https://core.ac.uk/download/18295271.pdf (accessed on 29th October 2025).

2. Aggarwal, M., Berio, E., Coudé du Foresto, V., Diener, P., Eisenhauer, F., Errmann, R., Froehly, C., et al. (2016). Astronomical photonics in the context of infrared interferometry and high-resolution spectroscopy. http://arxiv.org/abs/1608.02640 (accessed on 29th October 2025).

3. Ahmad, M. S., Ali, S., Hanif, M., Hussain, S., et al. (2008). ATR-FTIR spectroscopy detects alterations induced by organotin (IV) carboxylates in MCF-7 cells at sub-cytotoxic/-genotoxic concentrations. https://core.ac.uk/download/68737.pdf (accessed on 29th October 2025).

4. Aimar, P., Causserand, C., Ettori, A., Gaudichet-Maurin, E., et al. (2011). Permeability and chemical analysis of aromatic polyamide-based membranes exposed to sodium hypochlorite. https://core.ac.uk/download/12043412.pdf (accessed on 29th October 2025).

5. Alabastri, A., Calandrini, E., De Angelis, F., Garoli, D., et al. (2018). Thermoplasmonic effect of surface enhanced infrared absorption in vertical nanoantenna arrays. https://core.ac.uk/download/158822794.pdf (accessed on 29th October 2025).

6. Armăşelu, A. (2017). New Spectral Applications of The Fourier Transforms in Medicine, Biological and Biomedical Fields. https://core.ac.uk/download/322429245.pdf (accessed on 29th October 2025).

7. Aussel, H., Bergeron, J., Lutz, D., Moorwood, A. F. M., Genzel, R., Robinson, A., Veilleux, S., et al. (1999). Mid-infrared and optical spectroscopy of ultra luminous infrared galaxies: A comparison. https://core.ac.uk/download/25255719.pdf (accessed on 29th October 2025).

8. Avouris, P., & Low, T. (2014). Graphene plasmonic for terahertz to mid-infrared applications. http://arxiv.org/abs/1403.2799 (accessed on 29th October 2025).

9. Baden, N., Kebukawa, Y., Kobayashi, H., Kobayashi, K., et al. (2019). Nanoscale infrared imaging analysis of carbonaceous chondrites to understand organic-mineral interactions during aqueous alteration. https://core.ac.uk/download/pdf/286393583.pdf (accessed on 29th October 2025).

10. Baianu, I. C., & Guo, J. (2011). NIR calibrations for soybean seeds and soy food composition analysis: Total carbohydrates, oil, proteins and water contents. https://core.ac.uk/download/pdf/290467.pdf (accessed on 29th October 2025).

11. Bauke, S., Ctistis, G., Golibrzuch, K., Müller, M., et al. (2018). Quantitative measurement of combustion gases in harsh environments using NDIR spectroscopy. http://arxiv.org/abs/1810.09239 (accessed on 29th October 2025).

12. Beattie, R. J., Downey, G., Fearon, A. M., O'Donnell, C. P., et al. (2015). Prediction of naturally-occurring, industrially-induced and total trans fatty acids in butter, dairy spreads and Cheddar cheese using vibrational spectroscopy and multivariate data analysis. https://core.ac.uk/download/45656662.pdf (accessed on 29th October 2025).

13. Bellatreccia, F., Della Ventura, G., & Marcelli, A. (2013). SR-FTiR microscopy and FTIR imaging in the earth sciences. http://arxiv.org/abs/1312.5750 (accessed on 29th October 2025).

14. Birkner, J. P., Chang, G., Cheatum, C. M., Chung, H. S., Liang, C., Corry, B., Ganim, Z., et al. (2013). Vibrational spectra of a mechanosensitive channel. https://core.ac.uk/download/pdf/12961810.pdf (accessed on 29th October 2025).

15. Brownlee, D., Defrees, D., Irvine, W., Klein, H., et al. (2025). Exobiology in Earth orbit: The results of science workshops held at NASA, Ames Research Center. https://core.ac.uk/download/pdf/42820522.pdf (accessed on 29th October 2025).

16. Burbine, T. H., Dunn, T. L., Moskovitz, N. A., Reddy, V., et al. (2015). Mineralogy and surface composition of asteroids. http://arxiv.org/abs/1502.05008 (accessed on 29th October 2025).

17. Caldwell, J. D., Edgar, J. H., Folland, T. G., Liu, X., et al. (2018). Probing hyperbolic polaritons using infrared attenuated total reflectance micro-spectroscopy. http://arxiv.org/abs/1810.00732 (accessed on 29th October 2025).

18. Chen, Z. P., Jin, J. W., Li, L. M., et al. (2011). Quantitative spectroscopic analysis of heterogeneous mixtures: The correction of multiplicative effects caused by variations in physical properties of samples. https://core.ac.uk/download/9037183.pdf (accessed on 29th October 2025).

19. Cheng, G., Kuang, H., Qian, X., et al. (2018). Rapid determination of saponins in the honey-fried processing of Rhizoma Cimicifugae by near infrared diffuse reflectance spectroscopy. https://core.ac.uk/download/323080465.pdf (accessed on 29th October 2025).

20. Dengo, N., Gross, S., & Vittadini, A. (2017). Functionalization of colloidal transition metal sulphides nanocrystals: A fascinating and challenging playground for the chemist. https://core.ac.uk/download/141699569.pdf (accessed on 29th October 2025).
21. Dirac, P. A. M., Eid, R., Erdelyi, A., Greiner, W., Herrmann, R., Herzberg, G., et al. (2012). Infrared spectroscopy of diatomic molecules–A fractional calculus approach. http://arxiv.org/abs/1209.1630 (accessed on 29th October 2025).
22. Donais, M. K., & Vandenabeele, P. (2016). Mobile spectroscopic instrumentation in archaeometry research. https://core.ac.uk/download/55846882.pdf (accessed on 29th October 2025).
23. Frost, R., & Xi, Y. (2012). A vibrational spectroscopic study of plancheite $Cu_8Si_8O_{22}(OH)_4 \cdot H_2O$. https://core.ac.uk/download/10909530.pdf (accessed on 29th October 2025).
24. Gerlach, J. C., Parsons, C. L., & Whitehurst, M. (2025). A Fourier transform spectrometer for visible and near ultra-violet measurements of atmospheric absorption. https://core.ac.uk/download/pdf/42855919.pdf (accessed on 29th October 2025).
25. Grady, M. M., Jones, G. C., & Morlok, A. (2004). FT-IR microanalysis of mineral separates from primitive meteorites: Techniques, problems and solutions. https://core.ac.uk/download/41354.pdf (accessed on 29th October 2025).
26. Hanel, R. A., & Kunde, V. G. (2025). Fourier spectroscopy and planetary research. https://core.ac.uk/download/pdf/42892895.pdf (accessed on 29th October 2025).
27. Holmes-Smith, A. S., Lennard, F., & Smith, M. J. (2019). Development of non-destructive methodology using ATR-FTIR with PCA to differentiate between historical Pacific barkcloth. https://core.ac.uk/download/293865553.pdf (accessed on 29th October 2025).
28. Johnston, S. F. (2001). In search of space: Fourier spectroscopy, 1950–1970. https://core.ac.uk/download/223236192.pdf (accessed on 29th October 2025).
29. Jones, A. C., & Weddell, J. B. (2025). Observation requirements for unmanned planetary missions, part 2. https://core.ac.uk/download/pdf/85237609.pdf (accessed on 29th October 2025).
30. Laubereau, A., Mantel, K. H., & Zinth, W. (2007). Proceedings of the Thirteenth International Conference on Time-Resolved Vibrational Spectroscopy. https://core.ac.uk/download/12162307.pdf (accessed on 29th October 2025).

31. Lennard, F., Smith, M., & Thompson, K. (2017). A literature review of analytical techniques for materials characterization of painted textiles–Part 2: Spectroscopic and chromatographic analytical instrumentation. https://core.ac.uk/download/84591645.pdf (accessed on 29th October 2025).
32. Lestander, T. (2003). Multivariate NIR studies of seed–water interaction in Scots pine seeds (Pinus sylvestris L.). https://core.ac.uk/download/11694649.pdf (accessed on 29th October 2025).
33. McCord, T. B. (2025). Reflectance spectroscopy in planetary science: Review and strategy for the future. https://core.ac.uk/download/pdf/42832233.pdf (accessed on 29th October 2025).
34. McCrow, A. (2011). Laser based mid-infrared spectroscopic imaging–Exploring a novel method for application in cancer diagnosis. https://core.ac.uk/download/1592070.pdf (accessed on 29th October 2025).
35. Mitra, C. (2018). Mid-infrared spectroscopy and challenges in industrial environment. https://core.ac.uk/download/322439067.pdf (accessed on 29th October 2025).
36. Mouazen, A. M., Stenberg, B., Viscarra Rossel, R. A., et al. (2010). Visible and near infrared spectroscopy in soil science. https://core.ac.uk/download/11697073.pdf (accessed on 29th October 2025).
37. Pieters, C. M. (2025). The quickest, lowest-cost lunar resource assessment program: Integrated high-tech Earth-based astronomy. https://core.ac.uk/download/pdf/10447868.pdf (accessed on 29th October 2025).

CHAPTER 6

NUCLEAR MAGNETIC RESONANCE (NMR) SPECTROSCOPY

Contents

INTRODUCTION

The field of Nuclear Magnetic Resonance (NMR) Spectroscopy is a very complicated and useful field in both chemistry and physics. It is an important tool for figuring out molecular structure and doing dynamic analysis. It works on the principle of nuclear magnetic resonance, which says that atomic nuclei resonate when they are in a magnetic field. Scientists can learn about the molecular environment around the nuclei of interest by using electromagnetic radiation to interact with the nuclei. This method is very important for many things, including figuring out the structures of complicated organic molecules and studying biological macromolecules like proteins and nucleic acids. NMR is an essential method in materials science, pharmaceuticals, and biochemistry because it can give us detailed information about molecular dynamics and interactions. This greatly improves our understanding of chemical compounds at the molecular level. Furthermore, the various uses of NMR go beyond just determining the structure of molecules to include studying the kinetics and thermodynamics of chemical reactions. NMR spectroscopy allows for the real-time observation of molecular dynamics, which makes it easier to study reaction pathways, intermolecular interactions, and the changes in shape of biomolecules.

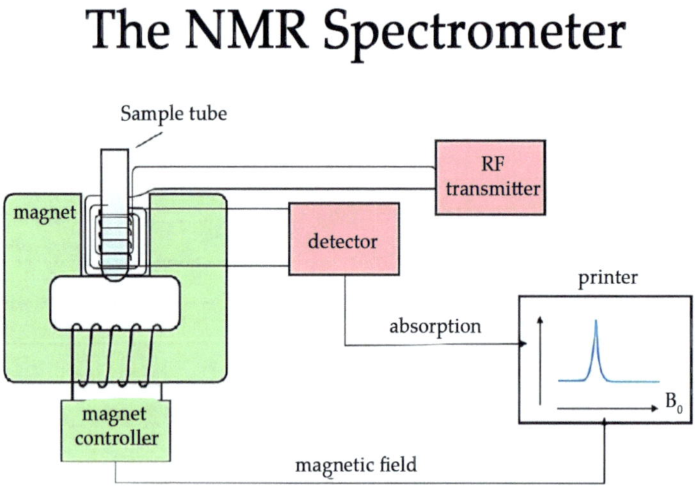

Figure 6.1. *Nuclear magnetic resonance (NMR) spectroscopy.*

Source: https://www.researchgate.net/profile/Saqib-Ali-5/publication/333657432/figure/fig1/AS:767821248200704@1560074259150/Schematic-presentation-of-a-typical-nuclear-magnetic-resonance-spectrometer-showing-the.png.

This adaptability is especially evident in the investigation of *Feshbach* resonances, where precise data is essential for comprehending the scattering phenomena relevant to ultracold atomic systems. Specifically, improvements in nuclear magnetic resonance (NMR) methods have made it possible to learn more about interactions such as those in sodium systems, where resonances can develop into complex behavior due to intricate nuclear properties, as recent research has shown (Trautmann *et al.,* 2011). NMR is at the crossroads of theory and practice, and it plays a big role in modern scientific research. Moreover, NMR's sensitivity to nuclear characteristics enhances its significance in uncovering subtle effects arising from atomic structure that can significantly impact experimental outcomes. For example, NMR can give us information about the mass, spin, and charge distribution of nuclei, which can help us understand nuclear physics through atomic spectra. This interaction is illustrated in the examination of isotope shifts and hyperfine structures, where alterations in nuclear properties directly influence atomic transitions, thereby deepening the comprehension of the nucleus itself. The investigation of these phenomena highlights the significance of NMR spectroscopy not only as a diagnostic instrument but also as a conduit connecting atomic and nuclear realms of physics, as evidenced by recent reviews in the discipline (Adriana *et al.,* 2010). NMR continues to develop in this way, combining theoretical predictions with experimental validation. This makes it an important part of modern scientific research (Figure 6.1).

Definition of Nuclear Magnetic Resonance (NMR)

Nuclear Magnetic Resonance (NMR) is a very complex analytical technique for investigating the magnetic properties of atomic nuclei. NMR uses the magnetic field generated by the nuclei when they are placed in an external magnetic field. In other words, certain nuclei will resonate at certain frequencies if placed in a strong magnetic field. These frequencies are determined by the strength of the magnetic field and the local environment of the nuclei. This resonance is measured and converted into spectra which contain a wealth of information about the structure and dynamics of molecules. Nuclear spin is the fundamental property on which NMR is based. It allows nuclei such as hydrogen, carbon, and phosphorus to behave like miniature magnets that point to or against the outside field, producing distinct resonance frequencies that can be analyzed further. Nuclear Magnetic Resonance spectroscopy (NMR) is a technique that has many applications in the field of science, which is why it is so versatile, for example, in chemistry, biochemistry, and medicine. NMR spectrometry is a technique used by chemists to determine the structure of molecules by allowing them to determine the arrangement of the atoms in complex compounds. In medicine, NMR is the principle behind Magnetic Resonance Imaging (MRI) which is a non-destructive method of imaging soft tissues in the body. NMR is not only useful for doctors to determine what is wrong with a person, but also what is occurring as a disease progresses. This dual utility highlights the relevance of NMRs, since they combine basic scientific research with useful applications directly to health and technology, thus pointing out its strategic importance not only to the development of knowledge but also to the enhancement of the quality of life. The sensitivity and resolution of NMR are continuously increasing thanks to new technique developments, thereby offering new prospects for discoveries in the field of materials science and nanotechnology. For instance, new methods such as optimal control theory-based methods have been demonstrated for facilitating high resolution nanoscale magnetic resonance imaging. These developments reflect the evolution of NMR from traditional applications to new applications that can identify materials on the nanoscale as reflected by significant reduction of spin resonance linewidth and dramatic improvements in spatial resolution (Budakian *et al.,* 2017). Moreover, the study of the optimal control theory opens a path towards the design of pulse sequences which can tolerate experimental imperfections. This innovation continuum is emphasized here to show that NMR has been significant not just in fundamental research but also in solving real-world applications in a wide variety of fields, making it a technology that will remain relevant to the scientific community (Glaser *et al.,* 2012).

Historical Background and Development of NMR

Nuclear Magnetic Resonance (NMR) spectroscopy is an experimental technique that is based on fundamental discoveries in quantum mechanics and nuclear physics during the early 1900s. NMR made its first appearance in the 1930s by physicist Isidor Rabi. He received the Nobel Prize in 1944 for his studies of molecular beams and for his development of magnetic resonance methods. During this time, NMR transformed from an idea to a real-life method that paved the way for future applications. As these early instruments were further developed, scientists came to see their usefulness in atomic and molecular characterization, which would prove to be very important for future discovery in chemistry and physics. In the years after World War II, the advent of radiofrequency technology made the use of NMR much easier. This paved the way for more sophisticated and sensitive applications in a wide range of science. During the next decades, the joint application of NMR and advances in computational methods of molecular structures made the technique more accessible in science. In the 1950s and 1960s, there was a great development in the field of NMR spectroscopy in terms of its sensitivity and resolutions. Also, two-dimensional NMR techniques were developed in this period. These methods provided scientists with new ways of visualizing the interaction of molecules with one another. Major contributions to the understanding of the use of these multidimensional techniques were made by Martin Karplus and Richard R. Ernst. As a consequence, NMR became not only a physical phenomenon, but also a useful tool in many branches of science, including chemistry, biology, and materials science. This development of NMR as an analytical tool proved crucial, as scientists were able to determine the structure and dynamics of molecules. Today, NMR spectroscopy is applied in various applications ranging from fundamental research to clinical diagnostics applications. In recent years, NMR has been further used in combination with other imaging and analytical approaches in the study of complex biological systems and samples from cultural heritage. The Nuclear Physics Division of the EPS, for example, does an excellent job of demonstrating the applications of nuclear and related techniques such as NMR in protecting cultural artifacts. This is cross-disciplinary evidence of modern scientific research (ANONYMOUS, 2016). Furthermore, improvements in pseudocontact shift analysis have increased the capabilities of NMR to study atomic-level interactions, particularly in systems with delocalized electrons (Charnock *et al.,* 2014). These developments not only highlight the importance of NMR in modern scientific research but also exemplify its evolution into a tool of fundamental importance in basic and applied science.

6.1. PRINCIPLES OF NMR SPECTROSCOPY

Nuclear Magnetic Resonance (NMR) Spectroscopy is based on interaction between atomic nuclei and external magnetic field. Nuclei which have an odd number of protons or neutrons, such as hydrogen-1 or carbon-13, have a magnetic moment. This means they can resonate when they are in a magnetic field and surrounded with the radiofrequency (RF) radiation. This resonance occurs at specific frequencies that are determined by the strength of the magnetic field applied and the presence of chemical neighbors around the nuclei. The spectral data obtained from this affords information on molecular structure, dynamics, and interactions. Chemists can determine the molecular structure, such as bond lengths and angles, based on the shift and splitting of the resonance signals. This is good for teaching them the better behavior of chemicals. With the development of new NMR methods for NMR, it has become increasingly useful in many scientific disciplines, including biology and materials science. The latter can be studied in detail in complex molecular systems using modern techniques, like two-dimensional NMR or diffusion NMR. These techniques are very important for determining the structure of proteins, nucleic acids, and polymers, which is essential for the improvement of drugs and materials. Furthermore, pulse sequence design based on the optimal control theory has enabled overcoming the experimental limitations, providing increased sensitivity and resolution. These developments ensure that NMR remains a powerful analytical tool, expanding the boundaries of what is possible in both research and practice in the current scientific environment (Glaser *et al.,* 2012). In addition, the application of combined new imaging techniques and NMR concepts, especially in magnetic resonance imaging (MRI) demonstrates how versatile the method is. In this paper, a new imaging framework in which the spectral-scanning magnetic resonance imaging (MRI) is incorporated in a unified way is presented for diverse applications, such as small-body imaging at non-uniform magnetic fields. This system can generate custom excitation signals and receive related MR responses and hence has considerable flexibility to meet various research needs (Babakhani *et al.,* 2008). Both broad and narrow-band experiments can be performed, and the tunable operation frequency enables a more detailed spectral analysis. Thus, these gains not only demonstrate the importance of NMR Spectroscopy in classic chemistry but also demonstrate the importance of NMR Spectroscopy in interdisciplinary applications at the interface between science and engineering (Figure 6.2).

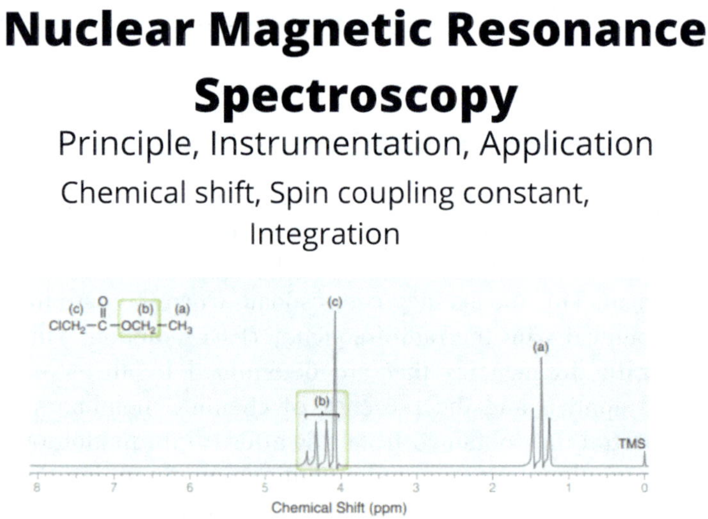

Figure 6.2. *Principles of NMR spectroscopy.*

Source: https://chemistnotes.com/wp-content/uploads/2021/11/Nuclear-Magnetic-Resonance-Spectroscopy.png.

6.1.1. Basic Concepts of Nuclear Spin and Magnetic Fields

To be able to understand Nuclear Magnetic Resonance (NMR) Spectroscopy you need to at least be familiar with the basic ideas behind nuclear spin and magnetic fields. Nuclear spin is the effect caused by the intrinsic angular momentum of the nuclei of atoms. When these nuclei are placed in a magnetic field, they can assume some quantized orientations. This phenomenon is based on the quantum mechanics, which says that nuclei with an odd number of protons and/or neutrons have a spin that is not zero. When an outside magnetic field is applied, these nuclear spins align in the same direction or opposite direction as the field. This causes energy levels to split, and this is known as Zeeman splitting. When the nucleus is exposed to some radiofrequency radiation, it moves between these states, which leads to a variety of possible energy states which can be used in NMR studies. This provides us with useful spectroscopic information on molecular structure and dynamics. Also, magnetization is an important part of nuclear spins which makes it possible to change and find signals in NMR spectroscopy. Without an external magnetic field, the spins of the nuclei are randomly oriented and that the net magnetization is zero. But when a magnetic field is applied, a net magnetization vector gets created because the number of spins with the field is more than those against the field. This gives rise to a macroscopic magnetization phenomenon. Pulse sequences in NMR use this magnetization to generate measurable signals by applying radiofrequency pulses to the spin system. The detection of these signals is fundamentally based on the interactions of the nuclear spins with the magnetic field, the resonant frequencies that are detected in NMR experiments correspond to the energy differences between the nuclear spin states and these energy differences can be affected by local chemical environments and interactions with electrons within the molecule, providing a wealth of information about molecular structure and interactions. The connection between nuclear spins and magnetic fields also applies to more complicated things such as spin relaxation and decoherence. Nuclear spins may experience various relaxation processes in response to their interaction with the environment. These processes can alter the stability and coherence of the nuclear magnetization. Interactions with surrounding spins or phononic excitations, which are found in the structure of the lattice, may lead to decoherence, particularly in atomistic and molecular systems. This can result in the loss of information over the course of time. For example, nuclear spins can be used as qubits in quantum technologies, when they are used in nanostructured materials. This means that we need to have a deep understanding of how decoherence works in a bid to make coherence times longer (Liu *et al.,* 2016). The coupling of nuclear spins to resonant cavities can also be used to increase the relaxation rates and this demonstrates the interaction of magnetic fields and quantum effects to control the dynamics of nuclear relaxation (Abragam *et al.,* 2015). These insights promote the importance of nuclear spin dynamics in both traditional NMR fields and new technologies in quantum information science.

6.1.2. Resonance Condition and Energy Transitions

The resonance condition is a general concept in Nuclear Magnetic Resonance (NMR) spectroscopy. It explains the orientation of nuclei spins in an external magnetic field. When nuclei are exposed to a particular frequency of radio frequency (RF) radiation, they resonate and are thus caused to move between energy levels. This is due to quantum mechanics which states that the energy difference between the state of the nuclear spins must be equal to the energy of the radiation striking the nuclei for the interaction to occur. This condition is very important as when it is satisfied the spins can absorb the maximum energy which is manifested as signals in the NMR spectrum. "It's not only the ability to control these energy changes, but we can also now see them to get a better understanding of molecular structure, as well as how things move and interact in complex samples." In practice, NMR spectroscopy uses these energy changes to investigate the molecular surroundings. Fine control of the RF frequency allows excitation of a particular transition in the nuclei, which is essential for obtaining high resolution spectra. When the spins transition from one energy state to another, they emit energy in the form of radio waves that can be detected and analyzed to gather more information about chemical environments, molecular structures and reactions that are taking place.

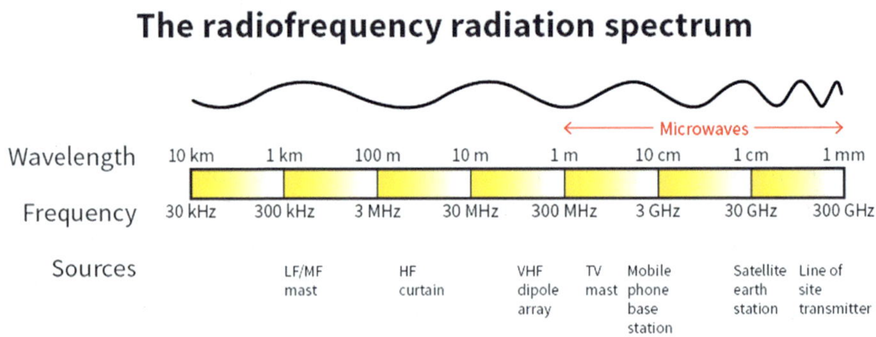

Figure 6.3. *Radiofrequency radiation.*

Source: https://www.arpansa.gov.au/sites/default/files/legacy/images/basics/rf-spec.jpg.

There are many experiments that use the resonance condition with experiments like spin echo or selective excitation pulses. Also, advances in NMR technology have enabled finding small differences in the magnetic environments of nuclei. This has increased the spectral resolution and the sensitivity which are very important for the characterization of complex biomolecules and materials. Recent findings in resonance techniques point toward possible interdisciplinary use beyond the traditional NMR spectroscopy. For instance, resonance is being used in new ways in developments in the control of the quantum states of spins in quantum dots. Coherent control of electron spins using continuous wave oscillating magnetic fields has been realized (Greilich *et al.,* 2006) and is analogous to the resonance condition found in NMR. Moreover, it has been shown that multiphoton transitions can create temporal modulation effects, which would be beneficial for experimental sensitivity in hydrogen-like ions, thus making an interesting connection between NMR techniques and quantum computer techniques (Zaretskii *et al.,* 2010). In addition to enhancing the existing theoretical picture of NMR spectroscopy, the resonance condition identifies new applications of quantum mechanics in the future (Figure 6.3).

6.1.3. Role of Radiofrequency (RF) Radiation in NMR

Nuclear Magnetic Resonance (NMR) spectroscopy is an important analytical tool in chemistry and biology, and a small amount of radiofrequency (RF) radiation is essential for NMR measurements. Nuclei of some isotopes resonate at a given frequency when exposed to RF radiation in a magnetic field. This resonance effect is directly proportional to both the magnetic characteristics of the nuclei, specifically their spin states, and the magnetic field around the nuclei that influences them. The RF radiation causes the nuclei to jump from a lower energy state to a higher energy state, from which we can obtain useful spectral data. By observing the signals that are emitted, scientists are able to obtain information about the molecular structure, dynamics, and interactions of the compounds under study. Such knowledge is critically important to advance in a variety of fields, such as pharmacology and materials science, and highlights the importance of RF radiation in elucidating complex molecular scenarios. Besides their excitation, the nuclei are also surrounded by an electronic environment, which can influence their fine structure and their interaction. This is directly connected with the role of RF radiation in NMR. The covalent electrons that surround the nuclei are sensitive to such things as the nuclear mass distribution and the nuclear charge distribution. This alters the resonant frequency of the nuclei when excited with RF. This relationship between nuclear parameters and the electronic properties is very important as it helps us to understand NMR spectra. Different chemical environments lead to different chemical shifts giving us structural information that otherwise would be difficult to obtain. This fact stresses the delicate interplay between electronic structure and nuclear magnetic properties and the key aspect of RF radiation to uncover complex molecular information in NMR spectroscopy, thereby being an indispensable tool for scientific analysis (Adriana *et al.,* 2010). Moreover, incorporation of RF radiation in the field of NMR spectroscopy has increased the usefulness of the NMR technique beyond structural elucidation, allowing advancement in biomedical research, particularly in the field of oxidative stress studies. For example, because reactive oxygen and nitrogen species (ROS and RNS) have the potential to affect biological systems by causing changes in nuclear spin, the NMR is susceptible to these changes. NMR is an exceptional method among the established approaches for the quantification of these radicals due to its non-destructive nature and the wide molecular elucidation capabilities which can be applied to a multitude of substances. As free radical research continues to expand, quite clearly, NMR techniques are playing a larger role in the investigation of oxidative damage. This reflects the changing role of RF radiation. Based on these findings, any RF radiation plays a significant role in the development of many scientific disciplines by combining the physical chemistry with biological implications (Gurer-Orhan *et al.,* 2017).

6.2. TYPES OF NMR SPECTROSCOPY

Nuclear magnetic resonance (NMR) spectroscopy is a wide-ranging field with numerous types of NMR spectrographs, each with a special use in while molecular analysis. One of the most common types of NMR spectroscopy is proton (^{1}H) NMR spectroscopy which is commonly used. It is mostly used for organic compounds as it is very sensitive to hydrogen atoms. This method provides researchers with important information about the hydrogen environment, which is important in order for them to determine structural and molecular dynamics. By examining chemical shifts, splitting patterns and integration, scientists can determine the location of hydrogen atoms in a molecule. This provides them with important information about functional groups and how molecules are joined. So, ^{1}H NMR is an important tool for identification of organic compounds, as well as drug discovery and development in the pharmaceutical industry. Carbon-13 (^{13}C) spectroscopy is another type of NMR that is important. It looks at the carbon nuclei in organic molecules. ^{1}H NMR mainly gives information about hydrogen atoms, while ^{13}C NMR gives more information in detail about the structure of carbon skeleton. This method is especially useful in the discovery of the different carbon environments in complicated organic molecules. Additionally, improvements in probe technology and relaxation enhancement have made it possible to obtain high-quality ^{13}C spectra, making it even more useful in fields such as metabolomics and natural product chemistry. The ability to acquire both quantitative and qualitative information from ^{13}C NMR strengthens its use as a complementary tool to ^{1}H NMR to better understand the molecular structures and behaviors of complex mixtures. Advancements in hyperpolarized NMR spectroscopy have brought a wide extension of the applications of this analytical tool, particularly in biomedical studies and material science. Hyperpolarized gases such as xenon-129 (^{129}Xe) and helium-3 (^{3}He) have unique advantages in terms of signal sensitivity to examine certain low-concentration species and dynamic processes that cannot easily be examined using standard NMR techniques. For example, researchers are looking into hyperpolarized xenon as a way to see how well the lungs are ventilating without having to cut into the body. This could be a big step forward for medical imaging. Recent advances in the generation and application of hyperpolarized gases prove their importance in non-invasive imaging and the study of microporous materials and bear witness to the growing interest in interdisciplinary approaches in NMR research (ANONYMOUS, 2016; Rychnovský *et al.,* 2009). As a result, the introduction of the hyperpolarized NMR represents a revolutionary phase in the study of complex systems in various areas of science (Figure 6.4).

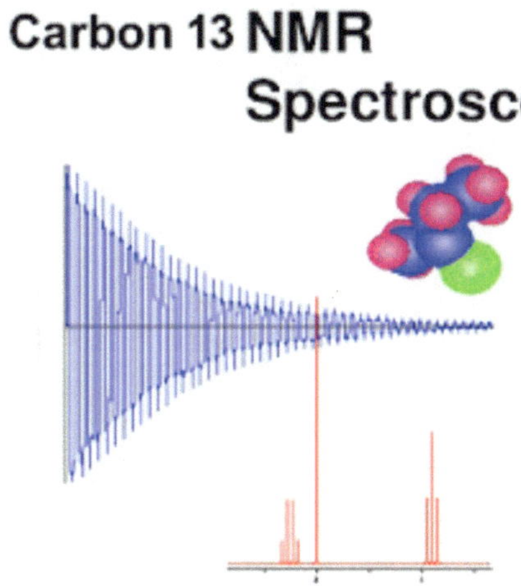

Figure 6.4. *Carbon-13 NMR.*

Source: https://2.bp.blogspot.com/-S5JAhKl0dAE/WBrWtahID0I/AAAAAAAAAnQ/35_47s9P-EIUAwsVzqYbKDQvVf4GTjhiQCLcB/s400/nmrspectroscopy-151116135705-lva1-app6892-thumbnail-4%2Bcopy.jpg.

6.2.1. Proton NMR (^{1}H NMR)

Proton NMR (^{1}H NMR) is a very important part of nuclear magnetic resonance (NMR) spectroscopy, since it looks at the magnetic properties of hydrogen nuclei. This is a very useful method to figure out the structures of organic compounds because hydrogen is so common in organic molecules. ^{1}H NMR can tell us a lot about the structure and dynamics of molecules by measuring the resonance frequency of hydrogen nuclei in different chemical environments. Technological advancements have made this method even more sensitive, making high-resolution imaging and analysis possible. For example, high-resolution nanoscale magnetic resonance imaging (nano-MRI) has shown big improvements in proton spin resonance linewidth reduction, getting amazing spatial resolution in polystyrene samples. This reveals how NMR could change the way it is used in many fields of science (Budakian *et al.,* 2017). In addition to helping us understand structures, ^{1}H NMR is very important in chemistry and biochemistry as well, especially when it comes to analyzing complex mixtures and biological samples.

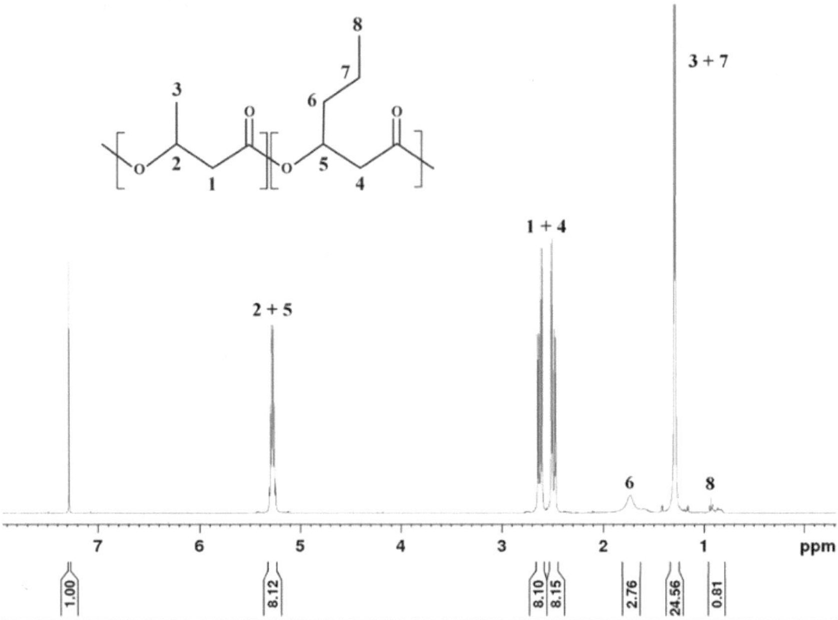

Figure 6.5. *Proton nuclear magnetic resonance spectroscopy.*

Source: https://www.researchgate.net/publication/367188633/figure/fig2/AS:11431281119992877@1676343186764/Proton-nuclear-magnetic-resonance-spectroscopy-H-NMR-spectrum-of-P3HB-co-3HHx.png.

The technique can provide signals directly proportional to the number of protons existing, so it is possible to analyze quantitatively. In addition, advancements in techniques such as algorithmic cooling have made ^{1}H NMR even better through the increase of spin polarization. This improvement allows a stronger signal, which is very important for finding small changes in how molecules behave and interact. Researchers have been able to surpass traditional limits for polarizations of proton ensembles through the use of selective reset techniques. This has accelerated the development of new pulse sequences in NMR (Brassard *et al.,* 2014). ^{1}H NMR has more uses than just the determination of the structure of a substance. It can also be applied in materials science, medicine, and pharmaceuticals. In terms of materials science, the ability to see the distribution of protons at nanoscale makes for new insights into molecular interactions and properties. In the field of medicine, ^{1}H NMR is very useful in the formation of diagnostic tools. For example, magnetic resonance imaging (MRI) uses NMR principles to make detailed pictures of body tissues. As research progresses, so does the technique, incorporating the old NMR applications and innovative technologies to solve today's scientific challenges. The versatility of ^{1}H NMR supports the use of this technique in many different disciplines and demonstrates the role this powerful analytical instrument plays in many areas (Figure 6.5).

6.2.2. Carbon-13 NMR (^{13}C NMR)

The use of Carbon-13 NMR (^{13}C NMR) in Nuclear Magnetic Resonance (NMR) Spectroscopy adds to the knowledge about molecular structures and dynamics. Carbon-13 is a naturally occurring isotope of carbon that occurs in about 1.1% of the earth's crust. This is a rare but significant part of organic compounds. The unique nuclear spin of Carbon-13 can be used to observe when molecular motions are inhibited and to determine how complex structures are formed in various materials. Pulse sequences and labeling techniques offer great possibilities for enhancing the sensitivity of ^{13}C NMR acquisitions. The latter allows studying molecular environments in high detail. This skill is particularly important in materials science and organic chemistry, where understanding the interaction between carbon-containing compounds can help to optimize the properties and functionalities of materials. Also, new theoretical approaches, particularly new cooling techniques for algorithms, have contributed to an improved understanding of ^{13}C NMR signals. This technique causes the Carbon-13 nuclei to be more polarized and the NMR spectra to be clearer and stronger. Algorithmic cooling produces highly polarized groups of spins. Application of NMR pulse sequences to these groups results in more resolved and meaningful spectra. In addition to making ^{13}C NMR even more useful in real life, these improvements surpass the knowledge limit specified by Shannon entropy, and the team was able to explore areas of molecular structure and dynamics that were never explored before (Brassard et al, 2014). This symbiosis of experiments and theory reveals the significance of ^{13}C NMR in learning more about the chemical and physical properties of carbon-rich materials. The information provided by ^{13}C NMR goes beyond the limits of organic chemistry, and is relevant to a wide range of fields, such as materials science and nanotechnology. Recent studies have examined hyperfine magnetic fields created at Carbon-13 nuclei in ferromagnetic graphite and have revealed large atomic scale interactions. These experiments provide evidence that ^{13}C NMR can be applied to magnetic properties investigation in complex systems such as systems with atomic vacancies and edge positions with a significant impact on their electronic and magnetic properties (Araujo-Moreira *et al.*, 2014). With the increasing number of applications of ^{13}C NMR it has become a valuable analytical tool that will provide us with important information on the complex relationships between molecular structure, magnetic properties, and materials behavior in non-metallic, carbon-based systems.

6.2.3. Two-Dimensional NMR Techniques

The recent introduction of two-dimensional NMR techniques has significantly contributed to the field of Nuclear Magnetic Resonance (NMR) Spectroscopy, enabling researchers to gain an in-depth understanding of molecular structures and dynamics. Whereas one-dimensional NMR provides limited information on the interactions between different nuclei within the molecule, two-dimensional NMR can be used to record complex interactions. This additional dimension helps us resolve overlapping signals, so we are able to more easily assign resonances to individual atoms or functional groups in larger molecules. Furthermore, chemoinformatics approaches that rely on the correlations between chemical shifts and coupling constants in multiple dimensions effectively extend the information of spectral data and lead to a deeper understanding of molecular structure. Two-dimensional methods have become indispensable in a broad range of areas including organic chemistry, biochemistry, and structural biology where the knowledge of the way molecules interact is crucial to understanding their function and their reactivity. Two-dimensional NMR techniques have been used to design new approaches that use more sophisticated detection schemes to increase sensitivity and resolution. For instance, the recent development of using nitrogen-vacancy (NV) defects in diamonds as sensors has demonstrated that they are potentially much more sensitive than ever before in NMR detection. Nuclear spins can be controlled and measured in a much faster and more robust way by exploiting the strong coupling capabilities of NV centers to the nuclei of interest. As demonstrated in recent studies, the results of NV centers in such samples with very low nuclear spin polarization show that they can be used to identify relevant spectral signatures. This technique has improved our understanding of molecular systems and broadened the use of NMR techniques in a variety of scientific fields, such as materials science and biophysics where accuracy is important (Cai *et al.,* 2015). Furthermore, two-dimensional NMR imaging represents a major development of function in NMR spectroscopy, specifically, for nanostructured materials and biological systems. Recent work has shown that single nitrogen-vacancy centers can be used for two-dimensional imaging with sub-12 nm spatial resolutions. As such, this capability is very important for the study of complex structures on the nanoscale, such as proteins and other molecules that are important to biology. The precessing protons produce oscillating magnetic fields that can be detected by sweeping samples over NV centers. This provides them with high resolution pictures that are generally beyond the scope of NMR sensitivity. As both these imaging techniques are technologically maturing further, they are likely to become valuable tools in the study of intricate molecular structures, enabling new research directions in both fundamental science and applied areas (Awschalom *et al.,* 2014) (Figure 6.6).

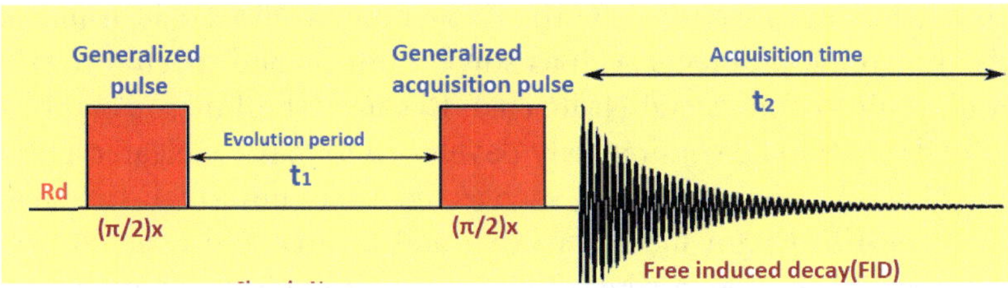

Figure 6.6. *Two-dimensional NMR (2D NMR) spectroscopy.*

Source: https://chemistnotes.com/wp-content/uploads/2023/01/Pulse-sequences-for-2D-nmr.png.

6.3. INSTRUMENTATION AND TECHNIQUES

Nuclear Magnetic Resonance (NMR) Spectroscopy is one of the best techniques for the analysis of molecular structures. NMR employs a lot of complicated equipment, such as a magnet, a radiofrequency (RF) transmitter, and a receiver. The strong magnetic field is what enables some isotopes, such as hydrogen 1 or carbon 13, to align their nuclear spins. Once the RF transmitter is aligned, it beams out pulses which throw off this alignment. The receiver then receives the nuclei's relaxation. This process produces a spectrum which reveals important information about the identity, structure, and behavior of molecules. It is very important to know how this complex function because it is the basis for interpretation of NMR results. This illustrates how these tools are used to further the development of the world from many branches of science like chemistry and biochemistry, down to medical diagnostics with the use of magnetic resonance imaging (MRI) (ANONYMOUS, 2016). NMR involves a plethora of techniques which can be varied depending on the needs of various research projects. For example, two-dimensional NMR helps to improve resolution by providing researchers with more information about the dimensions of molecules, which help them see more complicated interactions. In order to determine the relationship among the atoms of multi-component systems, scientists employ techniques such as COSY (Correlated Spectroscopy) and HSQC (Heteronuclear Single Quantum Coherence). Additionally, the pulse sequences can be optimized to investigate dynamic processes such as molecular motions or interactions in real time. Because it can be used for so many things, NMR spectroscopy is not only a way to figure out the structure of a molecule, but also a way to study how fast reactions happen and how they work. The inclusion of these sophisticated techniques into NMR has led to a huge increase in its use in interdisciplinary research from the material sciences to biophysical chemistry (McKinney *et al.,* 2000). Even though NMR's instruments and methods are strong, problems such as solvent effects and spectral overlap can make it difficult to understand NMR data. Because of all these complications, new methods have to be continuously developed to make it clearer and reduce the noise. New ideas in cryogenics, such as the use of ultra-high-field magnets, have made NMR spectra more sensitive and clearer and enabled scientists to see low-concentration analytes that were difficult to study previously. Moreover, the combination of NMR and other analytical techniques, such as mass spectrometry or chromatography, has become popular for in-depth molecular characterization. Such hybrid approaches are representative of how modern improvements in instrumentation and methodology are extending the usefulness of NMR spectroscopy and confirms its central role in scientific research and applications beyond simple structural analysis (ANONYMOUS, 2016) (Figure 6.7).

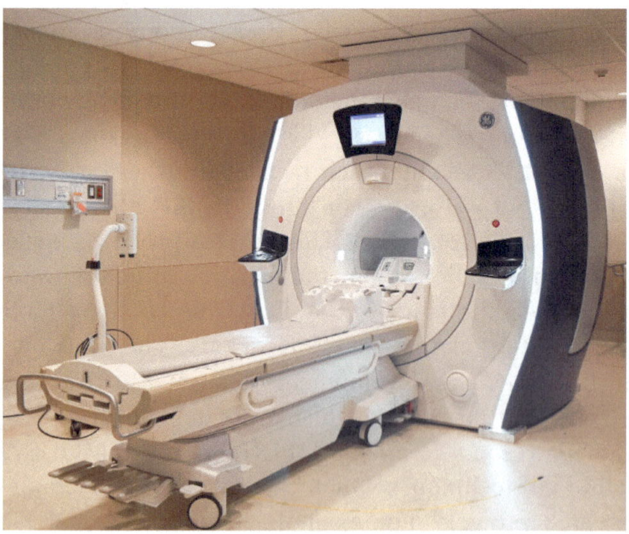

Figure 6.7. *MRI (magnetic resonance imaging).*

Source: https://www.hrh.ca/wp-content/uploads/2023/03/MRI.jpg.

6.3.1. Components of an NMR Spectrometer

The nuclear magnetic resonance (NMR) spectrometer is a complex tool with many important parts that work together to get detailed information about the structures of molecules. The NMR spectrometer has a powerful magnet in its center. This magnet creates a uniform magnetic field that is required to polarize the nuclei of interest, which mainly are hydrogen or carbon. When subjected to this magnetic field, nuclei align themselves with the magnetic field and a higher energy state is achieved. The strength of the magnetic field has a direct effect on the resolution and sensitivity of the spectroscopic measurements. As a result, it becomes an important part of the NMR system. The magnet must also work consistently, since even small changes can change the spectral outputs and lower the quality of the data. This indicates a higher level of control systems are required for the spectrometer setup. The radiofrequency (RF) transmitter is another important part of an NMR spectrometer. It sends energy pulses into the sample. This RF energy disrupts the alignment of the magnetically polarized nuclei, causing them to precess or wobble. When this RF pulse is switched off, the nuclei return to their original state and emit energy in the form of free induction decay (FID) signals which are picked up by the receiver coil. The design and accuracy of the RF pulse sequences is very important for getting the best NMR data. Recent progress in optimal control theory has led to improved sequences of pulses which support accurate measurements even when there are flaws in the experiment or distortions in the resonator. These improvements make the spectrometers work better, which is demonstrated by the fact that they provide a higher signal-to-noise ratio, which is important for picking up on small differences in chemical environments in complicated samples (Borneman *et al.,* 2012). Lastly, the detection system which has a receiver coil, is very important as in this system, it picks up the FID signals that are sent out and are turned into useful data. The signal is typically weak so has to be amplified and digitized before it can be analyzed. High-quality amplification and filtering systems help get more accurate spectral data, which helps solve the problem of overlapping signals that often happen in complex mixtures. Modern NMR spectrometers also include advanced software for processing and interpreting data. This software uses algorithms to deconstruct the spectra into useful structural information. Recent work has demonstrated that the use of sophisticated pulse design and signal analysis using advanced numerical optimization algorithms renders spectrometers useful in a wide range of disciplines from chemistry to medical imaging (Glaser *et al.,* 2012).

6.3.2. Sample Preparation and Handling

NMR spectroscopy is quite sensitive to the way in which the samples are prepared and handled, and the procedure is very sensitive to errors. Since the spectral data is a direct representation of the sample, the quality of the data is directly influenced by the quality of the preservation. It is necessary to prepare the sample correctly by removing any impurities that would interfere with the signal of interest. Grinding and sieving are two techniques which can be employed to make solid samples more uniform.

For the same reason, liquid samples should be carefully filtered to remove particles that might scatter RF radiation and obscure the signal. In addition, the selection of an appropriate solvent is extremely important because the NMR chemical shifts and peak resolution can be affected by the choice of solvent. By emphasizing the importance of these steps, you will be able to convey how they are the foundation for obtaining NMR data that can be reproduced and interpreted. Processing samples is labor intensive and has to be handled with care so as not to cause contamination and damage. This includes using clean and chemical resistant containers and storing the chemicals separate from air and moisture whenever possible. These substances have the power to alter the chemical structure of the chemicals. The treatment of the sample protocols should also allow discussion of the influence of temperature on samples as variations in temperature may alter the behavior of the molecules and, consequently, the NMR spectra.

One should take the utmost care in designing tools that transfer samples, in a way that the quality of the specimen should be maintained throughout its life. In addition, by using special techniques such as "keystones" that enable small sample sizes to be taken from larger samples whilst minimizing damage, the estimates you get may also be more robust than those Brennan et al. (2003) did. This new method shows how sample handling methods have evolved in modern analytical practices. In addition, advances in sample preparation techniques have made the interpretation of the use of NMR in a wide variety of fields, including biomedical research, easier. For example, the use of silicon-based micro and nanoparticles has demonstrated the ability to adapt the sample preparation to meet the requirements for specific applications such as medical imaging. Not only are these materials biocompatible, but they can also be chemically modified on their surfaces to make them even more ideal for delivery of drugs or imaging agents.

Purification techniques that enable *in-vivo* imaging of hyperpolarized nuclei represent a great step in the way samples are treated. The flexibility of NMR spectroscopy is demonstrated in applications such as gastrointestinal and tumor perfusion imaging. The progress made through advanced sample preparation has made it an integral part of achieving better spectral resolution as well as expanding the scope of NMR techniques in a variety of science, as suggested by Bhattacharya et al. (2013) (Figure 6.8).

Figure 6.8. *Chemical-resistant containers.*

Source: https://www.imectechnologies.com/wp-content/uploads/2019/04/Haz-Drums.jpeg.

6.3.3. Data Acquisition and Processing

Nuclear Magnetic Resonance (NMR) Spectroscopy is a powerful technique for providing structural and dynamic information on molecules based on a complex procedure of data acquisition and processing. The first step to obtain data is to excite nuclear spins with radiofrequency pulses. This excitation induces the precession of the nuclei, and the nuclei return to their normal state while releasing energy that is measured as free induction decay (FID). The quality and resolution of the data you obtain is highly dependent on the settings you choose during the pulse sequence, such as the length of the pulse, the inter-pulse delay, and the repetition rate. Optimized parameters increase the signal-to-noise ratio and allow for more easily obtainable chemical information. In the field of NMR analysis, precise configuration of the acquisition parameters is very important, since it directly influences the quality and quality of the resulting spectrum. Data processing is an important part of NMR Spectroscopy that takes raw FID signals and converts them into spectra that can be understood. The most important tool in this phase is the Fourier transformation, which converts data from time domain to frequency domain spectra. This change is very important because it allows chemists to observe resonances attributed to different environments of different nuclei in the sample. Further advanced data processing techniques such as baseline correction, phase correction, and apodization make the spectra even cleaner by making the peaks more visible and improving the resolution. Complicated multidimensional NMR experiments have been made faster recently, due to improvements in the data acquisition techniques, such as Hadamard spectroscopy. These new techniques are making it easier for scientists to rapidly investigate complex molecular systems. This is demonstrated by the recent development of fast quantum computing schemes based on Hadamard operations in NMR (Anderson *et al.,* 2006). The introduction of modern computational algorithms into data processing has revolutionized NMR Spectroscopy, making it even more useful for many different fields. For instance, applications of machine learning methods provide new ways of viewing large data sets and allow for the exploration of chemical landscapes that were previously inaccessible. This development not only makes data easier to manage but also creates new applications for NMR in areas such as agriculture and food science. NIR spectroscopy is an instrument, which is based on the principle of NMR, and it has been developed in several fields due to the fastness of chemical tests, such as the control of food quality (Baianu *et al.,* 2011). NMR Spectroscopy is still at the cutting edge of analytical chemistry and continues to offer powerful information on molecular structures and dynamics. This is because of the constant advances in the way data are acquired and processed.

6.4. CHEMICAL SHIFTS AND SPLITTING PATTERNS

Chemical shifts and splitting patterns are elements of fundamental study of nuclear magnetic resonance (NMR) spectroscopy. Chemical shifts occur as a result of changes in the electronic environment around nuclei which affects the resonance frequency of an atom. In a nutshell, nuclei in a different chemical state vibrate at different frequencies and this enables differentiation between similar atoms in a molecule. The ability to visualize these changes is very important in determining molecular structure and the interactions between molecules. In practice, such as in the case of potassium aluminum tetrahydride ($KAlH_4$), the NMR methods provide precise isotropic chemical shift data that is very significant in the analysis of complex spectra for describing materials with varying electronic environments (Felderhoff *et al.,* 2019). Due to spin-spin coupling, that is, when adjacent nuclei influence the magnetic environment of adjacent nuclei, splitting patterns are observed, also known as multiplicity. This is due to the interaction of spins of neighboring nuclei to each other leading to a series of signals in the spectrum that represent the relative positions and interactions of the nuclei. The simplest example of splitting is that of two next-nearest adjacent protons which give rise to a triplet character. As the number of protons next to each other increases, the splitting becomes more complex. This is called the $n+1$ rule, where n is the number of protons adjacent to each other. Understanding such dynamics is necessary to explain the shape of organic molecules, providing insights into connectivity and spatial structures that are not easily determined from the chemical shifts alone (Felderhoff *et al.,* 2019). Furthermore, methodological advances in NMR have helped to increase the understanding of the splitting pattern, especially in the near-zero-field NMR, in which conventional assumptions on the spin interactions are questioned. In this regime, the interactions may be considered as perturbations to electron-mediated scalar interactions, which yield unique insights into the molecular behavior at reduced magnetic fields. The weak magnetic fields in these environments are responsible for splitting of NMR lines even more, which provides us with more information from the spectra. The development of rules for interpreting these splitting profiles can be used in the interpretation of spectroscopic data to further advance the study of complex molecules and materials (Appelt *et al.,* 2011). As shown, chemical shifts and splitting patterns are both very valuable tools for NMR spectroscopists to determine the structure of molecules in great detail (Figure 6.9).

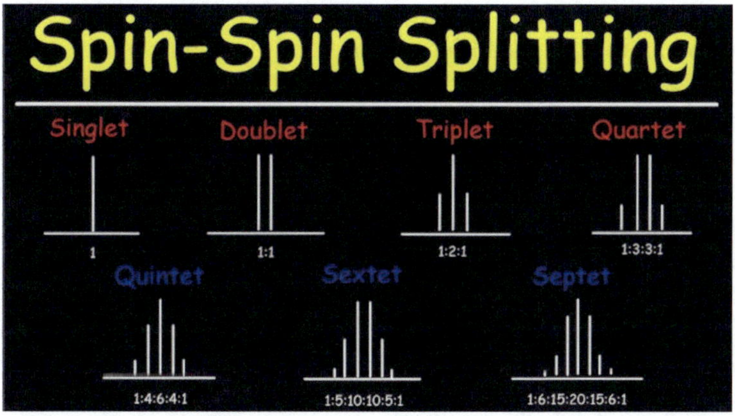

Figure 6.9. *Spin-spin splitting.*

Source: https://i.ytimg.com/vi/WJKfl3w52Mk/hq720.jpg?sqp=-oaymwEhCK4FEI-IDSFryq4qpAxMIARUAAAAAGAElAADIQj0AgKJD&rs=AOn4CLBjaY4FO-mh8r8Di0tgjBiw-DGkPQ.

6.4.1. Explanation of Chemical Shifts in NMR Spectra

To comprehend the complexities of nuclear magnetic resonance (NMR) spectroscopy, it is necessary to comprehend the concept of chemical shifts.

Chemical shifts, measured in parts per million (ppm), are caused by the magnetic environment (the magnetic field surrounding the nuclei) particularly hydrogen (protons) in organic compounds. Placing these nuclei in a magnetic field can alter their resonance frequency by shielding or deshielding the electronic environment around them. This phenomenon is necessary to distinguish between a variety of chemical environments in a molecule. For example, protons that are attached to electronegative atoms (like oxygen or nitrogen) show big downfield shifts due to deshielding effects, while protons that are in very shielded environments show upfield shifts. Chemists can determine the structural information about compounds by examining these shifts. This helps them to understand how molecules fit together and how they interact in more complicated systems. Chemical shifts in NMR spectra are affected by a number of things, such as how the molecules are shaped and how they interact with each other. This makes things further complicated. Two things can change the positions of peaks, and these are hydrogen bonding and steric hindrance, which change the expected chemical shifts. The resolution of the NMR spectrum can also provide us with information about dynamic processes, such as changes in shape or rotations of molecules, which gives us an indication of the flexibility of chemical environments. The interaction of different nuclei can lead to the occurrence of complicated splitting patterns and signal intensities, which makes the data of NMR experiments even more valuable. This highlights the need for a detailed analysis involving both chemical shifts and coupling constants (which are in general indicative of the interactions between neighboring nuclei). Such multifaceted evaluations are necessary in the characterization of biomolecules and complex organic compounds, as well as in the understanding of their functional dynamics in biological systems. Recent improvements in NMR spectroscopy have helped us to understand chemical shifts better, which lets them show us how biological processes work. For example, researchers have used *in vivo* NMR methods to look at changes in metabolism in living things, using chemical shifts to look at biochemical processes in real time. This method has been especially helpful in the field of medical research, whereby looking at shifts, signs of disease or problems with metabolism can be found. The inclusion of *ex vivo* analyses helps to improve interpretation by providing a relative framework for assessing abnormal changes in diseased tissues. By correlating NMR spectroscopic data with biochemical data from tissue homogenates, we can get a better understanding of how metabolic pathways change under different conditions. These extensive studies emphasize the important role of chemical shifts in NMR spectroscopy and increase its applicability in different areas such as organic chemistry and clinical diagnostics (Agrawala *et al.,* 2009; Atkinson *et al.,* 1991).

6.4.2. Understanding Splitting Patterns and Coupling Constants

Splitting patterns and coupling constants are of fundamental importance in nuclear magnetic resonance (NMR) spectroscopy for elucidating the structure of organic compounds. Splitting occurs due to the coupling of nuclear spins of neighboring atoms, resulting in unique peaks in the NMR spectrum. Each peak represents a different environment in which the nuclei in a molecule find themselves; this is dependent on the number of neighboring protons, as well as the effect of their magnetic fields on each other. The basis of this splitting is the $n+1$ rule, n being the number of equivalent neighboring nuclei. These types of interactions not only provide us with information about the connectivity of molecules, but it also discloses the stereochemical relationships, which are important for determination of the three-dimensional structures of complex molecules. Coupling constants, also referred to as J-values, measure the strength of the spin-spin coupling between nuclei, and they can thus be used to better understand these interactions.

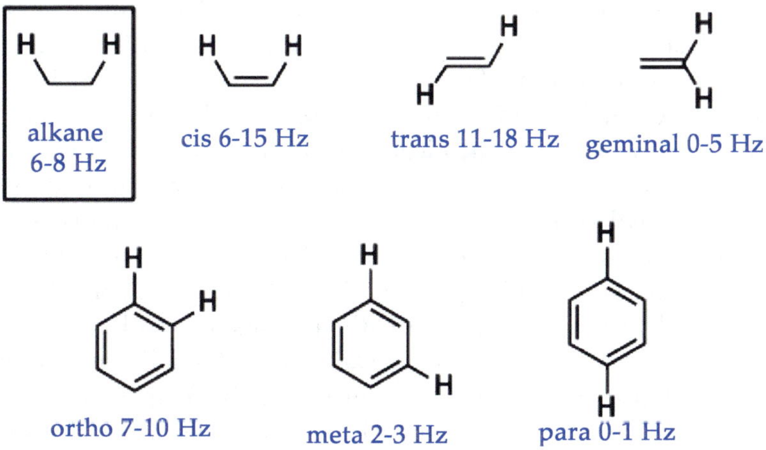

Figure 6.10. *Coupling constants.*

Source: https://www.chem.ucalgary.ca/courses/350/Carey5th/Ch13/coupling04.gif.

These constants are dependent on such factors as the nature of coupling, the nuclear distance between atoms and the electronic nature of the atoms involved. Chemists use different J-values to identify possible isomers because different types of coupling, such as long-range or vicinal coupling, produce different J-values. This knowledge at the level of detail provides much more insight into the NMR data for making informed guesses about the structure of the molecules. Hence, precise measurements of J-values simplify the mapping out of complex biochemical pathways and interactions which is of great importance in drug design and molecular biology. The analysis of splitting patterns and coupling constants is also compatible with modern NMR techniques such as two-dimensional NMR (2D NMR). These methods enable a more detailed description of relationships between different spins and can give us a better understanding of the manner in which molecules move. For example, coupling constants have been shown to be sensitive to the interaction between excited electronic states and nuclear spins when studying ultracold molecules and their spectroscopic properties (Bouloufa-Maafa *et al.,* 2017; Park *et al.,* 2015). These new techniques allow for the observation of the vibrational and rotational parameters of molecular species, providing insight into the complex molecular behavior. This is of theoretical and practical importance for chemistry and medicine (Figure 6.10).

6.4.3. Factors Affecting Chemical Shifts

The electronic environment of the nuclei of interest has a large impact on chemical shifts in nuclear magnetic resonance (NMR) spectroscopy. Electronegativity is very important; the higher the electronegativity, the more electric field is created which protects the nuclei nearby and results in downfield shifts. You can see this happen when comparing the chemical shift of protons in different functional groups. For instance, electronegative atoms such as fluorine or oxygen have a larger shift than hydrocarbons. Molecular geometry also gives rise to changes in chemical shifts. This is due to the fact that steric hindrance may alter the way that molecules are arranged in space and how electrons are spread out across the space. The hybridization state of carbon atoms also influences the shifts of protons. For example, *sp* hybridized carbons exhibit shifts at low magnetic fields compared to sp^2 or sp^3. Chemists can accurately interpret NMR spectra if they are aware of these characteristics. This helps them to figure out structures of unknown compounds and understanding the complex relationships between molecules. The solvent in which NMR experiments are conducted has a big effect on chemical shifts. The polarity of the solvent used could alter the environment surrounding the nuclei, and this could in turn alter how well they are shielded electronically. For example, in polar solvents, there may be stronger deshielding effects on the solute molecules due to stronger dipole interactions, leading to some noticeable downfield shifts. On the other hand, nonpolar solvents may not have the same effect which can make it easier to see certain compounds in the spectrum. Also, the concentration of the solute can alter the strength of these effects, as these intermolecular interactions are greater at higher concentrations. Also, impurities or additives in the solvent can cause changes in the chemical shifts that need to be taken into account when analyzing the data. So, the correct selection of a solvent is very important for the correct interpretation of NMR data and the determination of the chemical structure of analytes. Lastly, outside factors, such as temperature and pH, can also have a big effect on chemical shifts. Changes in temperature can alter molecular dynamics and the amount of molecular motion, which alters the shielding of the nuclei and results in chemical shifts which vary. Lower temperatures can slow motion of molecules, which can lead to sharper peaks, and potentially more defined chemical shifts as the averaging effects are reduced. Changes in pH can be responsible for changes in biological and environmental systems, particularly for labile protons that are components of ionization equilibria. For example, protonated and deprotonated species exhibit different chemical environments which may cause large shifts. This is especially important in research on signaling molecules, where changes in pH can change how molecules interact and how biological systems work. Studies examining the impact of the oceanic pH on organisms in the seas through methods such as NMR spectroscopy demonstrate this (Adamo *et al.,* 2016; Forse *et al.,* 2015).

6.5. APPLICATIONS OF NMR SPECTROSCOPY

Nuclear magnetic resonance (NMR) spectroscopy has a wide range of applications in many areas, but chemistry, biology, and medicine are the most significant. Nuclear magnetic resonance spectroscopy (NMR) is an important chemical technique for determining the structures of organic compounds. With amazing accuracy, researchers can determine the chemical shifts of different nuclei in a molecule and infer the molecular structures and functional groups. NMR can also be used to analyze complex mixtures since it leaves the chemical structure of the studied samples unaltered. The NMR is also very significant for the detection and quantification of small metabolites in biological systems. This allows us to know the metabolism pathways and the mechanism of diseases, which can promote the development of biochemical research and pharmacology. The application of NMR spectroscopy to allow Magnetic Resonance Imaging (MRI) has had the most impact on medicine. This high-resolution imaging method applies the principles of NMR imaging to obtain high-resolution images of soft tissue structures. It is a crucial investigation for diagnosis of a wide spectrum of medical problems including tumors and neurological disorders. MRI does not involve the use of ionizing radiation like X-rays or CT scans therefore making it a safer option for patients. MRIs have become even better with the recent advances in pulse sequence design based on optimal control theory. They now allow doctors to have clearer images and more precise diagnoses. MRI has revolutionized the care of the patient by enabling the treatment planning and monitoring that was previously not possible. In addition to applications in chemistry and medicine, NMR spectroscopy has new applications in the study of free radicals and oxidative stress as relevant to the understanding of a number of biological processes. Whereas the dynamics of reactive species have been assessed in the past by techniques such as Electron Paramagnetic Resonance (EPR), NMR is a more powerful tool to extend this knowledge to their behavior in living systems. "With NMR, we can study how these transient species react and interact in real time, so we can develop insights as to how they influence biochemistry and human health." It is concluded that NMR studies aimed at interrogating the oxidative status of tissues will contribute to the increasing basic and applied scientific knowledge on the mechanism of diseases such as cancer and the technique will be a powerful tool for these studies (Gurer-Orhan *et al.,* 2017). Thus, the versatility of NMR spectroscopy still impacts on various fields, highlighting its importance as an essential part of modern scientific research (Glaser *et al.,* 2012) (Figure 6.11).

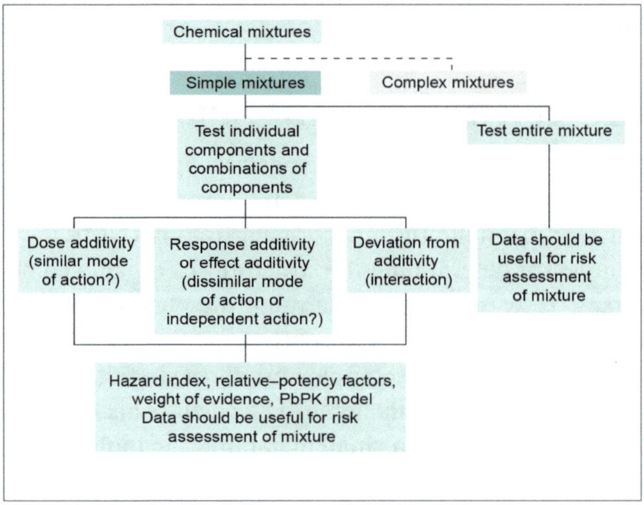

Figure 6.11. *Toxicology of simple and complex mixtures.*

Source: https://ars.els-cdn.com/content/image/1-s2.0-S016561470001720X-gr1.jpg.

6.5.1. Use of NMR in Organic Chemistry for Structure Elucidation

In organic chemistry, nuclear magnetic resonance (NMR) spectroscopy is a technique that is mostly used to figure out the structure of molecules. This allows chemists to determine molecular frameworks to great accuracy. Researchers may employ NMR methods such as the proton (^1H) and carbon (^{13}C) NMR to observe the difference in resonance frequencies of the nuclei. This provides them with information about the number and types of atoms that exist near them. For example, the specific chemical shifts seen in the spectrum give us detailed information about the electronic environment around each atom which is important for figuring out functional groups and how they are connected. When looking at complicated organic molecules NMR is very important since traditional methods may not work. This makes NMR a necessary tool for organic chemists who want to figure out structures accurately. One of the most important applications of NMR spectroscopy is for using two-dimensional NMR techniques such as HSQC and TOCSY. These methods make it much easier to assign both ^1H and ^{13}C resonances.

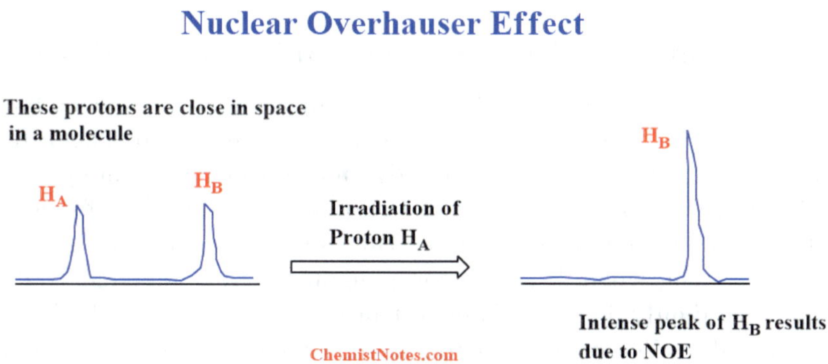

Figure 6.12. *Nuclear Overhauser effect (NOE).*

Source: https://chemistnotes.com/wp-content/uploads/2023/03/NOE-effect.png.

These multidimensional NMR techniques help connect spins and help provide a picture of the molecular structure in its entirety. HSQC (Heteronuclear Single Quantum Coherence) is particularly useful, because it allows you to visualize the relationship of proton and carbon atoms to each other in space. Recent studies have shown that the use of advanced NMR techniques is handy in figuring out the structure of carbohydrates, where more common methods may not work as well. The ability to initiate the process of structure elucidation from hydroxyl (–OH) groups highlights the power of these novel NMR methods, showing that they are applicable to the analysis of larger and more complex molecules, which can be a challenge because they are in very low concentrations in biological samples (Bauer *et al.,* 2018). Moreover, the use of specialized NMR techniques such as Nuclear Overhauser Effect Spectroscopy (NOESY) and Rotating-Frame Overhauser Effect Spectroscopy (ROESY) is of great use for structural characterization in organic chemistry. These methods are great at giving us a lot of information about how atoms are organized in space through interactions between and in molecules. This helps us to better understand how molecules are shaped and arranged. As shown in the literature, these methods are gaining popularity not only in organic chemistry, but also in the areas combining different areas of study, such as food science, where it is very important to understand the structure of complex ingredients. Researchers can better know the three-dimensional spatial structures of organic compounds, using NOESY and ROESY. This is important when studying reaction mechanisms, as well as intermolecular interactions (Akbari *et al.,* 2019). So, NMR spectroscopy is still an important tool for organic chemists, and it is helping to come up with new ways to figure out structures (Figure 6.12).

6.5.2. Applications in Biochemistry and Pharmaceuticals

Nuclear magnetic resonance (NMR) spectroscopy has revolutionized the way in which we think about molecular structures and dynamics in biochemistry and pharmaceuticals. NMR is a nondestructive method of investigating complex biomolecules. This will allow scientists to explore the intricacies of protein and ligand interactions and enzyme action. This ability is particularly relevant to drug development, where being able to pinpoint where compound candidates bind and how tightly they bind is important for scientists in developing better drugs. As well as allowing us to analyze samples in their natural environments, NMR provides a more complete understanding of biological processes which may open up new ways of treating diseases for which the conventional methods prove less effective. NMR applications have also been used for metabolomics, a developing field of biochemistry. By mapping the metabolic profiles in cells and tissues, researchers can compare healthy and sick states. This is useful for diagnosis and prevention of diseases. In addition, studies which emphasize the application of NMR in human diseases, toxicology, and pharmacology, have demonstrated the capability of NMR to provide rich information about the metabolome. By combining NMR with other advanced methods, scientists will be able to discover more about the many metabolites that affect health and disease. This paves the way for specialized programs of personalized medicine targeting particular metabolic pathways (Busi *et al.,* 2013). Moreover, the structural analysis of glycosaminoglycans such as heparin and heparan sulfate highlights the critical role played by NMR in elucidating the complexity of biomolecular interactions. The structural microheterogeneity of these molecules classes is responsible for their diverse biological functions, including cell signaling and inflammation. NMR is therefore used to get valuable information about the structural details that regulate these interactions. This enables them to design drugs that will alter their biological activities in useful ways. These findings can be applied to develop targeted therapies that benefit patient outcome while minimizing side effects (Bisio *et al.,* 2012).

6.5.3. Role of NMR in Materials Science and Engineering

More and more people in materials science and engineering are getting aware of the importance of nuclear magnetic resonance (NMR) spectroscopy in determining the structure and dynamics of molecules at the atomic level. NMR techniques help us to understand how atoms are arranged in materials which is important to understanding their physical and chemical properties. For example, improvements in high-resolution nanoscale magnetic resonance imaging have enabled scientists to obtain very high spatial resolutions, with proton spins with an accuracy of less than 2 nm. By combining NMR pulses optimized based on control theory, the proton spin resonance line width has been decreased greatly. This makes it easier to study complex systems of materials. These breakthroughs show the importance of NMR in material characterization in the development of advanced substances with customized functionality that is vital for new technologies (Budakian *et al.,* 2017). NMR is important for more than just basic material characterization; it is also important in checking the strength and performance of different engineering materials. Because the techniques don't damage the samples they can be used to do thorough tests and that's very useful in fields like aerospace and civil engineering. For example, NMR is commonly used for testing composite materials to assess how polymers break down or fail to ensure that structures remain safe and reliable during the time they are in use. NMR can also help us in understanding how materials diffuse which has a direct effect on how good a product works as well as how long it lives. These features make NMR an important analytical tool for engineers and material scientists who want to come up with new ideas and make existing materials better (ANONYMOUS, 2016). Additionally, the role that NMR plays in the development of new materials and nanostructures is substantial. As scientists study materials at the nanoscale, it becomes very important to be able to accurately probe and change atomic structures. Because NMRs are sensitive to changes in the local environment, they can help to find new uses for new materials, such as nanomaterials and biomaterials. This results in advances in a host of fields, such as energy storage where NMR can help researchers find out more about how electrolytes and electrodes interact in batteries, which can then aid in designing more efficient energy systems. The union of materials science and NMR technology not only sparks innovation but fosters interdisciplinary cooperation which has the potential to reshape the future of engineering applications in a variety of domains (Budakian *et al.,* 2017; ANONYMOUS, 2016).

6.6. ADVANTAGES AND LIMITATIONS OF NMR

One of the best things about Nuclear Magnetic Resonance or NMR spectroscopy is that they can analyze things without destroying them. This makes it ideal for investigating organic compounds, biomolecules, and complex mixtures without altering their structure. NMR is unlike traditional methods of chemical analysis because it can be expensive and time consuming and it can also destroy samples. NMR on the other hand, allows you to obtain detailed information about molecules without consuming the sample. The method gives researchers a lot of information about molecular structures, dynamics, and interactions, which allows them to work out how the atoms in a compound are arranged in three dimensions. This is particularly useful in the biochemistry and pharmaceuticals industries, where it is important to know how molecules interact with one another and change shape, in order to make and design drugs. Furthermore, there are improvements in techniques like algorithmic cooling which may enhance the sensitivity of NMR that can potentially expand its use in high resolution analyses of intricate systems (Fernandez *et al.,* 2004). NMR spectroscopy has so many advantages, but it has certain disadvantages. One of the biggest problems is that it's not as sensitive as other analytical methods, such as mass spectrometry. This restriction comes from the fact that there isn't much difference in the population of spin states at thermal equilibrium, which usually means that you need a lot of samples or a long time to get detectable signals. In addition, NMR equipment typically requires very costly and sophisticated technology, such as superconducting magnets, radiofrequency coils, and sophisticated computer systems. This makes it more difficult for smaller labs and research institutions to get. This cost may make it more difficult for NMR to be used widely in many fields of study, particularly in situations where budgets are tight. So, despite the fact that NMR is a very useful tool in analytical chemistry, because of low sensitivity and availability it is difficult for researchers to take full advantage of this potential. Another thing that needs to be considered when using NMR is the technical knowledge and experience required to correctly read spectra. To adequately analyze NMR data, you typically have to know a lot about chemistry, as well as how to use advanced software tools to be able to interpret spectra. People not experienced with the field can misinterpret results, leading to incorrect conclusions concerning molecular structures or dynamics and make research results less reliable. NMR provides us with valuable information on how molecules change shape and interact with each other on a basic level. However, some systems are so complicated that the spectra are difficult to read.

This situation illustrates how important it is to keep learning and training in the field and to make better computer algorithms to help people to understand data better. As NMR continues to change, it will be important to find new ways to deal with these problems and make technological advances in order to keep it as an important tool for modern scientific research (Baianu *et al.,* 2011).

6.6.1. Advantages of NMR Over Other Spectroscopic Techniques

Nuclear magnetic resonance (NMR) spectroscopy has some definite advantages over other spectroscopy methods, especially in the information provided about molecular structure and dynamics. NMR is superior to another type of spectroscopy, ultraviolet-visible (UV-Vis) spectroscopy, in that UV-Vis spectroscopy primarily reveals electronic transitions, whereas NMR provides information regarding atomic environments and molecular interactions. For instance, UV-Vis methods are only sensitive to chromophores, and NMR is sensitive to nuclei spins which reveals molecular arrangement in space based on chemical shift and coupling constants. NMR can reveal very small details of complex molecules and thus it is very useful in many fields such as organic chemistry, biochemistry, and pharmaceuticals where it is important to know the structure of molecules in order to produce drugs and materials. NMR spectroscopy is well known as a method for determining the structure of molecules but is also known for being non-destructive which differentiates it from other methods like infrared (IR) spectroscopy which can alter or destroy samples during an analysis. Because NMR can be used on biological samples without harming them, it can be applied to rare or valuable samples, without concern that they might be damaged or lost. Also, NMR can provide very precise quantitative data on sample concentrations, important for quality control and following the rules. For example, being able to analyze liquid samples directly without pre-treatment is a huge plus in food chemistry and other fields where samples integrity is very important. Due to these features, NMRs can be utilized for a wide variety of chemical investigations, which shows that it is the best characterization tool. Also, the fact that NMR is particularly well suited to the study of complex mixtures without the need for a lot of separation methods makes it even better. Nuclear Magnetic Resonance (NMR) can tell you a lot about the components of a mixture and traditional methods such as gas chromatography (GC) or high-performance liquid chromatography (HPLC) require the sample to be cleaned first. This capability is particularly useful for metabolomics, in which determining which metabolites are present in biological samples and their relative abundances can provide insights into metabolic pathways. Near Infrared (NIR) spectroscopy is a quick technique of analysis but it does not provide much structural information and might not be as suitable for more complex matrices (Baianu *et al.,* 2011). Nuclear magnetic resonance (NMR), however, is suitable in a wide variety of circumstances. This makes it a critical tool for researchers and businesses that require robust analytical techniques. This is indicated by the ever-increasing potential of new techniques such as terahertz spectroscopy which are being applied to niche applications (Abbott *et al.,* 2017) (Figure 6.13).

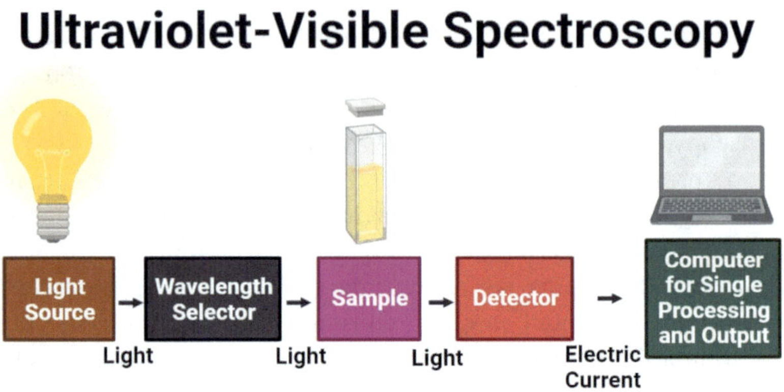

Figure 6.13. *UV-Vis spectroscopy.*

Source: https://microbenotes.com/wp-content/uploads/2023/11/UV-Vis-Spectroscopy.jpeg.

6.6.2. Limitations and Challenges in NMR Spectroscopy

Nuclear magnetic resonance (NMR) spectroscopy is a very valuable technique for molecular structural investigation, but it has some problems and drawbacks. One major problem is that NMR techniques are very sensitive in nature. Conventional NMR techniques are hard for low concentration samples. This can result in weak signals that are difficult to detect, resulting in low resolution and unreliable data. This problem is exacerbated by the presence of multiple signals within a mixture, as any one of them may mask important spectral information. These limitations need to be improved upon to make NMR more useful in a wider range of fields from drug discovery to materials science, by enhancing the detection sensitivity and resolution. Other novel ways, including using the nitrogen-vacancy centers in diamond for NMR imaging could improve the sensitivity and provide exceptional spatial resolution which could alleviate some of the most critical issues with conventional methods (Awschalom *et al.,* 2014). Environmental effects can also be applied to NMR spectroscopy, which can result in bad experimental results. Temperatures, magnetic field inhomogeneities, and sample viscosity are some of the variables that can cause data to be inconsistent. Relaxation times of the spins vary with the temperature and spectral lines may broaden, which complicates interpretation of spectral data. In addition, the resonance frequency can be distorted by unevenness of the magnetic field, which can result in distortion of the signal. Despite much progress in improving temperature control and magnetic field homogenization, these issues remain a challenge area and deserve continuous research and development in the field. To overcome these discrepancies, advanced techniques and algorithms are currently being explored such as heat-bath algorithmic cooling methods, in order to improve the signal-to-noise ratio and, as a result, the reliability and reproducibility of measurements in NMR (Brassard *et al.,* 2014). NMR spectroscopy is also difficult because samples are difficult to prepare, and they must be optimized before analysis. Sample preparation can become time consuming and laborious, and you need to be very careful not to contaminate or damage your samples. Also, experimental conditions such as pH, ionic strength, and solvent selection must be optimized to maximize the analysis, which adds to the complexity of the analysis, and makes it more difficult to duplicate the analysis. In some cases, the presence of paramagnetic species might cause signal quenching, which requires additional sample purification steps. We have to rely on concepts from chemistry, physics, and engineering to enhance our methods and techniques in order to solve these problems. There is also the possibility that the development of more sophisticated computer methods for data analysis might make spectra easier to interpret, which would further enhance the usefulness of NMR spectroscopy both in academic research and clinical settings.

6.6.3. Future Developments and Improvements in NMR Technology

Nuclear magnetic resonance (NMR) is here to stay – and new hardware and software will make the technology better. Since many of the current NMR instruments rely on the magnets for their sensitivity, these better designs, including high-temperature superconductors will likely lead to even more sensitive and higher resolution instruments. These magnets improve the homogeneity and strength of the magnetic field, which is beneficial for obtaining clearer spectra and resolving significantly overlapped resonances. In addition, enhanced pulse sequence algorithms will allow scientists to extract more information from complex samples of particular use in chemistry and biochemistry. Refinements in these algorithms can be used to make NMR analysis even more useful and efficient for a wider array of applications. This will reduce the time required to obtain data and improve the accuracy of measurements. In addition, the joint use of NMR technology with other spectroscopic techniques is one of the major areas for future research and development. When NMR is combined with other techniques such as mass spectrometry or infrared spectroscopy, it will provide more information about how molecules are structured and how they move around. Not only will this interdisciplinary approach enhance the data obtained from samples, but also it will allow complex biological systems to be analyzed in greater detail. For instance, integration of NMR with imaging techniques might produce major shifts in diagnostic medicine, enabling physicians to observe metabolism occurring in real time in living organisms. As a result, scientists of various disciplines are likely to collaborate to develop new ways to improve the application of NMR spectroscopy in research, as well as in a clinical environment. Also, advances in artificial intelligence and data interpretation are significantly influencing the future of NMR technology. NMR experiments generate a lot of data, and high-level algorithms are often needed to interpret complex spectral data. By using machine learning, researchers are able to create predictive models that allow them to make more accurate spectral assignments and can also be used to identify novel compounds. Further, automation of experimental protocols using AI (artificial intelligence) is revolutionary as it enables high-throughput NMR screening, which is very useful in drug discovery and materials science. Several recent strategic documents have argued that the nuclear physics community has to collaborate in order to gain more insight into the strong force and nuclear matter. The integration of these technologies may also play a very significant role in this work. These developments collectively herald a new golden age for NMR and make it more useful and effective for solving modern-day scientific problems.

SUMMARY

In conclusion, Nuclear Magnetic Resonance (NMR) Spectroscopy is a powerful analytical tool that provides us with a lot of information about the structure and dynamics of molecules. In addition to helping us to find compounds, it is also used to better understand biochemical interactions in fields such as medicine and biochemistry. NMR has the flexibility of *in vivo* and *ex vivo* techniques, allowing scientists to study metabolites and molecular environments in living tissues. This capability is most clearly demonstrated in research that has demonstrated quantitative changes in metabolism associated with disease states such as comparing normal tissues to those that possess pathological properties (Agrawala *et al.,* 2009). This ability to integrate these data in diagnostic methodology has increased the relevancy of NMR in the development of personalized medicine and targeted therapies. Furthermore, NMR findings are directly crucial for elucidating complex biological phenomena. For instance, the biochemical environment of migraine patients has been studied using proton NMR and major changes in energy metabolism were identified, which can open up new avenues for treatment (Achten *et al.,* 2012). The ability to define metabolic disorders improves understanding of disease mechanisms and highlights the need for continued research in NMR methodology to improve these results. Not only do the results fill in gaps in our understanding of medicine at the moment, but they also open up new ways of diagnosing diseases that could bring a shift in how we treat them. In conclusion, the continuous development of NMR Spectroscopy offers the potential to enhance scientific and clinical understanding of human health and disease. By continual advances in methodology and technology, NMR can now provide more definitive information on the biochemical changes and metabolic processes that occur as a result of various diseases. Recent studies have provided evidence that there is a close correlation of NMR findings with the clinical symptoms of diseases. This indicates that NMR has a great potential for use in research and clinical settings. As a result, the continued research on NMR applications will significantly contribute to the development of effective diagnostic and therapeutic methods for improved patient outcomes and deeper understanding of human biology.

As it is evident from the discussion of NMR Spectroscopy, there have been many important developments and applications in various fields of science. A major breakthrough is the use of in-cell NMR to investigate proteins in their natural environments, which leads to a better understanding of protein dynamics and structural properties. For instance, in the case of a-synuclein in *E. coli*, the NMR represents a powerful method for rendering disordered protein dynamics in biologically relevant conditions. With this work, the structural differences of proteins become more clear and new methods are highlighted to resolve the spectral overlap in high-populated cellular environments: a deconvolution method was developed to overcome the measurement noise (Binolfi *et al.,* 2013). "This approach has further increased the significance of NMR and fostered a deeper understanding of molecular interactions occurring in biological systems." NMR spectroscopy is also a powerful tool in materials science particularly in the investigation of defect structures in various materials including graphite. It is concluded that NMR is successful in revealing the intricate relationship between structure and magnetic properties, seen in investigations of magnetic properties of monovacancy defects in neutron-irradiated graphite. It is shown that the bulk paramagnetism and its correlation with the temperature dependent NMR properties are a manifestation of the microscopic dynamical characteristics of the material observed using NMR. The sensitivity of NMR to molecular interactions can be seen by the increase in relaxation rate of the nucleus due to variation of the hyperfine coupling involving defects. This allows scientists to find out more about how defection has an effect on the properties of materials (Dmytriieva *et al.,* 2017). These results improve our understanding of fundamental physics and possible applications of carbon-based materials for technology. Finally, NMR spectroscopy's role in biological and material sciences highlights its versatility and analytical power. The analysis of dynamic protein behavior and defect properties in materials shows how NMR can be used to improve and revolutionize the scientific understanding in many different areas. As NMR techniques continue to be refined, the impact on drug discovery, the development of new materials, and the study of complex biological processes is enormous. The continued evolution of NMR techniques is illuminating the lack of clarity on these domains and expanding the horizon of knowledge and enabling new applications in diverse disciplines. Despite major developments in both protein dynamics and materials science, NMR spectroscopy remains an important and revolutionary tool for research today.

REFERENCES

1. Abbott, D., Dinovitser, A., & Valchev, D. G. (2017). Terahertz time-domain spectroscopy of edible oils. https://core.ac.uk/download/151170691.pdf (accessed on 29th October 2025).
2. Abragam, A., Bienfait, A., Lo, C. C., Weis, C. D., Wood, C. J., Esteve, D., Vion, D., et al. (2015). Controlling spin relaxation with a cavity. https://core.ac.uk/download/79501662.pdf (accessed on 29th October 2025).
3. Achten, E., Paemeleire, K., & Reyngoudt, H. (2012). Magnetic resonance spectroscopy in migraine: What have we learned so far? https://core.ac.uk/download/55801062.pdf (accessed on 29th October 2025).
4. Adamo, B., Bopp, Briffa, Brown, Browne, Clements, et al. (2016). Ocean acidification affects marine chemical communication by changing structure and function of peptide signaling molecules. https://core.ac.uk/download/151161191.pdf (accessed on 29th October 2025).
5. Agrawala, A., Rehani, M., & Sharma, R. (2009). Detection of metabolites by proton *ex vivo* NMR, *in vivo* MR spectroscopy peaks and tissue content analysis: Biochemical-magnetic resonance correlation: Preliminary results. https://core.ac.uk/download/pdf/288964.pdf (accessed on 29th October 2025).
6. Akbari, M., Setiowati, D., Arima, Sedaghat Doost, A., et al. (2019). A review on nuclear Overhauser enhancement (NOE) and rotating-frame Overhauser effect (ROE) NMR techniques in food science: Basic principles and applications. https://core.ac.uk/download/188644933.pdf (accessed on 29th October 2025).
7. Anderson, A. K., et al. (2006). Hadamard NMR spectroscopy for two-dimensional quantum information processing and parallel search algorithms. https://core.ac.uk/download/44137998.pdf (accessed on 29th October 2025).
8. Appelt, S., Blanchard, J., Bluemich, B., et al. (2011). Near-zero-field nuclear magnetic resonance. http://arxiv.org/abs/1107.1706 (accessed on 29th October 2025).
9. Araújo-Moreira, F. M., Bernardes, L. V., Cunha-Filho, F. E., et al. (2014). Hyperfine magnetic field in ferromagnetic graphite. http://arxiv.org/abs/1406.1119 (accessed on 29th October 2025).
10. Atkinson, K., Avison, M. J., Baldassarri, S. R., Bode, J., Bäckman, L., Calne, D. B., Cohen, J. S., et al. (1991). Effect of cyclosporine on hepatic energy status and on fructose metabolism after portacaval shunt in dog as monitored by phosphorus-31 nuclear magnetic resonance spectroscopy

in vivo. https://core.ac.uk/download/12204641.pdf (accessed on 29th October 2025).

11. Awschalom, D. D., Kim, M., Mamin, H. J., et al. (2014). Proton magnetic resonance imaging with a nitrogen-vacancy spin sensor. http://arxiv.org/abs/1406.2983 (accessed on 29th October 2025).
12. Babakhani, A., Hajimiri, A., & Hassibi, A. (2008). A spectral-scanning magnetic resonance imaging (MRI) integrated system. https://core.ac.uk/download/4886088.pdf (accessed on 29th October 2025).
13. Baianu, I. C., & Guo, J. (2011). NIR calibrations for soybean seeds and soy food composition analysis: Total carbohydrates, oil, proteins and water contents. https://core.ac.uk/download/pdf/290467.pdf (accessed on 29th October 2025).
14. Bauer, J., Brown, G. D., Kuemmerle, R., et al. (2018). A solution NMR approach to determine the chemical structures of carbohydrates using the hydroxyl groups as starting points. https://core.ac.uk/download/187087209.pdf (accessed on 29th October 2025).
15. Bhattacharya, P. K., Cassidy, M. C., Chan, H. R., et al. (2013). *In-vivo* magnetic resonance imaging of hyperpolarized silicon particles. http://arxiv.org/abs/1305.3332 (accessed on 29th October 2025).
16. Binolfi, A., Fitzpatrick, A. W. P., Maltsev, A. S., Morar, A. S., Fauvet, B., Fauvet, B., McNulty, B. C., et al. (2013). In-cell NMR characterization of the secondary structure populations of a disordered conformation of α-synuclein within E. coli cells. https://core.ac.uk/download/16467855.pdf (accessed on 29th October 2025).
17. Bisio, A., Canales, A., Rhomberg, A. J., Casu, B., Casu, B., Kuberan, B., Kuberan, B., et al. (2012). Heparin and Heparan Sulfate: Analyzing Structure and Microheterogeneity [Chapter]. https://core.ac.uk/download/78054335.pdf (accessed on 29th October 2025).
18. Borneman, T. W., & Cory, D. G. (2012). Bandwidth-limited control and ringdown suppression in high-Q resonators. http://arxiv.org/abs/1207.1139 (accessed on 29th October 2025).
19. Bouloufa-Maafa, N., Dulieu, O., Guo, M., Lụ, et al. (2017). High resolution molecular spectroscopy for producing ultracold absolute ground-state ^{23}Na ^{87}Rb molecules. http://arxiv.org/abs/1710.04883 (accessed on 29th October 2025).
20. Bracco, A., Chomaz, P., Gaardhøje, J. J., et al. (2010). Perspectives of nuclear physics in Europe: NuPECC long range plan 2010. https://core.ac.uk/download/19502483.pdf (accessed on 29th October 2025).

21. Brassard, G., Elias, Y., Fernandez, J. M., et al. (2014). Experimental heat-bath cooling of spins. https://core.ac.uk/download/pdf/81759419.pdf (accessed on 29th October 2025).

22. Brennan, F., et al. (2003). Aerogel keystones: Extraction of complete hypervelocity impact events from aerogel collectors. http://arxiv.org/abs/astro-ph/0312460 (accessed on 29th October 2025).

23. Budakian, R., Chen, A. Q., Cory, D. G., et al. (2017). High-resolution nanoscale solid-state nuclear magnetic resonance spectroscopy. http://arxiv.org/abs/1707.01062 (accessed on 29th October 2025).

24. Busi, M. V., Gomez Casati, D. F., & Zanor, M. I. (2013). Metabolomics in plants and humans: Applications in the prevention and diagnosis of diseases. https://core.ac.uk/download/52479380.pdf (accessed on 29th October 2025).

25. Cai, J., Kost, M., & Plenio, M. B. (2015). Resolving single molecule structures with nitrogen-vacancy centers in diamond. https://core.ac.uk/download/77002999.pdf (accessed on 29th October 2025).

26. Charnock, G. T. P., & Kuprov, I. (2014). A partial differential equation for pseudo contact's shift. *Physical Chemistry Chemical Physics, 16*(42), 22695–22701. https://pubs.rsc.org/en/content/articlelanding/2014/cp/c4cp03106g (accessed on 29th October 2025).

27. Detmold, W., Dudek, J. J., Edwards, R. G., et al. (2019). Hadrons and nuclei. http://arxiv.org/abs/1904.09512 (accessed on 29th October 2025).

28. Dmytriieva, D., Helm, M., Kühne, H., et al. (2017). Monovacancy paramagnetic in neutron-irradiated graphite probed by ^{13}C NMR. http://arxiv.org/abs/1707.03748 (accessed on 29th October 2025).

29. Felderhoff, M., & Zibrowius, B. (2019). On the preparation and NMR spectroscopic characterization of potassium aluminum tetrahydride $KAlH_4$. http://arxiv.org/abs/1902.08540 (accessed on 29th October 2025).

30. Fernandez, J. M., Lloyd, S., Mor, T., et al. (2004). Algorithmic cooling of spins: A practicable method for increasing polarization. http://arxiv.org/abs/quant-ph/0401135 (accessed on 29th October 2025).

31. Forse, A. C., Grey, C. P., Griffin, J. M., et al. (2015). *In situ* NMR and electrochemical quartz crystal microbalance techniques reveal the structure of the electrical double layer in supercapacitors. https://core.ac.uk/download/42338971.pdf (accessed on 29th October 2025).

32. Glaser, S. J., Janich, M., Lapert, M., et al. (2012). Exploring the physical limits of saturation contrast in magnetic resonance imaging. http://arxiv.org/abs/1208.5865 (accessed on 29th October 2025).

33. Greilich, A., Imamoglu, A., Kogan, A., Johnson, A. C., Johnson, A. C., Holleitner, A. W., Buizert, C., et al. (2006). Driven coherent oscillations of a single electron spin in a quantum dot. http://arxiv.org/abs/cond-mat/0608459 (accessed on 29th October 2025).

34. Gurer-Orhan, H., Saso, L., & Suzen, S. (2017). Detection of reactive oxygen and nitrogen species by electron paramagnetic resonance (EPR) technique. https://core.ac.uk/download/154947312.pdf (accessed on 29th October 2025).

35. Liu, R. B., Ma, W. L., & Yang, W. (2016). Quantum many-body theory for electron spin decoherence in nanoscale nuclear spin baths. http://arxiv.org/abs/1607.03993 (accessed on 29th October 2025).

36. McKinney, M. A., Uhl, F. M., & Wilkie, C. A. (2000). Polybutadiene cross-linked with various diols – Effect on thermal stability. https://core.ac.uk/download/67756474.pdf (accessed on 29th October 2025).

37. ANONYMOUS. (2016). The Nuclear Physics Division of the European Physical Society. Nuclear physics for cultural heritage. https://core.ac.uk/download/46663971.pdf (accessed on 29th October 2025).

38. Pálffy, A., Borchert, G. L., Bouchiat, M. A., Bransden, B., Demtröder, W., Goldanskii, V., Otten, E. W., et al. (2010). Nuclear effects in atomic transitions. http://arxiv.org/abs/1106.3218 (accessed on 29th October 2025).

39. Park, J. W., Will, S. A., & Zwierlein, M. W. (2015). Two-photon pathway to ultracold ground state molecules of ^{23}Na ^{40}K. http://arxiv.org/abs/1505.01835 (accessed on 29th October 2025).

40. Rychnovský, J. (2009). Study of the properties of hyperpolarized xenon-129 for magnetic resonance imaging. https://core.ac.uk/download/30310956.pdf (accessed on 29th October 2025).

41. Trautmann, A., Smirnov, B. M., Tiemann, E., Tiesinga, E., Tiesinga, E., Mies, F. H., Appmeier, J., et al. (2011). Feshbach spectroscopy and analysis of the interaction potentials of ultracold sodium. https://core.ac.uk/download/pdf/15474382.pdf (accessed on 29th October 2025).

42. Zaretskii, D. F., Zaretskii, D. F., Sobelman, I. I., Pavlichenkov, I. M., & Landau, L. D. (2010). Time modulation of K-electron capture decay of hydrogen-like ions with multiphoton resonance transitions. http://arxiv.org/abs/1002.0075 (accessed on 29th October 2025).

CHAPTER 7

MASS SPECTROMETRY

Contents

INTRODUCTION

Mass spectrometry (MS) is an important analytical method that many scientists use to find and measure chemicals. The basis of mass spectrometry is its ability to turn samples into ions, sort them by their mass-to-charge ratio (m/z) and find them. The process mainly has three steps: ionization, mass analysis, and detection. To understand how mass spectrometry has become an important tool in fields like biochemistry, environmental science, and pharmaceuticals, you need to know these basic ideas. Mass spectrometry also makes it possible to study complex mixtures, which helps fields like proteomics and metabolomics move forward. As the technique becomes more advanced, it is more important than ever to understand both the basic principles and the more advanced methods that make it work better in different situations (Gunzer *et al.,* 2018). One important change in mass spectrometry is the use of different ionization methods, which affect the quality and sensitivity of the results. Electrospray ionization (ESI) and matrix-assisted laser desorption ionization (MALDI) are two methods that are praised for being able to make ions with little fragmentation. This makes them perfect for biological samples.

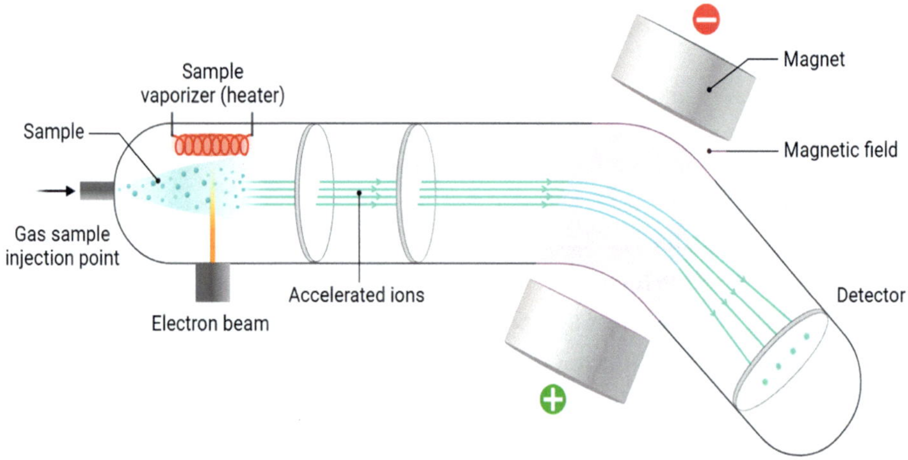

Figure 7.1. *Mass spectrometry (MS).*

Source: https://microbenotes.com/wp-content/uploads/2018/12/Mass-Spectrometry-MS.jpeg.

On the other hand, techniques that produce rich fragmentation patterns, like electron capture dissociation (ECD) and electron transfer dissociation (ETD), are very helpful for figuring out the structure of proteins and looking at post-translational modifications. It is very important to choose the right ionization methods because they have a direct effect on how well the mass spectra can be interpreted. Understanding these methods not only improves the dependability of data, but it also helps researchers deal with difficult analytical problems in natural product chemistry and other fields (Bugni, 2017). Mass spectrometry is used in many areas of science and gives us new information about both basic and applied research. Mass spectrometry has been very useful for finding unknown substances and measuring organic compounds, from monitoring the environment to diagnosing medical conditions. The wide range of uses is also shown by how it is used to study the chemical makeup of biological samples, where even small changes in chemical structure can have big effects on biology. Recent improvements in data interpretation methods, like advanced algorithms for fragmentation analysis, have also made mass spectrometry better at analyzing complicated samples. In the end, researchers who want to use mass spectrometry as a powerful analytical tool in modern scientific research need to have a full understanding of it, especially its principles, techniques, and uses (Gunzer *et al.,* 2018) (Figure 7.1).

Definition and Significance of Mass Spectrometry (MS)

Mass spectrometry is a powerful way to find and measure compounds by looking at the mass-to-charge ratio of their ions. This technique gives researchers a lot of information about the molecular makeup of complicated mixtures, which lets them figure out chemical structures and reactions in a way that has never been done before. The process includes ionizing chemical species and separating the ions that come out based on their mass-to-charge ratios. This makes it possible to do both qualitative and quantitative analysis. This ability is very important in fields like biochemistry, pharmacology, and environmental science, where knowing how molecules interact, and change is necessary for gaining new knowledge and coming up with new uses. The importance of mass spectrometry is further highlighted by its capacity to examine post-translational modifications in proteins, which are crucial for cellular functions. For example, the discovery of more than 300 known modifications, such as N- and O-glycosylation, demonstrates the technique's ability to reveal intricate biological phenomena. Studies have demonstrated that this glycosylation exist not only in eukaryotes but also in significant human pathogens, such as *Campylobacter jejuni* and *Mycobacterium tuberculosis*. Mass spectrometry has been instrumental in clarifying these modifications, significantly enhancing our comprehension of disease mechanisms, and facilitating the identification of potential therapeutic targets, thereby illustrating its transformative potential in biomedical research (Graham *et al.,* 2010). In proteomics and metabolomics, mass spectrometry is becoming more and more common to use with liquid chromatography techniques, like Liquid Chromatography Time-of-Flight Mass Spectrometry (LC-TOFMS). This platform enables the thorough examination of biological mixtures, yielding extensive datasets that necessitate advanced statistical techniques for analysis. Researchers are working on building strong probabilistic models to help them make sense of LC-TOFMS data. This helps them find unknown metabolites and makes the conclusions they draw from the data more reliable. These improvements in method development not only make analyses more accurate, but they also open the door to new discoveries in many fields of science. This shows how important mass spectrometry is to modern research (Ipsen *et al.,* 2011).

Historical Development of Mass Spectrometry

In the early 1900s, scientists were looking for better ways to study chemical compounds, which is when mass spectrometry first came about. J. J. Thomson laid the groundwork when he invented the first mass spectrometer in 1912. This made it possible to measure the masses of ionized particles. This early model used a magnetic field to move ions out of the way, which made it possible to separate them based on their mass-to-charge ratios. Thomson's groundbreaking research not only improved our understanding of atomic structure, but it also paved the way for future progress in the field. Even though the Thomson mass spectrometer had some problems at first, it showed how useful mass spectrometry could be as an analytical tool, which interested both chemists and physicists. This early stage was very important for the later creation of more advanced mass spectrometric methods that would change analytical chemistry in many areas. In the middle of the 20th century, mass spectrometry changed a lot because of new technologies that made it possible to use it in more ways and made it more sensitive. The time-of-flight (TOF) mass spectrometer was a major breakthrough in the 1940s. It made it possible to analyze heavier ions with better resolution.

Also, the introduction of electron impact ionization in the 1950s made it easier to study complicated organic molecules, which broadened the scope of mass spectrometry beyond just simple elemental analyses. The advent of tandem mass spectrometry (MS/MS) further enhanced the technique, facilitating fragmentation pattern analysis that established the foundation for contemporary proteomics and metabolomics research. The growing sophistication of mass spectrometric techniques has enabled the characterization of larger biomolecules, rendering the system indispensable in various applications, including pharmaceutical development, environmental analysis, and clinical diagnostics. Mass spectrometry has become an important part of research in many fields in the last few years, especially in life sciences and materials characterization. The emergence of plant metabolomics exemplifies this trend, with researchers utilizing mass spectrometry, frequently in tandem with chromatography, to examine intricate metabolite profiles in plant systems. This combination not only helps find and measure low molecular weight compounds, but it also helps us understand how growth and development are controlled in different species. Improvements in mass spectrometry methods, like secondary ion mass spectrometry and surface-enhanced techniques, show how innovative the field is, especially when it comes to preserving and studying historical objects, like painted textiles. These improvements show how important it is for scientists and conservators to work together to improve analytical skills and protect cultural heritage (Gullberg *et al.,* 2005; Lennard *et al.,* 2017).

Overview of Mass Spectrometry Applications

Mass spectrometry has become an essential technology in many areas, providing deep insights into the makeup and interactions of molecules. One of its main uses is in metabolomics, where scientists try to figure out the metabolic profiles of biological samples. Scientists can use mass spectrometry to find and measure a wide range of low molecular weight metabolites. This helps us understand metabolic pathways and how they are controlled in different organisms. Combining mass spectrometry with chromatographic methods increases sensitivity and specificity, allowing for the analysis of complicated mixtures. To accurately identify metabolites, it is essential to develop tandem mass spectral databases and software tools (Blaženović *et al.,* 2018). These improvements make annotations more reliable, which is important for understanding the complexities of systems in clinical and translational research. Mass spectrometry is also very useful in proteomics, where it helps scientists study the structures and functions of proteins.

The capacity to discern proteins within intricate mixtures has transformed our comprehension of numerous biological processes, particularly those associated with disease mechanisms. Mass spectrometry can give you important information about the identities of proteins, post-translational changes, and how proteins interact with each other. Researchers can use methods like liquid chromatography coupled with mass spectrometry (LC-MS) to do large-scale protein profiling, which is very helpful for finding biomarkers and developing new drugs. Furthermore, the careful processing of the resulting data necessitates strong statistical and analytical techniques to guarantee significant interpretations of the proteome (Gullberg *et al.,* 2005). This combination of generating a lot of data and doing advanced analysis shows how important mass spectrometry is to the life sciences. Mass spectrometry is important in environmental science as well as biology. It is used to find and measure pollutants, which helps researchers and policymakers figure out how safe and healthy the environment is. Mass spectrometry is very sensitive and can look at a lot of different types of samples, which makes it possible to accurately keep track of pollutants like pesticides, heavy metals, and new pollutants in air, soil, and water. A key feature of these kinds of applications is the ability to do high-throughput analyses, which is necessary for quickly evaluating environmental data. As new ways to analyze data are created, mass spectrometry keeps getting better, making it even more useful for getting important information for environmental risk assessments and regulatory frameworks. This adaptability underscores the essential function of mass spectrometry in various scientific fields.

7.1. BASIC PRINCIPLES OF MASS SPECTROMETRY (MS)

Mass spectrometry is based on basic ideas that help scientists figure out what chemical compounds are and how much of them there are by looking at their mass-to-charge ratios. The first step is to ionize a sample, which changes the molecules into charged particles, which can be either positive or negative. There are different ways to ionize something, like electron impact, chemical ionization, or electrospray ionization. Each method changes the properties of the ions that are created. After ionization, an electric field speeds up the ions and sends them into a mass analyzer, where they are sorted by their mass-to-charge ratios. The basic relationship that governs this separation is very important for getting the right mass, because it lets ions be classified in a way that makes it easier to find and analyze the sample's parts. The accuracy of these mass measurements is what makes mass spectrometry such an important analytical tool in many scientific fields. There are many kinds of mass spectrometry, and each one is made for a different use or analytical need. Time-of-Flight (TOF), Quadrupole, and Orbitrap mass spectrometry are some of the most common methods. TOF mass spectrometry, for example, measures how long it takes for ions to move a certain distance, which lets you figure out their mass. In contrast, Quadrupole mass spectrometry uses electric fields to dynamically filter ions based on their mass-to-charge ratios. This method works especially well for figuring out how much of something is in a complex mixture. Researchers need to think about the pros and cons of each of these methods based on the type of sample and the desired result. To choose the right method for a specific analytical problem, you need to know about the different types of mass spectrometry. This will make the experiment more effective. Fragmentation patterns are very important in mass spectrometry because they give information about the structural properties of the compounds being studied. During analysis, ions can break apart, creating fragment ions that are then looked at to see how their mass-to-charge ratios compare. These fragmentation patterns are typical of the molecular structure, which helps figure out what unknown compounds are and makes it easier to identify them. Additionally, advanced software tools have been created to help link observed fragmentation patterns to known structures in databases, which makes it easier to identify small molecules. For example, new methods like in silico fragmentation tools are being used more and more to guess how things will break apart based on their structure. As research progresses, the enhancement of spectral databases and the incorporation of computational techniques lead to more dependable metabolomic studies, demonstrating the significant influence of fragmentation patterns on mass spectrometry applications across diverse scientific and medical domains (Blaženović *et al.,* 2018). Mass spectrometry has become a powerful tool in clinical diagnostics, environmental monitoring, and pharmaceutical research, showing how useful and important it is in modern analytical chemistry (Bolton *et al.,* 2014) (Figure 7.2).

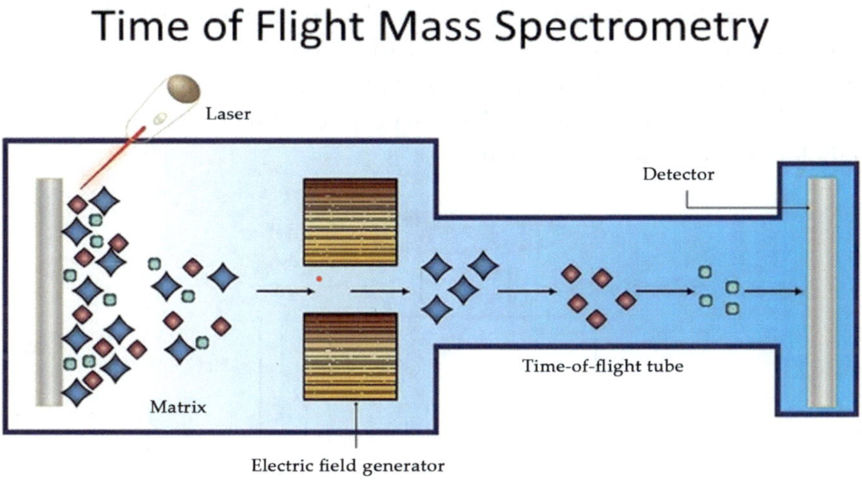

Figure 7.2. *Laser used for TOF-MS.*

Source: https://www.rpmclasers.com/wp-content/uploads/2022/02/Image-Time-of-Flight-ToF-Mass-Spectrometry-Application-Diagram-Social.png.

7.1.1. Ionization Methods and Their Importance

Ionization is a key part of mass spectrometry because it determines the quality and reliability of the analyses that come after it. Different types of ionization methods, like Electron Ionization (EI), Electrospray Ionization (ESI), and Matrix-Assisted Laser Desorption/Ionization (MALDI), work for different types of compounds. The effectiveness of each technique is significantly determined by the physical and chemical characteristics of the analyte, which affect the types of ions generated and their fragmentation patterns during spectral analysis. Notably, EI is often better for volatile and thermally stable compounds, while ESI is necessary for polar and larger molecules like proteins and nucleic acids. This flexibility shows how important it is to choose the right ionization methods for mass spectrometric analysis to get accurate molecular characterization and identification. This will help us better understand complex biological systems. Different ionization methods are important for more than just finding analytes;

they are also necessary for correctly figuring out chemical structures and composition information. For instance, the production of particular fragment ions during the ionization process can facilitate the clarification of molecular structures, including the identification of tautomeric forms, as demonstrated by research examining the influence of mass spectral data on tautomerism in diverse organic compounds. Ionization causes small changes in the stability or reactivity of tautomer, which makes it possible to do a more detailed analysis that links the ratios of fragment abundance to the keto/enol content. So, learning about these things not only makes analysis more accurate, but it also gives us a better understanding of chemistry that can be used in other areas of science, like metabolomics and organic chemistry.

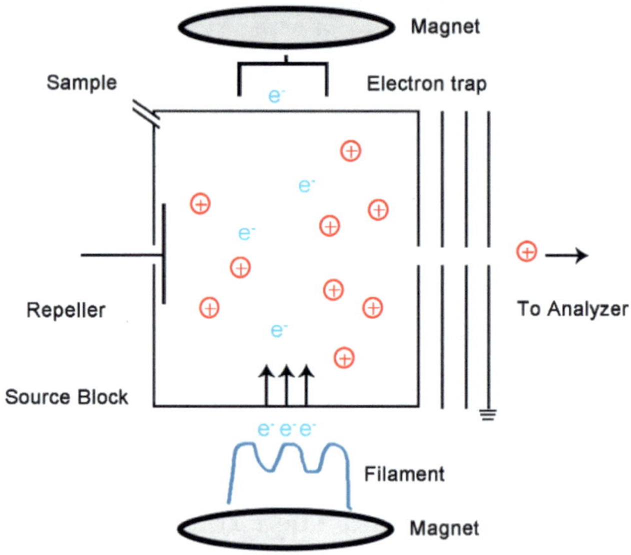

Figure 7.3. *Electron ionization.*

Source: https://www.creative-proteomics.com/images/Electron-Ionization-1.png.

This helps both theoretical knowledge and practical uses of mass spectrometry. The development of software and computational tools to support ionization methods has also changed the way mass spectrometric analysis is done. New tools like in silico fragmentation have made it easier to annotate complex mixtures, which has helped researchers figure out molecular structures from tandem mass spectral data. These improvements help with the problems that come up when trying to annotate unknown compounds, which is an important part of untargeted metabolomics. For example, tools like MS-

FINDER and MetFrag help identify metabolites by comparing experimental data to large databases. This makes predictive metabolic models more reliable in translational research. This combination of computational methods not only makes ionization methods better at analyzing things, but it also helps us understand chemicals and their roles in biological systems better. This makes ionization methods more important in the overall context of mass spectrometry (Blaženović *et al.,* 2018; Eduardo *et al.,* 2008) (Figure 7.3).

7.1.2. Mass-to-Charge Ratio (m/z) and Its Role

The mass-to-charge ratio (m/z) is a key part of mass spectrometry because it helps identify and describe ions based on their mass and charge. A mass spectrometer makes ions, speeds them up, and then sorts them by their m/z values. The m/z of each ion directly affects how it moves through the mass analyzer, which is what finally determines whether it is detected. This feature lets scientists figure out the molecular weight of compounds, which makes it easier to find different analytes in a sample. It is very important to measure the m/z ratio accurately because even small errors can cause compounds to be misidentified, which can make mass spectrometric data much less reliable in fields like proteomics and metabolomics. The function of m/z transcends simple identification; it is crucial for comprehending fragmentation patterns of molecules during MS/MS analysis. When ions hit neutral gas molecules, they break apart into smaller ions. The m/z values of these fragment ions show how intense and spread out they are, which shows how the precursor ions are structured. Scientists have started to use the intensity patterns of fragmentation spectra to learn more about how peptides break down, which means that m/z values are not just numbers that tell you what they are, but also numbers that tell you how they act when they break down. The analysis conducted using Bayesian neural networks has illustrated how the m/z values of fragment ions can improve algorithms for high-throughput protein identification, thus optimizing the utility of the information obtained from MS/MS spectra (Bowler *et al.,* 2008). Moreover, improvements in methods for processing and analyzing data have highlighted the importance of m/z in metabolomics, where complex datasets require advanced algorithms to make sense of them. Tools that analyze m/z data let you annotate and see metabolites, which are very important for understanding biochemical pathways and interactions. As researchers keep working on open-source software for untargeted metabolomics workflows, the importance of m/z as a basic building block cannot be overstated. As mass spectrometry datasets become more complex, especially in metabolomic studies, it is important to have complete resources that explain the latest advances in m/z analysis tools. These tools not only make it easier to process data, but they also improve the accuracy of identifying and measuring metabolites, which helps the field of mass spectrometry as a whole (Misra *et al.,* 2016).

7.1.3. Detection and Analysis of Ions

Finding and studying ions is a key part of mass spectrometry because it connects theory to practice. Mass spectrometry makes ions, studies them, and finds them to give important information about what chemicals are made of. Different ways to ionize molecules, like Electrospray Ionization (ESI) and Matrix-Assisted Laser Desorption/Ionization (MALDI), make it possible to turn neutral molecules into charged ones. These processes facilitate the identification of numerous compounds, especially within intricate biological matrices.

The capacity to analyze and dissect ion characteristics, including mass-to-charge ratio (m/z) and fragmentation patterns, improves the specificity and sensitivity of analyses, which is essential for untargeted metabolomics studies seeking to clarify the complex biochemistry of organisms (Blaženović *et al.,* 2018). As researchers investigate ions more deeply, the significance of tandem mass spectrometry (MS/MS) becomes evident, especially in structural elucidation and the verification of chemical identity. MS/MS is a powerful way to break ions into smaller pieces in a controlled way, which lets you look at the pieces in detail. This analytical method is very important for labeling unknown compounds. It helps create complete databases that can be used in many fields, such as pharmacology and environmental science. Additionally, sophisticated computational tools like MS-FINDER and CFM-ID automate the analysis of mass spectral data, giving researchers accurate structural information about complicated mixtures and making it easier to find new metabolites (Blaženović *et al.,* 2018).

The combination of these tools shows how technology has changed mass spectrometry's ability to analyze ions in a big way. The detection and analysis of ions are not limited to small organic molecules; they encompass various biological modifications, including glycosylation, which are essential for comprehending microbial pathogenesis. Mass spectrometry has become an important tool for studying post-translational modifications, like N- and O-glycosylation in bacterial proteins, which were once thought to only happen in eukaryotic systems. This has important effects on research into human pathogens like *Campylobacter jejuni* and *Mycobacterium tuberculosis*, where glycosylation affects how dangerous they are and how well surface proteins work as a whole. Researchers can find and map these changes using mass spectrometry. This helps them understand their biological significance and come up with targeted treatment plans. The ongoing advancements in this area illustrate the crucial role mass spectrometry plays in expanding our understanding of the molecular landscape within complex biological systems (Figure 7.4).

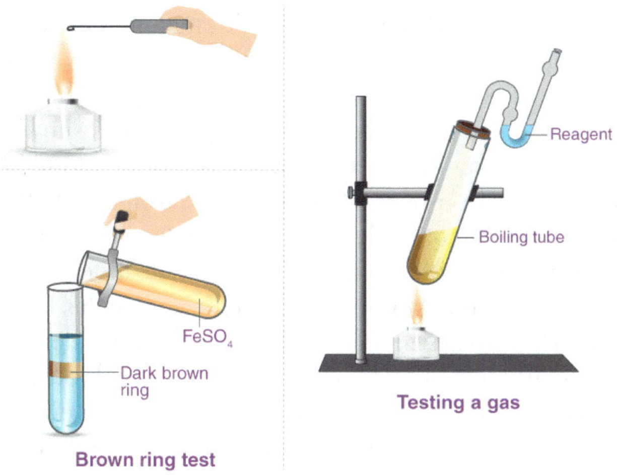

Figure 7.4. *Systematic analysis of anions.*

Source: https://cdn1.byjus.com/wp-content/uploads/2023/01/Systematic-Analysis-of-Anions.png.

7.2. TYPES OF MASS SPECTROMETRY

Mass spectrometry is an important analytical method that can be broken down into different types based on how they work and what they are used for. Electrospray Ionization (ESI) mass spectrometry is a well-known type that is great for studying big biomolecules like proteins and nucleic acids. ESI is unique because it can make ions directly from a solution. This makes it great for studying complicated mixtures without having to separate them first. This method makes things more sensitive and can give you detailed information about molecular weight and structure. Matrix-Assisted Laser Desorption/ Ionization (MALDI) is another important type. It is especially useful for studying macromolecules and polymers. MALDI can make intact molecular ions using laser pulses, which makes it possible to study large and delicate biomolecules. This broadens the uses of mass spectrometry in genomics and proteomics. Gas Chromatography-Mass Spectrometry (GC-MS) and Liquid Chromatography-Mass Spectrometry (LC-MS) are two examples of mass spectrometry techniques that are different from each other. GC-MS is frequently employed for volatile compounds, utilizing gas chromatography to segregate analytes prior to mass spectrometric evaluation. This method is used a lot in food safety, forensic analysis, and environmental monitoring. On the other hand, LC-MS is good at analyzing compounds that are not volatile and are sensitive to heat. This is because liquid chromatography is so flexible. LC-MS has changed the way pharmaceutical and clinical labs work by making it possible to quickly analyze complex biological samples with high sensitivity and specificity. These different mass spectrometry setups show how adaptable the method is across many fields and how important it is to modern analytical chemistry. The introduction of tandem mass spectrometry (MS/MS) has also greatly improved the depth of analysis that can be done with mass spectrometry techniques. MS/MS allows for sequential mass selection and fragmentation, which gives us information about the structure of molecules by looking at how they break apart. This advanced technique is essential for identifying unknown compounds, especially in metabolomics, where complex biological samples are common. The utilization of standardized databases and in silico fragmentation tools, as outlined in the literature, is essential in the metabolite identification process (Blaženović *et al.,* 2018). These developments suggest an increasing necessity in clinical and translational research for precise compound annotation and pathway mapping, thereby highlighting the advancing role of mass spectrometry as an essential instrument in scientific research and diagnostics (Figure 7.5).

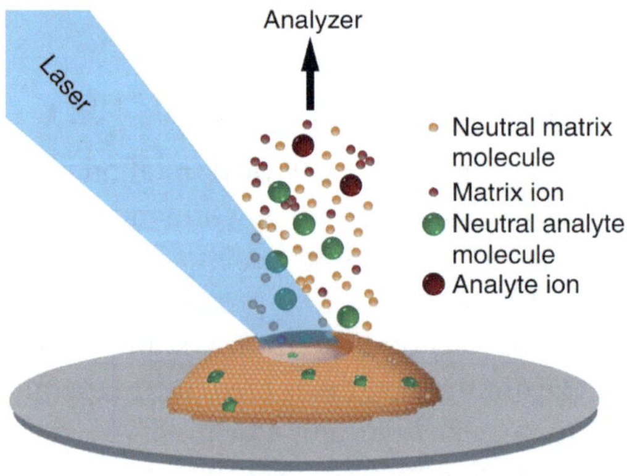

Figure 7.5. *Matrix assisted laser desorption ionization.*

Source: https://www.creative-proteomics.com/images/Matrix-assiste-Laser-Desorption-Ionization.jpg.

7.2.1. Electrospray Ionization (ESI)

Electrospray Ionization (ESI) has become a crucial method in mass spectrometry, enabling the examination of various biomolecules, especially large macromolecules like proteins and nucleic acids. The method uses a high-voltage electric field to turn a liquid sample into an aerosol of charged droplets, which then evaporates to make gas-phase ions. This method is especially useful because it can keep labile biomolecules intact, which lets for detailed analysis without too much fragmentation. ESI is a soft ionization technique that is very sensitive and can analyze complicated mixtures. This makes it very important in proteomics and metabolomics, where it is important to understand the structure and function of biomolecules in order to advance scientific knowledge.

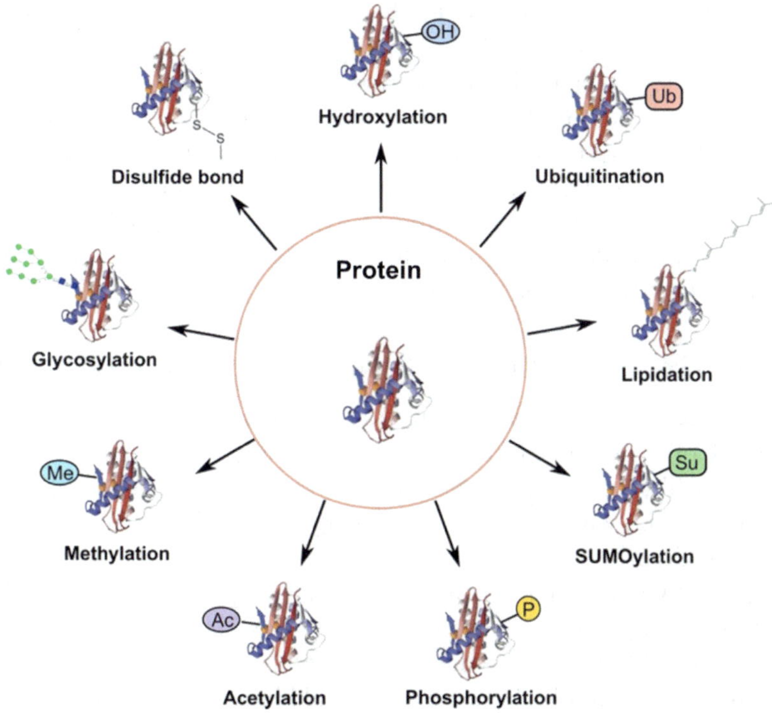

Figure 7.6. *Strategies for post-translational modifications (PTMs).*

Source: https://www.creative-proteomics.com/blog/wp-content/uploads/2018/06/Strategies-for-Post-Translational-Modifications-PTMs-0.png.

The importance of ESI is heightened by its utility in examining post-translational modifications (PTMs) of proteins, which are essential to cellular functions and signaling pathways. For example, mass spectrometry can find glycosylation patterns, which can help us understand how these changes affect how proteins work in living things. ESI facilitates the effective ionization of glycoconjugates and their subsequent analysis, elucidating their biological functions, especially in pathogens like *Campylobacter jejuni* and *Mycobacterium tuberculosis*. These results highlight the efficacy of ESI in clarifying the molecular mechanisms that contribute to microbial virulence, demonstrating how advancements in mass spectrometry are enhancing our comprehension of infectious diseases, thereby illustrating the convergence of biochemistry and public health. Also, ESIs can be used to analyze a wide range of biomolecules, not just proteins. These include lipids and carbohydrates. This wide range of uses comes from the fact that it works well with many different separation methods, like liquid chromatography. As mass spectrometry gets better, new ideas in ESI and other related fields have made it even more important for molecular diagnostics, especially for finding pathogens and sequencing genetic materials. These kinds of improvements are very important because they make it easier for us to find and deal with new infectious diseases and genetic mutations. Moreover, the incorporation of computational tools for the analysis of ESI-derived data is expected to enhance clinical information, facilitating novel diagnostic capabilities in microbiology and genomics, and fundamentally transforming methodologies in both research and clinical environments (Graham *et al.,* 2010; ANONYMOUS, 2025) (Figure 7.6).

7.2.2. Matrix-Assisted Laser Desorption/Ionization (MALDI)

Matrix-assisted laser desorption/ionization (MALDI) is a major step forward in the field of mass spectrometry. It is especially good at looking at big biomolecules like proteins and polymers. The process involves putting the sample in a matrix, which absorbs laser energy and makes it easier for nearby molecules to desorb and ionize. When a laser pulse hits the matrix, it turns into gas, taking the analyte with it as ions. Compared to other ionization methods, this one greatly reduces fragmentation, which keeps the molecules' structure intact. MALDI is an important tool for the thorough analysis of complex biological samples because it can find high-molecular-weight species. This lets researchers look at the structural composition and post-translational modifications of proteins in a wide range of fields, from proteomics to clinical diagnostics. The use of MALDI to study glycosylation events shows how important it is to modern biochemistry. Glycosylation, a common modification that happens after translation, has a big effect on many biological processes and ways that diseases spread. Recent research has revealed the presence of N- and O-glycosylation in bacteria, previously considered unique to eukaryotes, illustrating their role in processes such as virulence in human pathogens like *Campylobacter jejuni* and *Mycobacterium tuberculosis*. Researchers can use MALDI mass spectrometry to find and describe these glycosylation patterns with great accuracy. This shows how important they are to biology and how they make diseases worse. Combining MALDI with bioinformatics tools makes it even easier to look at complicated datasets from these kinds of studies, giving us a better understanding of how glycosylation affects health and disease

(Graham *et al.,* 2010). In the larger field of mass spectrometry, MALDI stands out because it can handle the complicated data that comes from biological systems with many factors. Diseases like diabetes have many different causes that need to be studied in depth and detail. Nevertheless, bioinformatics encounters considerable obstacles in managing and scrutinizing the redundancy inherent in datasets, exemplified by the representation of multiple peptides derived from a single protein. This complexity requires sophisticated analytical methods capable of concurrently addressing experimental and technological intricacies (Chadt *et al.,* 2011). MALDI mass spectrometry, when used with strong data analysis methods, helps researchers get around these problems by giving them clear, understandable data that can lead to important biological discoveries. The combination of MALDI with advanced bioinformatics not only helps to understand how diseases work, but it also makes it easier to find possible therapeutic targets. This makes mass spectrometry more useful in many fields of science.

7.2.3. Gas Chromatography-Mass Spectrometry (GC-MS)

Gas chromatography-mass spectrometry (GC-MS) is a powerful analytical tool that combines gas chromatography and mass spectrometry to create a strong platform for separating and identifying complex mixtures. The gas chromatography part helps separate volatile compounds by measuring how they vaporize and interact with the column's stationary phase. After the separation, mass spectrometry can find and describe the different parts by measuring their mass-to-charge ratios. Not only does this combination improve sensitivity and specificity, but it also lets you find compounds in very small amounts. This makes GC-MS very useful for many things, such as checking the safety of food and monitoring the environment. The usefulness of GC-MS in metabolomics shows how important it is to understand biological systems and pathways. In this context, identifying small molecules is often difficult because biological samples are so complicated. This means that annotation processes need to be quick and easy. Researchers have come up with methods like in silico fragmentation and database matching to help them make sense of mass spectrometry data, mainly to find out what unknown metabolites are. As outlined in the literature, these methodologies aid in the development of predictive metabolic models and enhance the biological interpretation of intricate systems, thereby affirming that GC-MS continues to be a pertinent and indispensable instrument in both translational and clinical research (Blaženović *et al.,* 2018). GC-MS is especially useful in forensic and security settings because it can look for certain compounds and additives that could be signs of illegal activity. For example, finding organic additives like diphenylamine and its derivatives in smokeless powders is very important for forensic investigations. Using thermal desorption with GC-MS, scientists can find and measure these compounds with amazing accuracy and sensitivity. The method, which has been tested on different additives, shows how well GC-MS works to get reliable data that is important for investigations into explosives and other forensic research areas (Jordan *et al.,* 2019). GC-MS continues to improve our ability to study and understand complicated chemical phenomena in many fields because it can be used in many different ways.

7.3. INSTRUMENTATION IN MASS SPECTROMETRY

The mass spectrometry instrumentation is an important part of its ability to analyze and identify a wide range of chemical compounds. Mass spectrometry is based on three main parts: the ion source, the mass analyzer, and the detector. The ion source turns the sample into ions, and the mass analyzer sorts these ions by their mass-to-charge ratios (m/z). There are different kinds of ion sources, like electron impact (EI) or electrospray ionization (ESI), that can be used to study different types of molecules, from small organic compounds to large biomolecules. Also, advanced mass analyzers like quadrupoles and time-of-flight (TOF) instruments have better resolution and sensitivity, which lets researchers find even tiny amounts of substances in complicated mixtures. This makes it easier for researchers to work on new ideas in many different fields. After ions are made and separated, the next step is to find them. This is where the detector comes in. Common detectors, like electron multipliers and ion traps, change the ion signal into an electrical signal that can be measured. This quantification ultimately helps to describe the sample and figure out what it's made of. The advancement of technology has facilitated the incorporation of additional methodologies, such as tandem mass spectrometry (MS/MS), which permits more extensive fragmentation analyses of ions. In this context, the capability to perform in silico fragmentation analysis, as outlined in contemporary research, improves the efficacy of mass spectrometry instruments in annotating unknown compounds. Researchers can use advanced mass spectrometry tools like MS-FINDER and CFM-ID to look at complicated biological systems and explain pathways clearly. This leads to better predictive metabolic models in clinical research, which shows how versatile advanced mass spectrometry tools can be (Blaženović *et al.,* 2018). Furthermore, contemporary advancements in mass spectrometry instrumentation have initiated a transformative era in numerous scientific fields, including proteomics and metabolomics. The integration of enhanced software and databases for identification and annotation optimizes the associated workflows, offering researchers more lucid insights into the molecular complexities of analyzed samples. The rigorous evaluation of small molecule identification strategies and databases, including those facilitated by MassBank repositories, is indispensable for comprehensive scientific research (Bracco *et al.,* 2010). As the need for accurate and dependable mass spectrometry analysis grows, future improvements will probably focus on making detection methods more sensitive and accurate, as well as making it easier to use computer resources to interpret data. This combination of advanced software and instruments leads to a better understanding of biological networks and chemical interactions, which opens the door to new uses in both research and industry (Figure 7.7).

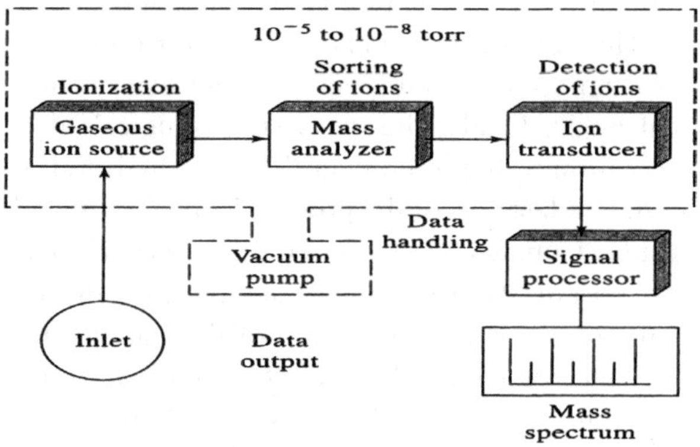

Figure 7.7. *Instrumentation for mass spectrometry.*

Source: https://www.expertsmind.com/CMSImages/963_Instrumentation%20for%20Mass%20Spectrometry.png.

7.3.1. Components of a Mass Spectrometer

The basic function of a mass spectrometer depends on its core parts, which work together to help analyze ions made from samples. The ion source is a key part of the spectrometer's function. It is where ionization happens, turning neutral molecules into charged particles that electric and magnetic fields can move around. This step is very important because it decides how effective and what kind of ionization will be used, which can change depending on the sample and the results you want. There are different types of ionization sources, such as Electron Ionization (EI), Electrospray Ionization (ESI), and Matrix-Assisted Laser Desorption/Ionization (MALDI). Each one is used for a different purpose. After the ions are made, they are sent to the mass analyzer, where they are sorted by their mass-to-charge (m/z) ratios. This separation is very important because it sets the stage for the next steps of analyzing and interpreting the data. This shows how important the ion source and mass analyzer are to the design of a mass spectrometer. The mass analyzer is very important in the next step of mass spectrometry because it determines how well the instrument works. There are many different types of analyzers, such as Quadrupole, Time-of-Flight (ToF), and Ion Trap analyzers. Each has its own benefits in terms of speed, resolution, and sensitivity. Quadrupole analyzers use oscillating electric fields to sort ions by their m/z ratios. This makes them great for targeted analyses and quantitation. On the other hand, Time-of-Flight analyzers measure how long it takes ions to travel a set distance. This gives you high-resolution data that works well for complicated mixtures. The efficiency of these analyzers affects how well you can see molecular patterns, such as fragmentation that happens during the mass analysis process. This fragmentation is necessary for figuring out the structure, which shows how important it is to pick the right mass analyzer based on the study's analytical goals. Lastly, the mass spectrometer's detection system is very important for turning the ionic data that was analyzed into a format that can be understood. The last part of the process is the detectors, like the Electron Multiplier or Faraday Cup, which measure the signals that ions make and record them as a mass spectrum. After that, these signals are amplified, which lets us measure and qualitatively analyze the parts of the sample. The resolution and sensitivity of the detector have a big effect on how accurate the results are. This is especially important in fields like metabolomics, where you need to figure out complex information about metabolites that are not very common. Recent studies have shown that mass spectrometry has become a key method for studying complex biological samples thanks to advances in bioinformatics. It has helped us understand things like glycosylation patterns that are important for cellular processes (Graham *et al.,* 2010) and plant metabolomes (Gullberg *et al.,* 2005). Through these interconnected parts, mass spectrometry becomes a powerful tool for modern analytical chemistry, giving us information about how molecules are structured and what they do.

7.3.2. Role of the Ion Source

The ion source in mass spectrometry is very important for figuring out how good and specific mass spectral analysis is. It is the first step in the process, where sample molecules are ionized so that they can be found and analyzed based on their mass-to-charge ratios. Different ionization methods, like Electron Impact (EI), Electrospray Ionization (ESI), and Matrix-Assisted Laser Desorption/Ionization (MALDI), show how important it is to be able to work with different types of samples. For example, ESI works especially well for polar and larger biomolecules because it makes ions directly from solution. On the other hand, MALDI works best with solid samples and large macromolecules because it uses a matrix to absorb laser energy, which helps with the desorption and ionization process. This variety of ion source methods makes it possible to use mass spectrometry in many different ways, which affects how it can help us understand complicated biological and chemical systems. When it comes to mass spectrometry applications, the choice of ion source can have a big effect on how the analysis breaks things down. The process of ionization not only creates ions, but it also affects how those ions break apart when they are analyzed further, like in tandem mass spectrometry (MS/MS). MS/MS uses methods like collision-induced dissociation (CID) that depend on the initial ion source's ability to make precursor ions that can be broken down easily. For instance, certain ion sources can cause fragmentation patterns that show the structure of molecules, which is very important in areas like metabolomics and proteomics. A well-optimized ion source can improve the analysis's sensitivity and specificity, making it easier to find and measure metabolites in complex biological matrices.

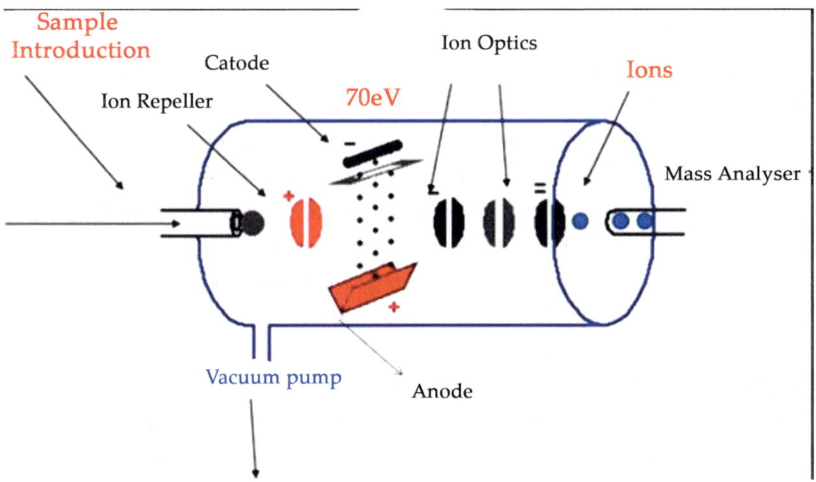

Figure 7.8. *Ion source.*

Source: https://ars.els-cdn.com/content/image/3-s2.0-B9780323850629000040-f04-04-9780323850629.jpg.

This can lead to important biological insights (Blaženović *et al.,* 2018). Also, improvements in ion source technology are very important for the continued growth of mass spectrometry applications. Researchers are looking into new ways to ionize samples, and adding new sources can make it easier to analyze samples that are more difficult to work with, such as those that are volatile, thermally labile, or not very common. For example, new technologies like laser-driven ion sources could improve spatial selectivity and dose accuracy in medicine, especially in cancer treatment. Although proton and heavy ion therapy are not yet standard in clinical practice, they show how advanced ion sources can target tumors more accurately than traditional methods. This is a big step forward from basic analysis to applied sciences that can affect how patients are treated (Bolton *et al.,* 2014). Advancements in ion source technology enhance mass spectrometry analysis and broaden its applicability across multiple disciplines, underscoring its significance as a fundamental element in the scientific research and application landscape (Figure 7.8).

7.3.3. Function of the Mass Analyzer

The mass analyzer is an important part of mass spectrometry because it separates ions based on their mass-to-charge (m/z) ratio. Mass analyzers help find and count molecules in a sample by using different analytical methods. After ionization, the separation process takes place. Ions are sent to the mass analyzer, where they are sorted and found. Different kinds of mass analyzers, like quadrupole, time-of-flight (TOF), and ion trap, use different methods to separate the particles. The effectiveness of a mass analyzer has a direct effect on the overall resolution and sensitivity of mass spectrometric data. This means that it is important to choose the right type for each use. The mass analyzer's job is more than just filtering ions; it also helps with complicated analyses that are important in many areas of research, such as plant metabolomics, where finding complex metabolite patterns is very important (Gullberg *et al.,* 2005). Mass analyzers do more than just separate things; they also help us understand how molecules break apart. Collision-induced dissociation (CID) is one way to do this. In CID, ions collide with gas molecules, which breaks them apart. We also look at the resulting fragment ions, which give us detailed information about the structure of the original molecules. This kind of fragmentation analysis is very important for figuring out what's in complex mixtures, especially in biological samples where there may be more than one metabolite. Each fragment corresponds to certain chemical bonds or structural features, which lets researchers figure out what the original compounds were. The mass analyzer's ability to separate these fragments shows how important it is in a wide range of fields, from pharmacokinetics to biochemical research, and how useful it is for understanding molecular complexities. Also, improvements in mass analyzer technology have made them work much better, which has led to high-throughput analyses that are becoming more and more important in modern research. Ultra-high-resolution mass spectrometry and other new technologies have made it possible to find small parts in complicated biological samples. This has helped us learn more about how different biochemical pathways and interactions work. This sensitivity is crucial in plant science, where the examination of low-abundance metabolites is fundamental to clarifying metabolic pathways and regulatory mechanisms. The advancement of resilient methodologies in mass spectrometry, especially when integrated with chromatographic techniques, has facilitated the analysis of extensive biological datasets and metabolomics. As mass spectrometry develops, the mass analyzer is still very important for finding and describing new compounds. This opens the door to new discoveries in many areas of science (Gullberg *et al.,* 2005).

7.4. FRAGMENTATION PATTERNS IN MASS SPECTROMETRY

The fragmentation patterns seen in mass spectrometry are very important for figuring out the structure of complicated molecules. Ionization of molecules frequently results in their fragmentation into smaller components; a phenomenon observed during tandem mass spectrometry (MS/MS) analysis. These fragmentation patterns provide significant insights into molecular architecture, enabling researchers to ascertain structural identity by analyzing the mass-to-charge ratios of the resultant ions. There are many things that affect fragmentation, such as the energy of the collision-induced dissociation process and the chemical properties of the analytes themselves. It is crucial to utilize these fragmentation pathways, as they can generate unique fingerprints for each compound, aiding in the differentiation of structurally analogous substances in complex mixtures, which is particularly significant in metabolomics and proteomics. In the field of mass spectrometry, software tools that can predict fragmentation patterns are very useful. In metabolomics, researchers frequently face difficulties in annotating small molecules due to the extensive diversity of metabolites found in biological samples. New computational methods, like those used in MS-FINDER, CFM-ID, and MetFrag, make it easier to guess how compounds will break down, which helps with the systematic process of labeling unknown compounds (Blaženović *et al.,* 2018). These tools use well-known spectral databases to compare theoretical fragmentation patterns with experimental results, which makes it easier to find what you're looking for. Researchers use these in silico fragmentation tools to learn more about metabolic pathways and biological systems. This helps us map out complicated biochemical networks. Furthermore, improvements in mass spectrometry methods and the addition of new technologies like ion mobility spectrometry have made the study of fragmentation patterns even better. Adding collision cross-section modeling makes it possible to better understand how ions behave in different situations. This multimodal method improves resolution and may be able to tell the difference between isomeric compounds that would otherwise give similar mass spectra (Bolton *et al.,* 2014). As new methods are created, researchers will learn more about how fragmentation patterns change with different ionization techniques. This will help them better understand mass spectral data. This information is very important not only for analytical chemistry but also for real-world uses like drug development and environmental monitoring. This shows how useful mass spectrometry is in modern scientific research (Figure 7.9).

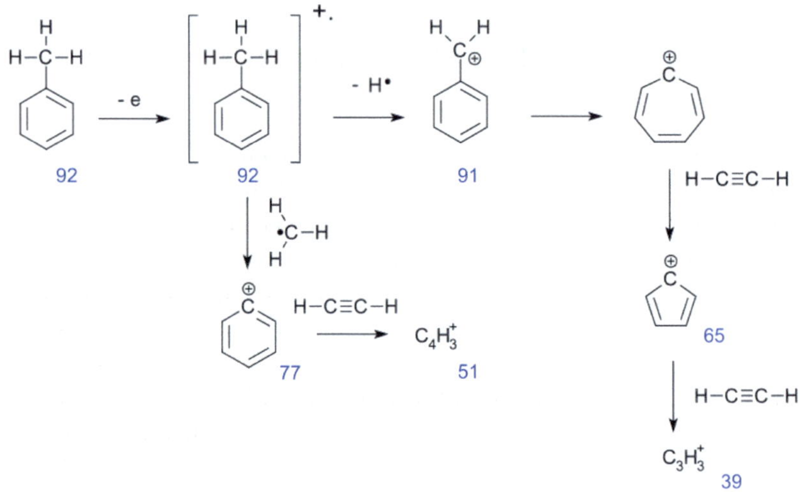

Figure 7.9. *Fragmentation pattern in mass spectrometry.*

Source: https://upload.wikimedia.org/wikipedia/commons/thumb/6/6c/Toluene-Fragmentation.svg/800px-TolueneFragmentation.svg.png.

7.4.1. Types of Fragmentation: Cleavage and Rearrangement

It is very important to know the different types of fragmentation effects in mass spectrometry because they have a big impact on how complex molecules are analyzed and understood. Cleavage and rearrangement are the two main ways that fragmentation happens. Cleavage fragmentation is the process of breaking certain chemical bonds in a molecule, which results in separate ions that come from the parent ion. The structure of the molecule and the bonds that are likely to break can often help us guess what will happen next. The resulting fragment ions yield significant structural information applicable for structural elucidation or compound identification, especially in organic compounds, where bond strengths and chemical environments can predictively influence fragmentation pathways. Rearrangement fragmentation adds complexity to mass spectrometric analyses in addition to cleavage. During rearrangement, fragments can form when atoms or groups move around within the molecular structure. This can create ions that are very different from the original parent ion. This phenomenon is especially noticeable in glycan studies, where fucose migration happens a lot. Experimental analysis indicates that proton mobility is essential for the migration of residues during fragmentation, implying that the energy barriers for these rearrangements may be less substantial than those for cleavage. This kind of fragmentation can make it harder to understand mass spectra because it can create unexpected ions that make it harder to use traditional methods to identify them. To figure out the structural details of the target molecules, more advanced analytical techniques are needed (Helden *et al.,* 2019). To fully understand mass spectrometry data and meet certain analytical needs, it is important to understand how cleavage and rearrangement fragmentation work together. For example, some applications may prefer cleavage to create simpler fragment patterns that are easier to analyze, while others may use rearrangement patterns to get a better look at complex molecular structures. Lipid analyses exemplify this complexity, as differing fragmentation behaviors under various conditions can distinctly identify lipid sources and their degradation pathways. Researchers can change their mass spectrometric methods to fit specific experimental goals because they can tell the difference between these types of fragmentation. Comprehending the functionality of these mechanisms augments the efficacy of mass spectrometry in chemistry, biochemistry, and biomolecular research, facilitating more precise and sophisticated analytical results (Bachratá *et al.,* 2009).

7.4.2. Factors Influencing Fragmentation

There are many things that affect fragmentation in mass spectrometry, including how well it works and how ions break up. The structure of the analyte itself, such as its molecular weight, functional groups, and overall shape, is one of the most important factors. Molecules with complicated or stiff structures may break apart in ways that are different from simpler molecules, which can lead to unique product ion spectra. Furthermore, the presence of particular functional groups can either promote or impede fragmentation by stabilizing certain ionic species during the dissociation process. This underscores the necessity of comprehending the molecular attributes that dictate fragmentation behaviors, which are crucial for precise identification and classification of compounds. As a result, having a deep understanding of these structural effects can make mass spectrometric analyses much more reliable in a wide range of uses. The ionization method used in mass spectrometry is another important factor that affects fragmentation. Different ionization methods, like electrospray ionization (ESI), matrix-assisted laser desorption/ionization (MALDI), or atmospheric pressure chemical ionization (APCI), make ions with different amounts of internal energy. When ions have more internal energy, they can break apart more easily when they enter the mass analyzer. Also, the choice of ionization method has a direct effect on the number of different fragmentation pathways that are available. For example, it can make certain bonds break more often or certain product ions form. To get the best results from mass spectrometry and get consistent analytical results from complex samples, it's important to know the subtleties of each ionization method and how they affect fragmentation patterns.

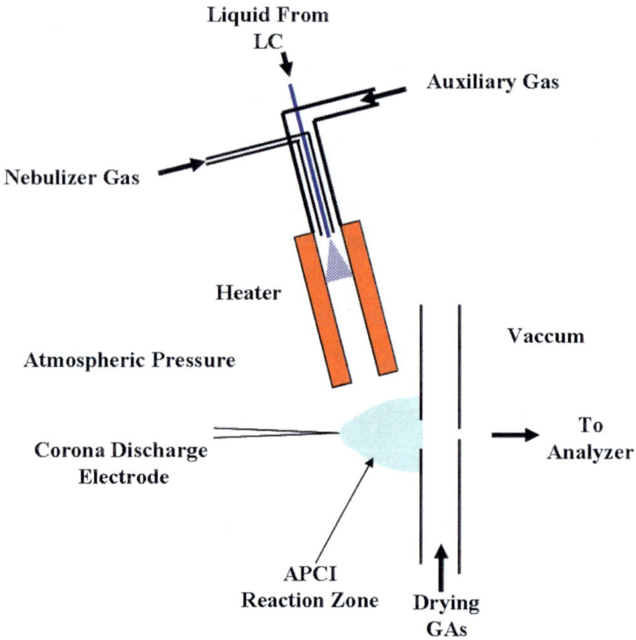

Figure 7.10. *Atmospheric pressure chemical ionization.*

Source: https://encyclopedia.pub/media/common/202212/mceclip0-63883df4c-b3a0.png.

Experimental conditions, including collision energy and the type of collision gas, are also very important in deciding how fragmentation will turn out. In tandem mass spectrometry (MS/MS) experiments, precise control over collision energy can enable the selective activation of ions, allowing researchers to investigate targeted fragmentation pathways. For example, changing the collision energy can make certain fragment ions more likely to form while making others less likely to form. This gives you a better idea of the molecular structure. The choice of collision gas can also affect how well energy is transferred during fragmentation, which in turn can change the spectra that come out. By thoroughly evaluating these experimental parameters, researchers can refine fragmentation models, as indicated by recent studies utilizing machine learning approaches to analyze large datasets and predict fragmentation behavior based on various influencing factors, reinforcing the utility of an evidence-based framework in mass spectrometry (Bowler *et al.,* 2008). Furthermore, certain biologically significant molecules, including lactoferrin, illustrate that even minor alterations in fragmentation resulting from glycosylation patterns can profoundly influence biological functions, highlighting the complex nature of fragmentation factors (Barile *et al.,* 2017) (Figure 7.10).

7.4.3. Interpretation of Fragmentation Patterns

Understanding fragmentation patterns in mass spectrometry is an important part of the analytical process that has a big impact on finding unknown compounds. Mass spectrometry analysis ionizes molecules and then breaks them down into smaller ions. Each fragmentation pattern serves as a molecular fingerprint. This unique pattern lets scientists figure out structural details about the compound, such as its functional groups and how its molecules are connected. As biological samples become more complex, like in metabolomics, it becomes very important to accurately interpret these patterns in order to identify and describe metabolites. Using established databases is often key to getting accurate results from mass spectrometry analyses (Armitage *et al.,* 2017). These databases give useful information about similar fragmentation patterns seen in known compounds. Researchers have created advanced algorithms and software tools to make it easier and more accurate to understand fragmentation patterns. These new features make it easier to compare experimental data to existing spectral libraries, which speeds up the process of identifying things. Researchers can now find small differences in fragmentation patterns that could mean differences in isomers or related compounds thanks to improvements in high-resolution mass spectrometry (HRMS). Moreover, multivariate statistical techniques facilitate the thorough examination of intricate data sets by discerning patterns and correlations among various metabolites. As mass spectrometry advances, improved data processing methods will enhance our comprehension of fragmentation patterns, ultimately facilitating superior identification and quantification of metabolites across diverse biological contexts (Gullberg *et al.,* 2005). In plant metabolomics, the analysis of fragmentation patterns is even more important because plant metabolites are so varied and complicated. The effective identification of these metabolites, numerous of which exhibit overlapping masses and analogous ionization characteristics, is significantly dependent on the capacity to analyze their distinct fragmentation profiles. Improvements in methods like Gas Chromatography-Mass Spectrometry (GC-MS) have made it possible to quickly analyze many plant samples at once. However, the data that comes from these tests is often very complicated and needs complicated deconvolution techniques to be understood. Researchers can improve their ability to make sense of the physiological effects of different growth conditions or treatments on plant metabolomes by using hierarchical methods to process GC/MS data. This expanding corpus of knowledge significantly enhances our comprehension of plant biology and fosters the advancement of applications in agriculture and pharmacology (Armitage *et al.,* 2017).

7.5. DATA ANALYSIS IN MASS SPECTROMETRY

Data analysis in mass spectrometry (MS) is a complex process that is very important for turning raw data into useful information. The first steps in data analysis are to carefully calibrate the mass spectrometers and check the instrumental parameters, which are very important for getting accurate mass-to-charge ratios. Then, peak detection and deconvolution are important steps to find analytes in complex spectra. More and more, advanced software tools are being used to make these tasks easier. This shows the shift from handling data by hand to handling it automatically. But there are still problems, especially when it comes to figuring out overlapping signals and handling large amounts of data, which needs strong analytical methods to make sure the results are accurate. As the samples being analyzed get more complicated, the need for more advanced data analysis methods grows. This shows how important it is to keep making progress in this field so that mass spectrometry can keep up with its changing uses. The annotation of metabolites and other compounds remains a significant challenge within mass spectrometry, especially in untargeted metabolomics, building on foundational data processing techniques. To make sure that unknown compounds are correctly identified, annotating small molecules requires structured elucidation methods and specialized software. Researchers can use tools like MS-FINDER and CFM-ID to help them with this task. These tools let researchers match fragmentation patterns to existing databases to add information about compounds. Building tandem mass spectral databases is very important for boosting annotation confidence, which in turn leads to more accurate biological interpretations. These advancements are essential for associating compounds with metabolic pathways and creating predictive models in translational and clinical research, thereby augmenting the applicability of mass spectrometry in extensive biological investigations (Blaženović *et al.,* 2018). Moreover, the interaction between data analysis and mass spectrometry encompasses plant metabolomics, where comprehending the regulatory mechanisms of growth and development is essential. Researchers are trying to measure and find low molecular weight metabolites in different plant tissues. To do this, they need to use a mix of analytical methods, such as gas chromatography-mass spectrometry (GC-MS). To handle the large amounts of data that these analyses create, you need good data processing workflows that may include things like baseline correction and multivariate analysis. This strong method not only makes it easier to find important metabolic changes, but it also makes it easier to compare different sample groups. The creation of semi-automated methods for managing data is a big step forward in solving the problem of overlapping peaks in high-throughput analyses. It shows how data analysis fits into the bigger picture of mass spectrometry applications (Gullberg *et al.,* 2005) (Figure 7.11).

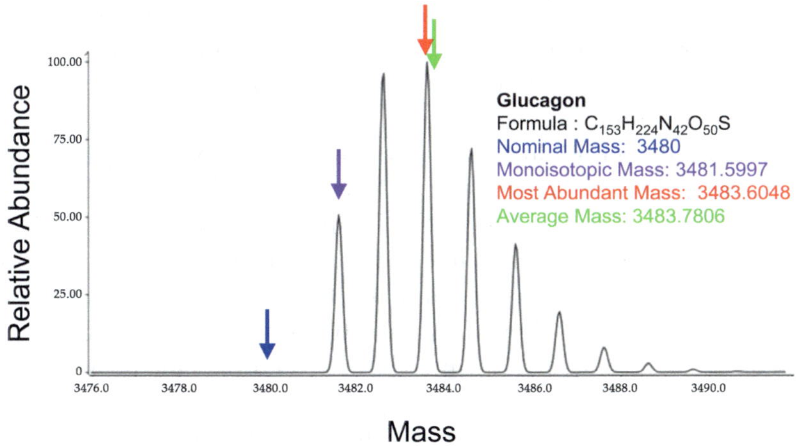

Figure 7.11. *Mass spectrometry data analysis.*

Source: https://cdn.prod.website-files.com/64f74bf861c5c5b812076920/65030f3a56d29b968f715890_f63927f8.png.

7.5.1. Mass Spectrum Interpretation

Mass spectrum interpretation is an important part of mass spectrometry. It helps scientists make sense of complicated data and get useful information from the results. The mass spectrum is the most important part of this process. It shows the amount of ions in a sample against their mass-to-charge (m/z) ratios, which shows that there are different types of molecules in the sample. Each peak stands for a different ion, usually a molecular ion or a fragment ion that comes from the breakdown of the original molecules. Researchers must be skilled at identifying the different peaks and linking them to the molecular structures in the sample in order to draw useful conclusions. Additionally, comparing spectra of similar compounds under different conditions can reveal significant physiological or biochemical phenomena, thereby improving the comprehension of the composition of the analyzed samples. Using different fragmentation patterns makes the interpretation process even better because they give important clues about the molecular structure of the analytes. When molecules are ionized during mass spectrometry, they can break apart in specific ways, creating patterns that are unique to certain functional groups or structural motifs. For example, certain ions may break apart in ways that can be predicted, creating ions that can be traced back to the molecule's original structure. This part of mass spectrum interpretation is very important for both qualitative and quantitative analyses. It helps researchers figure out the details of complicated biochemicals or metabolites. Combining fragmentation data with other analytical methods, like chromatography, is a synergistic way to separate complex mixtures and make analytical results clearer. Mass spectrum interpretation has applications in diverse domains, such as plant metabolomics and microbiology, where the clarification of molecular structures bears considerable consequences. Studies on Arabidopsis thaliana show that researchers in plant metabolomics often use gas chromatography-mass spectrometry (GC/MS) to look at complicated metabolite profiles. In this context, the capacity to interpret mass spectra is crucial, as researchers are required to analyze overlapping peaks and differentiate them to identify specific metabolites associated with growth regulation and developmental processes (Gullberg *et al.,* 2005). In microbiology, comprehending post-translational modifications, including glycosylation in human pathogens, necessitates proficient mass spectrum interpretation to delineate essential modifications that influence bacterial virulence (Graham *et al.,* 2010). Consequently, proficiency in mass spectrum interpretation is essential across various fields, enhancing the significance of mass spectrometry as a formidable analytical instrument.

7.5.2. Software Tools for Data Analysis

In the field of mass spectrometry (MS), specialized software tools that help with the problems that come up in high-throughput experiments are becoming more and more important for analyzing and interpreting data. These software solutions not only make it easier to work with raw data, but they also make it easier to understand the results. Tools created for metabolomics focus on handling large datasets that come from different analytical platforms. As shown in many studies, such as Misra et al. (2016), the lack of a complete overview of the latest software innovations has made it harder for researchers to fully use advanced methods in their work. As a result, the advancement of analytical capabilities requires a continuous dedication to assessing and incorporating these software tools, guaranteeing that researchers can proficiently traverse the complex data environments linked to mass spectrometry applications. Furthermore, the importance of data visualization and automation in mass spectrometry is immense. Traditional data management systems often can't handle the huge amounts of data that are created, so visualization tools are needed to show trends and insights. Software packages made for metabolomics often have tools for data annotation and visualization, which are very important for understanding complicated results. For example, some open-source tools have been created that make these tasks easier, which cuts down on the time researchers spend getting data ready and analyzing it. Misra et al. (2016) say that these kinds of tools are especially useful for untargeted metabolomics workflows because they help users find changes in metabolites more easily. By utilizing these advancements, researchers can gain a more comprehensive understanding of the biological implications of their findings, thereby enhancing decision-making in experimental designs.

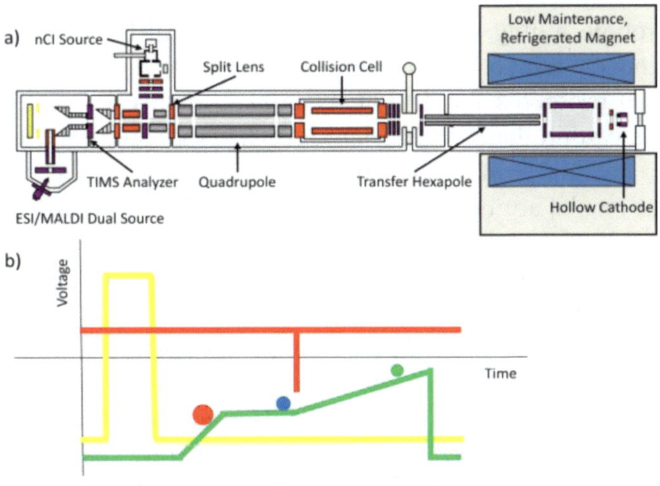

Figure 7.12. *Trapped ion mobility spectrometry.*

Source: https://ars.els-cdn.com/content/image/1-s2.0-S1387380617303299-gr5.jpg.

For detailed lipidomic studies, it is important to improve the sensitivity and specificity of mass spectrometry analyses through software integration, in addition to processing and visualizing data. Innovative workflows, like those that use trapped ion mobility spectrometry (TIMS) and parallel accumulation-serial fragmentation (PASEF), show how software tools can improve analytical performance. As shown in, these methods greatly increase the number of lipid species that can be identified, which expands the possibilities for lipid analysis in complex biological matrices. Using such advanced software makes it possible to get high-resolution information about lipids, which is important for figuring out how diseases work and how metabolism works. The combination of powerful software tools and advanced mass spectrometry methods not only makes data analysis better, but it also sets the stage for future research that will help us understand complex biochemical pathways (Figure 7.12).

7.5.3. Challenges in Data Interpretation

Interpreting data from mass spectrometry (MS) is a complicated process that has a big effect on how reliable and useful research results are. One of the main problems comes from how complicated the datasets are, especially in high-throughput experiments. These datasets frequently entail the concurrent analysis of multiple metabolites, generating extensive volumes of information that can exceed conventional processing capacities. Because of this, researchers often have trouble telling the difference between real biological signals and background noise, which can lead to mistakes. Additionally, the absence of standardized protocols for data analysis exacerbates the issue, as varying methodologies can produce inconsistent results across studies. This variability highlights the need for enhanced analytical tools to aid data interpretation and guarantee that results are both reliable and replicable across different contexts (Misra *et al.,* 2016). Another big problem is that mass spectrometry data needs to be annotated, which is very important for figuring out what compounds are in a biological sample. Assigning chemical identities to detected peaks is a difficult process because fragmentation patterns can vary widely and signals can overlap, which can make results less clear. Researchers must depend on extensive databases and in silico resources to correctly categorize metabolites; however, these repositories frequently lack completeness or exhibit inconsistencies. Recent literature has highlighted the urgent requirement for enhanced data processing and visualization tools that mitigate these constraints, facilitating researchers in deriving significant insights from intricate datasets more efficiently (Misra *et al.,* 2016). Without these kinds of improvements, it could still be hard to understand mass spectrometry data, which could slow down progress in the field. Moreover, the amalgamation of data from various sources, including metabolomics, introduces additional challenges in attaining a comprehensive understanding of biological systems. Different methods used to collect data, as well as different ways of preparing samples and analyzing them, can lead to results that don't match up and are hard to explain. For example, in studies designed to diagnose diseases such as tuberculosis, where swift and precise identification is essential, the analysis of intricate datasets is imperative. Research employing electrospray ionization tandem mass spectrometry to analyze mycolic acids has demonstrated that a targeted methodology can uncover distinct profiles for mycobacterial species, underscoring the significance of meticulous data analysis. Nonetheless, these focused analyses frequently undervalue the overarching difficulties associated with the integration of heterogeneous datasets and the attainment of uniform results, underscoring the continual need for methodological progress in data interpretation within mass spectrometry research.

7.6. APPLICATIONS OF MASS SPECTROMETRY (MS)

Mass spectrometry (MS) is used in a wide range of scientific fields, such as chemistry, biology, and environmental science. One of the most common uses is in metabolomics, where scientists look at biological samples to find and measure small molecules. This field is becoming more and more important for understanding metabolic pathways and how living things respond to changes in their environment or treatment plans. The combination of mass spectrometry and liquid chromatography has changed the way we can separate complex mixtures, making it possible to find metabolites that were not known before. The creation of extensive spectral databases and advanced software tools has improved the annotation of compounds from untargeted metabolomic studies, facilitating superior biological interpretations and applications in clinical research and personalized medicine (Blaženović *et al.,* 2018). In plant science, mass spectrometry is essential for clarifying metabolic profiles to comprehend the regulation of growth and development. Gas Chromatography-Mass Spectrometry (GC-MS) and other methods are used to get low molecular weight metabolites out of different plant tissues and study them (Gullberg *et al.,* 2005). This high-throughput method not only finds individual parts in complex mixtures, but it also measures their relative concentrations, making it easier to compare different plant species or conditions. The creation of strong methods, such as semi-automated data processing strategies, has made it much easier to analyze large sets of samples. These kinds of improvements are necessary to find out all the different ways that environmental factors or genetic changes affect plant metabolism and traits. This will help push the limits of agricultural and ecological research. Mass spectrometry has also become a key technology in proteomics, allowing scientists to look at protein expression, changes, and interactions. The capacity to elucidate the structure and sequence of proteins significantly impacts drug discovery, biomarker development, and the comprehension of disease mechanisms. Tandem mass spectrometry (MS/MS) is very important for these uses because it gives fragmentation patterns that help identify and describe protein complexes. Mass spectrometric techniques improve sensitivity and accuracy, making it possible to find proteins that are not very common. This opens up new areas of proteomic research. The ongoing improvements in mass spectrometry processes, such as better ionization methods, are always making the technology more useful for biological and medical research (Blaženović *et al.,* 2018).

7.6.1. Role in Pharmaceuticals and Drug Development

Mass spectrometry is an important part of drug development and pharmaceuticals because it makes it easier to find and measure compounds in complex biological matrices. Mass spectrometry enables researchers to effectively investigate the pharmacokinetics and pharmacodynamics of novel drug candidates through the utilization of high-sensitivity and high-selectivity methodologies. The n-in-one method, which combines and mixes samples before analysis, is an example of a game-changing way to quickly find promising compounds early in the drug discovery process. Using this method with mass spectrometry makes it much easier to find drugs that work by measuring their absorption and lipophilicity *in vitro*, like in caco-2 cell models. This makes it easier to find drugs that work by measuring their absorption and lipophilicity. This synergy between advanced analytical techniques and drug development shows how important mass spectrometry is for getting new drugs to market faster (Kangas *et al.,* 2010). Mass spectrometry's role in the detailed analysis of fragmentation patterns enables scientists to comprehend the stability and behavior of drug molecules during development. A significant study emphasized the influence of various substituents on the fragmentation of nitrophenoxy benzoate compounds, demonstrating that alterations in molecular structure directly affect fragmentation results. This relationship is not only theoretical; it has real-world consequences for evaluating the efficacy and safety of novel pharmaceuticals. Comprehending the behavior of molecules during fragmentation can yield insights into their potential interactions within biological systems, informing modifications to improve drug efficacy and mitigate toxicity. Mass spectrometry is an important tool for predicting how stable drug candidates will be, which is very important for successful pharmaceutical development (Kangas *et al.,* 2010). Mass spectrometry can be used to develop methods for analyzing drug samples in addition to identifying and characterizing compounds. Researchers have developed many new mass spectrometric methods that can make it easier to find and measure things. For instance, small mass spectrometric ionization techniques like DAPPI-MS let you quickly screen samples with little or no pretreatment. This efficiency is especially useful in bioanalysis and environmental monitoring, where time and accuracy are very important. Scientists can get more accurate and sensitive measurements of drug concentrations in complex matrices by using advanced instruments that combine chromatography and mass spectrometry, like GC-μAPCI-MS. This shows how important mass spectrometry is in the drug development process.

7.6.2. Use in Environmental Analysis

Mass spectrometry has become an essential instrument in environmental analysis, allowing researchers to identify and measure a diverse array of pollutants and naturally occurring substances across different matrices. Mass spectrometry makes it easier to look at air, soil, and water samples for volatile organic compounds, heavy metals, pesticides, and new contaminants by being very sensitive and specific. In environmental studies, where pollutants are often present in very small amounts among many other substances, it is especially important that the technology can tell the difference between them. This methodological strength enables precise environmental monitoring and evaluation, thereby facilitating regulatory adherence and policy formulation focused on safeguarding ecosystem integrity and human safety. A key part of environmental mass spectrometry is its use in metabolomics, which tries to figure out how organisms' biochemical profiles change when they are exposed to stressors or changes in their environment. Finding and measuring metabolites can help us understand how different organisms interact with each other and how biochemical pathways work, which can help us make better decisions about how to manage the environment. Advanced techniques, like tandem mass spectrometry (MS/MS) and different fragmentation patterns, make the analysis more sensitive and precise, which makes it possible to confidently identify unknown compounds. The incorporation of in silico fragmentation tools, as highlighted in recent studies, has enhanced the precision and efficacy of compound identification, facilitating more accurate interpretation of complex environmental samples (Blaženović *et al.,* 2018). Furthermore, the progress in mass spectrometry technologies and data analysis techniques provides unparalleled prospects for environmental research. Liquid chromatography coupled with mass spectrometry (LC-MS) is one method that can quickly separate and identify compounds based on their unique mass-to-charge ratios. This ability is essential for breaking down the complex chemical landscapes that can be found in polluted areas. Researchers can make predictive metabolic models that show how certain compounds affect ecosystems over time by using mass spectrometry along with databases and retention time models. These kinds of thorough studies not only help with cleanup efforts, but they also help businesses in many fields come up with ways to be more environmentally friendly (ANONYMOUS, 2010). The ongoing improvement of mass spectrometry methods will surely keep helping us learn more about environmental problems (Figure 7.13).

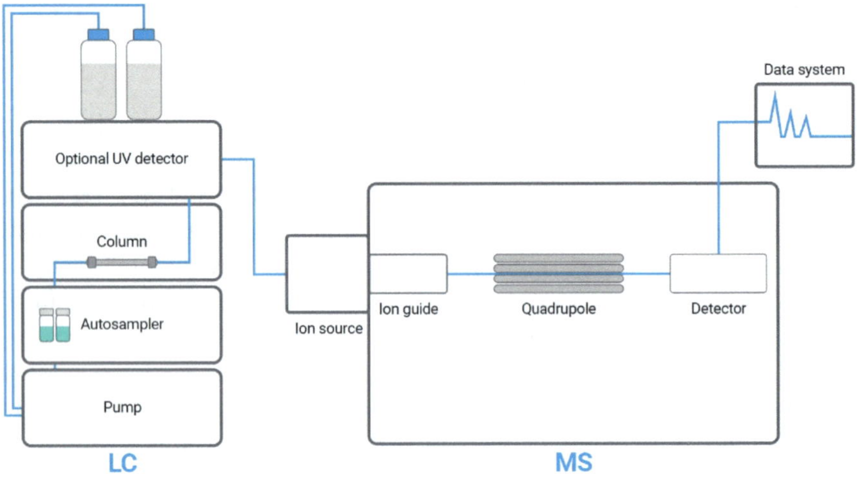

Figure 7.13. *Liquid chromatography coupled with mass spectrometry (LC-MS).*

Source: https://www.agilent.com/cs/promotions/images/lc-ms-fundamentals-basic-principles-1638x950-agilent.jpg.

7.6.3. Applications in Proteomics and Genomics

Mass spectrometry has become a fundamental technique in proteomics and genomics, offering essential information about the molecular composition and structure of biomolecules. Its ability to detect and measure proteins, peptides, and metabolites with great sensitivity helps us learn more about how biological systems work. In proteomics, mass spectrometry makes it easier to figure out how much protein is being expressed and how it has been modified after translation. This is important for understanding how proteins work and how they are controlled. Tandem mass spectrometry and other methods can analyze complex mixtures, which is important for finding biomarkers for diseases and figuring out how cells work. This ability to break down complex biological interactions is a big help for translational research, which aims to turn lab results into clinical applications that can help patients. In genomics, mass spectrometry provides researchers with non-invasive information about genetic variations and modifications that are responsible for heritable traits and diseases. The technology allows for the examination of genomic DNA and RNA, especially through methods such as mass spectrometry-based sequencing. This method has changed the way we can find genetic mutations linked to a wide range of conditions, from rare disorders to common diseases. This has helped us learn more about population genomics and personalized medicine. Also, combining mass spectrometry with bioinformatics tools makes it easier to understand data, which helps the analysis of sequencing results move toward a clear picture of how genes work and interact with each other. As genomic research moves forward, the combined work of mass spectrometry and computational biology is making it possible to make big strides in the treatment and prevention of genetic diseases. Mass spectrometry does more than just identify proteins and nucleic acids; it also moves research forward by using metabolomics to figure out complicated biochemical pathways. The large number of metabolites that cells make is a biochemical signature that shows physiological states, disease conditions, and environmental effects. Researchers can use untargeted metabolomics techniques to study and describe metabolites, which helps them learn more about metabolic diseases and how drugs interact with each other. The use of advanced in silico fragmentation tools and large data repositories to annotate small molecules has made it easier to find metabolites and connect them to biological pathways, as shown by Blaženović et al. (2018). Furthermore, the characterization of post-translational modifications, including glycosylation, is essential for comprehending proteomic dynamics and their effects on cellular functions, especially in human pathogens where these modifications can affect virulence, as outlined by Graham et al. (2010). Consequently, mass spectrometry is a fundamental technology in proteomics and genomics, facilitating a more profound comprehension of biological complexity.

SUMMARY

Mass spectrometry is a fundamental analytical technique that has revolutionized numerous scientific disciplines, including chemistry, biology, and medicine. Mass spectrometry helps us understand complex biological systems better by letting us accurately identify and measure compounds. The basic ideas behind this method, like ionization, mass analysis, and detection, have been carefully worked on and improved over time, resulting in many uses. Mass spectrometry has shown unmatched skill in making clear the complex biochemical pathways that make cells work, from metabolomics to proteomics. As noted by Blaženović et al. (2018), the development of software and databases has made it easier to annotate compounds, which has made metabolic analyses completer and more reliable. Additionally, the various types of mass spectrometry methods, such as time-of-flight (TOF) and ion trap methods, make this technology more flexible. These techniques facilitate the examination of a diverse array of substances, addressing the particular requirements of researchers from multiple fields. High-resolution mass spectrometry has gained prominence in clinical applications, elucidating disease mechanisms and therapeutic targets. The accuracy and sensitivity of mass spectrometry not only enable precise quantitative analysis but also pave the way for innovative discoveries in non-targeted metabolomics, as demonstrated through structured elucidation methodologies by Blaženović et al. (2018). Because of this, its role in advancing scientific knowledge keeps growing. The future of mass spectrometry looks bright, with even more new uses and improvements on the way. Ongoing research and development in tandem mass spectrometry and computational tools are poised to improve the precision of compound identification, thereby enabling novel advancements across various domains. The combination of new methods, like using collision cross-section modeling with ion mobility experiments, is an example of the cutting edge of modern mass spectrometry (Blaženović *et al.,* 2018). The potential of mass spectrometry is immense, affecting both academic research and clinical practices, particularly in cancer therapy, where the accuracy of targeted treatments is essential. Mass spectrometry will continue to be an important tool for helping us learn more about complex biochemical systems as the technology gets better.

To understand how mass spectrometry can be used in many different ways, you need to know its basic principles and types. The main idea behind mass spectrometry is to measure the mass-to-charge ratio of ions. This makes it possible to find and measure the amounts of different compounds in different samples. The method can be broken down into different types, such as electrospray ionization (ESI) and matrix-assisted laser desorption/ionization (MALDI). Each type works best with certain types of molecules. These different ways of ionizing make mass spectrometry useful in many areas, such as proteomics and environmental analysis, showing how flexible it is. As the principles of mass spectrometry continue to develop, researchers are working to improve these methods, which will make them more accurate and reliable when analyzing complex samples. This shows how important mass spectrometry is to modern scientific research. Fragmentation patterns are another important part of mass spectrometry that have a big effect on how mass spectra are read. After ions are created, they can break apart, creating daughter ions that give important information about their structure. Understanding these fragmentation patterns is important for figuring out what chemicals are in a mixture, especially if it is a complicated one. Different computer programs, like MS-FINDER and CFM-ID, make it easier to analyze these patterns. Researchers can be surer of finding unknown metabolites by combining in silico fragmentation with experimental data, as shown in important reviews of small molecule identification. Fragmentation patterns serve a purpose beyond just analysis; they are the basis for a better understanding of molecular structures and pathways, which helps progress in metabolomics and other areas (Blaženović *et al.,* 2018). Mass spectrometry has many uses, including in biomedical research, environmental sciences, and food safety. This method gives each of these fields its own unique benefits. In biomedical research, for example, mass spectrometry helps find metabolites that could be used as biomarkers for diseases. This could help doctors figure out how to diagnose and treat patients. In environmental science, it is used to find small amounts of pollutants in water and soil, which is important for pollution monitoring and ecological assessments. Mass spectrometry is also used in food safety to find contaminants and make sure that ingredients are real, which protects consumers. These different uses show how important it is to know the basic ideas and methods of mass spectrometry. Furthermore, as analytical techniques continue to improve, future innovations are likely to broaden its uses and make it more effective at solving difficult scientific problems (ANONYMOUS, 2010).

REFERENCES

1. Armitage, E. G., Barbas, C., Gil de la Fuente, A., et al. (2017). Differentiating signals to make biological sense–A guide through databases for MS-based non-targeted metabolomics. https://core.ac.uk/download/82893507.pdf (accessed on 29th October 2025).
2. Bachratá, R. (2009). Molecular study of lipids in humic acids by sequential chemical degradation. https://core.ac.uk/download/30295631.pdf (accessed on 29th October 2025).
3. Barile, D., German, J. B., Karav, S., et al. (2017). Studying lactoferrin N-glycosylation. https://core.ac.uk/download/323069288.pdf (accessed on 29th October 2025).
4. Barsch, A., Brunner, A., Mann, M., et al. (2020). Trapped ion mobility spectrometry and PASEF enable in-depth lipidomic from minimal sample amounts.
5. Blaženović, I., Fiehn, O., Ji, J., et al. (2018). Software tools and approaches for compound identification of LC-MS/MS data in metabolomics. https://core.ac.uk/download/323069688.pdf (accessed on 29th October 2025).
6. Bolton, P. R., Ledingham, K. W. D., Ma, C. M., et al. (2014). Towards laser driven hadron cancer radiotherapy: A review of progress. https://core.ac.uk/download/211698981.pdf (accessed on 29th October 2025).
7. Bowler, L. D., Feng, J., & Zhou, C. (2008). A machine learning approach to explore the spectra intensity pattern of peptides using tandem mass spectrometry data. https://core.ac.uk/download/48957.pdf (accessed on 29th October 2025).
8. Bracco, A., Chomaz, P., Gaardhøje, J. J., et al. (2010). Perspectives of nuclear physics in Europe: NuPECC long range plan 2010. https://core.ac.uk/download/19502483.pdf (accessed on 29th October 2025).
9. Bugni, T. (2017). Review of Mass Spectrometry: Instrumentation, Interpretation, and Applications. In R. Ekman, J. Silberring, A. M. Westman-Brinkmalm, & A. Kraj (Eds.), Journal of Natural Products. Wiley, New York. https://www.semanticscholar.org/paper/456bcfe8fba6affe526975c98540815e80cb56fa (accessed on 29th October 2025).
10. Castro, E. A., Furlong, J. J., Schiavonni, M. M., & Allegretti, P. E. (2008). Tautomeric equilibria studies by mass spectrometry. https://core.ac.uk/download/pdf/287803.pdf (accessed on 29th October 2025).
11. Chadt, A., Colorni, A., Gamez-Pozo, A., Rasche, A., Tiss, A., Tiss, A., Sauve, A. C., et al. (2011). Biomarker discovery and redundancy reduction towards classification using a multi-factorial MALDI-TOF MS T2DM

mouse model dataset. https://core.ac.uk/download/107845.pdf (accessed on 29th October 2025).

12. Graham, R. L. J., & Hess, S. (2010). Mass spectrometry in the elucidation of the glycoproteome of bacterial pathogens. https://core.ac.uk/download/4884362.pdf (accessed on 29th October 2025).
13. Gullberg, J. (2005). Metabolomics: A tool for studying plant biology. https://core.ac.uk/download/11694915.pdf (accessed on 29th October 2025).
14. Gunzer, F., Krüger, S., & Grotemeyer, J. (2018). Photoionization and photofragmentation in mass spectrometry with visible and UV lasers. Mass Spectrometry Reviews, 38(2), 202–217. https://www.semanticscholar.org/paper/7a43d8d1ca36fdf6a24c1b7addf3a16f900268f8 (accessed on 29th October 2025).
15. Ipsen, A., & Ipsen, A. (2011). Probabilistic modelling of liquid chromatography time-of-flight mass spectrometry.
16. Janiszewska-Drobinska, B., et al. (2013). Rapid method for Mycobacterium tuberculosis identification using electrospray ionization tandem mass spectrometry analysis of mycolic acids. https://core.ac.uk/download/71976389.pdf (accessed on 29th October 2025).
17. Jordan, A., Zeichner, A., Agarwal, B., Gilbert-López, B., Cruces-Blanco, C., Shen, C., West, C., et al. (2019). Applications of direct injection soft chemical ionization-mass spectrometry for the detection of pre-blast smokeless powder organic additives. https://core.ac.uk/download/228157014.pdf (accessed on 29th October 2025).
18. Lennard, F., Smith, M., & Thompson, K. (2017). A literature review of analytical techniques for materials characterization of painted textiles–Part 2: Spectroscopic and chromatographic analytical instrumentation. https://core.ac.uk/download/84591645.pdf (accessed on 29th October 2025).
19. Luosujärvi, L. (2010). Miniaturized mass spectrometric ionization techniques for environmental analysis and bioanalysis. https://core.ac.uk/download/14916573.pdf (accessed on 29th October 2025).
20. Misra, B. B., & van der Hooft, J. (2016). Updates in metabolomics tools and resources: 2014–2015. https://core.ac.uk/download/46559843.pdf (accessed on 29th October 2025).
21. von Helden, G., Lettow, M., Manz, C., et al. (2019). The role of the mobile proton in fucose migration. https://core.ac.uk/download/225124611.pdf (accessed on 29th October 2025).

CHAPTER 8

RAMAN SPECTROSCOPY

Contents

INTRODUCTION

Raman spectroscopy is a strong analytical method that has unique benefits for figuring out how molecules are structured and how they interact with each other. Raman spectroscopy uses the inelastic scattering of monochromatic light to find and measure vibrational modes in molecules. This gives each molecule a unique fingerprint of its chemical makeup. The method is especially noteworthy because it doesn't destroy samples, which means that their physical or chemical properties don't change when they are studied. Also, it can be used in many areas, such as chemistry, biology, and materials science. This makes it a useful tool for both researchers and professionals in the field. As industries put more and more emphasis on fast and accurate analytical methods, Raman spectroscopy stands out because it can give real-time information for a wide range of uses. Raman spectroscopy is not only flexible, but it is also perfect for samples with complicated matrices.

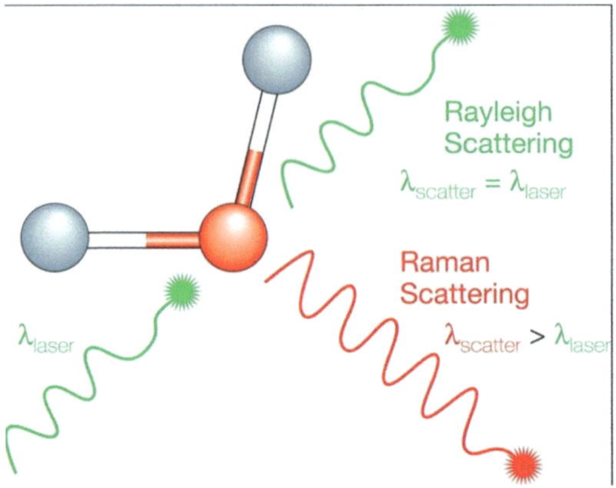

Figure 8.1. *Raman spectroscopy.*

Source: https://static.horiba.com/fileadmin/Horiba/_processed_/7/9/csm_fig_1_Raman_principle_01_94fb55ba09.jpg.

For example, it is an important tool in both academic and industrial research because it lets you look at biomaterials and environmental samples without having to do a lot of sample preparation. Applications in pharmaceutical sciences, like keeping an eye on the makeup of drug formulations, show how important Raman spectroscopy is for quality control and development. FT-Raman spectroscopy and other methods have been used successfully to chemotype plant sources. For example, Nickless et al. (2015) looked at nectar from *Leptospermum scoparium* plants and found that the amount of dihydroxyacetone was linked to possible health benefits in honey products. Researchers are still looking into how different compounds interact with each other, and Raman spectroscopy is very important for helping us understand how molecules behave. Technology improvements, like the creation of more sensitive detectors and the combination of different spectroscopic techniques, have made it easier for Raman spectroscopy to grow. The use of Raman spectroscopy with other methods, like hot-melt extrusion (HME) monitoring, shows how important this technique is becoming for process control and quality assurance. As pharmaceutical manufacturing moves to continuous processes, Raman spectroscopy's role in making sure products are of high quality becomes more important. Researchers can collect data on important factors during the manufacturing process by using real-time monitoring. This helps make the formulation of products more efficient and consistent (Beer *et al.,* 2013). The combination of Raman spectroscopy with new manufacturing methods shows how important it is for both improving analytical capabilities and meeting the needs of modern science and industry (Figure 8.1).

Definition of Raman Spectroscopy

Raman spectroscopy is a strong way to analyze things that is based on the inelastic scattering of monochromatic light, usually from a laser, by molecular vibrations in a sample. Raman scattering is the name of this process that lets you figure out what molecular structures are and what materials are like. Raman spectroscopy works differently than other types of spectroscopies, like infrared (IR) spectroscopy, which looks for light that is absorbed. Instead, Raman spectroscopy looks for energy that is shifted from the light that hits the sample as it interacts with the sample's molecular vibrations. The unique way that molecules scatter light gives them a "fingerprint" that is very useful for telling the difference between different chemical compounds. This makes it very useful in fields like chemistry, biology, and materials science. Raman spectroscopy has many uses in both basic and applied research. For example, in the field of cultural heritage, it helps researchers study painted textiles, which helps them learn more about how materials break down and how they are made. This part of the study shows how conservators and scientists work together to improve preservation techniques and methods based on detailed analytical results. Recent advancements have integrated complementary techniques such as Fourier Transform Infrared Spectroscopy (FTIR) and Surface-Enhanced Resonance Raman Spectroscopy (SERS) to enhance the efficacy of Raman Spectroscopy in tracing historical dye compositions and production methods, thereby elucidating both artistic intent and technological progress of the era (Lennard *et al.,* 2017). Furthermore, the fundamental principles of contemporary optics have facilitated substantial applications of Raman spectroscopy beyond conventional domains. The application of classical light principles to non-classical light phenomena, including squeezed and entangled states, has created novel opportunities in quantum communication and high-resolution imaging. These advancements augment the sensitivity of Raman techniques, facilitating the detection of minute concentrations of substances or even individual molecules. Researchers are still looking into and coming up with new ways to use Raman spectroscopy. Its role in cutting-edge research and technology is still very important, showing how adaptable and useful the technique is in a changing scientific world (Ghatak *et al.,* 2017).

Historical Background and Development

Raman spectroscopy, a powerful analytical tool, came from early 20th century research on how light scatters, especially work done by physicists like C. V. Raman and his peers. The journey to its practical use started in 1928, when C. V. Raman and K. S. Krishnan first saw the inelastic scattering of light. This is now known as the Raman effect. This important discovery led to new ideas about the molecular and atomic structure of materials and was the beginning of Raman spectroscopy. At first, it was only used for basic research, but its features quickly drew attention for material characterization, which led to a rapid change in analytical chemistry. Raman spectroscopy made a lot of progress in the 1960s and 1970s thanks to new technologies, especially better lasers, and more sensitive detectors. These advancements facilitated enhanced resolution and precision, permitting researchers to utilize Raman spectroscopy across diverse disciplines, including biology, chemistry, and materials science. Furthermore, the integration of Fourier Transform techniques transformed data acquisition and processing, augmenting the analytical capacity of the method. Raman spectroscopy's versatility was further demonstrated in the conservation of cultural heritage artifacts, offering essential insights into the composition of pigments and materials utilized in historical items, as evidenced by the SERS and Raman analyses of the *Hasti Afunei* sarcophagus (Bianco *et al.,* 2019). The function of cooperative endeavors between scientists and conservators is essential for the successful utilization of Raman spectroscopy in the preservation of painted textiles and analogous artifacts. Comprehensive literature reviews show that combining vibrational spectroscopic techniques like Raman and Fourier Transform infrared spectroscopy gives us a deeper understanding of materials, techniques, and how things break down. This kind of collaboration between different fields has been very helpful in learning about both old and new ways of preserving things (Lennard *et al.,* 2017). As the field continues to grow, Raman spectroscopy's ability to analyze complicated materials will surely help us learn more about heritage artifacts. This will help protect them for future generations and help people appreciate their historical importance.

8.1. PRINCIPLES OF RAMAN SPECTROSCOPY

The principles of Raman Spectroscopy are based on how light interacts with molecular vibrations. This lets us describe materials at the molecular level. Rayleigh scattering is the name for the process that happens when monochromatic light, usually from a laser, hits a sample and most of it scatters elastically. However, some of this light scatters inelastically, which changes the energy of the molecules in the sample in a way that corresponds to their vibrational modes. The Raman spectrum is a unique molecular fingerprint that comes from this difference in energy. Raman Spectroscopy is very useful because it can identify and measure the molecular makeup of a wide range of samples, from gases to solids. This method has become essential in many areas of science, such as materials science and chemistry, where knowing how molecules interact is important for making progress in both theory and practice. Raman Spectroscopy is a unique analytical method because it doesn't destroy samples and doesn't need much preparation. This trait is especially useful for analyzing delicate materials, like painted textiles, where keeping the sample's integrity is very important. The capacity to perform measurements *in situ* augments its utility, particularly in conservation science.

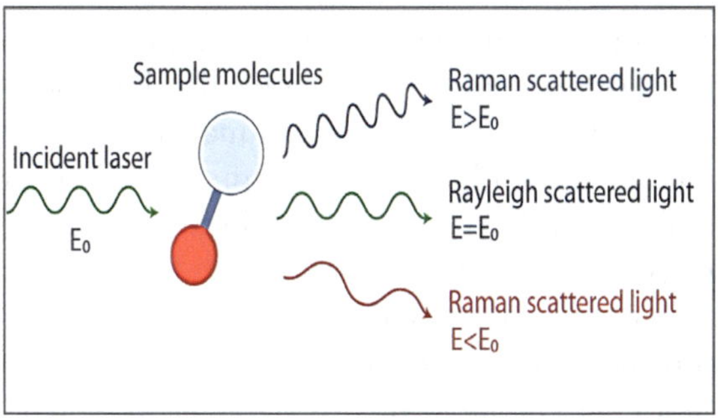

Figure 8.2. *Principles of Raman spectroscopy.*

Source: https://www.smacgigworld.com/assets/blog-img/Raman-Scattering-Basics.webp.

Furthermore, improvements in surface-enhanced Raman spectroscopy (SERS) have greatly increased the sensitivity of the technique, allowing it to find even small amounts of analytes. Through the collaboration of conservators and scientists, the incorporation of Raman Spectroscopy into the examination of painted textiles enhances comprehension of the materials employed and their deterioration mechanisms. This method not only has analytical uses, but it also helps to protect cultural heritage in a big way, showing that it has real-world uses beyond just research. Moreover, improvements in Raman Spectroscopy technologies keep making it more useful and applicable in both science and industry. The use of mapping techniques, which allow for spatially resolved analysis of samples, has been especially helpful in characterizing materials. Localized measurements give us important information about heterogeneous samples, which helps us better understand the properties of materials. Recent developments have also included improvements in instruments, which have made Raman Spectroscopy easier to use and more available for a wide range of uses. This evolution is aided by the growing collaboration between fields, which makes it easier to understand the data from Raman spectral analysis. All of these things show how important Raman Spectroscopy is, not just as a powerful analytical tool, but also as a growing field that deals with difficult problems in science and industry. It promises to keep making progress in material analysis and conservation efforts (Baianu, 2011; Lennard *et al.,* 2017) (Figure 8.2).

8.1.1. Basic Principles of Light Scattering

Light scattering is a basic physical process that happens when light hits matter and changes the direction, intensity, or wavelength of the light. In Raman spectroscopy, this phenomenon is mainly about inelastic scattering, which happens when the energy of incoming photons changes because they interact with the vibrational modes of molecules. When you shine light on a sample, most of it bounces off in an elastic way, keeping its original energy and wavelength. But a small amount of the light is affected by inelastic scattering, which changes the energy of the light in a way that matches the vibrational states of the molecules in the sample. This change in energy is what makes Raman spectroscopy able to give unique molecular fingerprints. This lets scientists' study different materials or look into complicated molecular interactions. The Raman effect is directly related to the vibrational energy levels of molecules, which depend on their atomic mass and how stiff their bonds are. First-order Raman scattering, in particular, involves single transitions between vibrational energy levels. This gives information about specific phonon modes in a material. Recent research has shown that higher-order processes can also affect the Raman spectra, making it harder to understand the results. In higher-order scattering, multiple phonons are excited at the same time. This shows how phonons interact across the entire spectrum, which gives more detailed information about how the material vibrates. So, to fully understand the properties of materials, it is important to separate these contributions, as shown by the rules of energy conservation during scattering events (Draxl *et al.,* 2017). Improvements in computer methods have made it easier to understand Raman scattering processes, especially when it comes to exciton-phonon coupling. This interaction can have a big effect on the strength and shape of Raman peaks, especially when resonance conditions are present, when certain excitation wavelengths interact strongly with the electronic structure of the material. These improvements are very important for understanding the complex details in the Raman spectra, and they give us a better idea of how materials behave at the atomic level. Furthermore, advancements like spin noise spectroscopy (SNS) contest conventional limits in optical methodologies by focusing on nonperturbative characteristics and offering novel approaches for material characterization in unparalleled manners (Glazov *et al.,* 2015). As these methods improve, they add to the conversation about the basic ideas behind light scattering and what they mean for Raman spectroscopy in both theory and practice.

8.1.2. Stokes and Anti-Stokes Scattering

Stokes and anti-Stokes scattering are two different results of the interaction between incoming photons and molecular vibrations. They are at the heart of Raman spectroscopy. When photons lose energy while transferring it to a molecule's vibrational modes, this is called Stokes scattering. This makes the emission have a longer wavelength. This process is very important for accurately detecting molecular structures because the Raman shift that corresponds to Stokes lines is usually stronger because the Boltzmann distribution favors lower energy states at room temperature. In real life, the strength of Stokes lines can often tell us a lot about the molecular composition and dynamics, which is why they are so important in vibrational spectroscopy. On the other hand, anti-Stokes scattering happens when the incoming photons gain energy, which makes the emission have a shorter wavelength. This process is less common because it depends on the population of excited vibrational states, which is activated by heat. So, anti-Stokes lines are usually weaker than Stokes lines, but they are very important for showing information about high-energy vibrational states and giving us information about the thermal population of molecular systems. The difference between Stokes and anti-Stokes scattering can also be very helpful when looking at how temperature changes in samples. This is because the relative strengths of these two types of scattering show how molecular populations change under certain thermal conditions. Recent improvements in laser technology have made Raman spectroscopy even better at seeing both Stokes and anti-Stokes scattering. The advent of ultrafast laser sources has enabled real-time observation of molecular dynamics, permitting researchers to investigate phenomena that were previously unattainable via conventional spectroscopic methods. Furthermore, the utilization of broadband laser pulses in conjunction with innovative methodologies may result in improved spectral resolution and resilience to environmental fluctuations, thereby enhancing the precision of spectral measurements. This new idea fits with recent research that shows coherent anti-Stokes Raman line shapes show strange behavior when you take into account the fact that modern laser pulse durations are not symmetrical (Bengtsson *et al.,* 2014). This suggests a promising new area in the ongoing development of Raman spectroscopy. Moreover, integrating noise into laser excitation has developed as a technique to enhance resolution and efficiency, providing considerable benefits in remote sensing and biological imaging applications (Weiner *et al.,* 2007) (Figure 8.3).

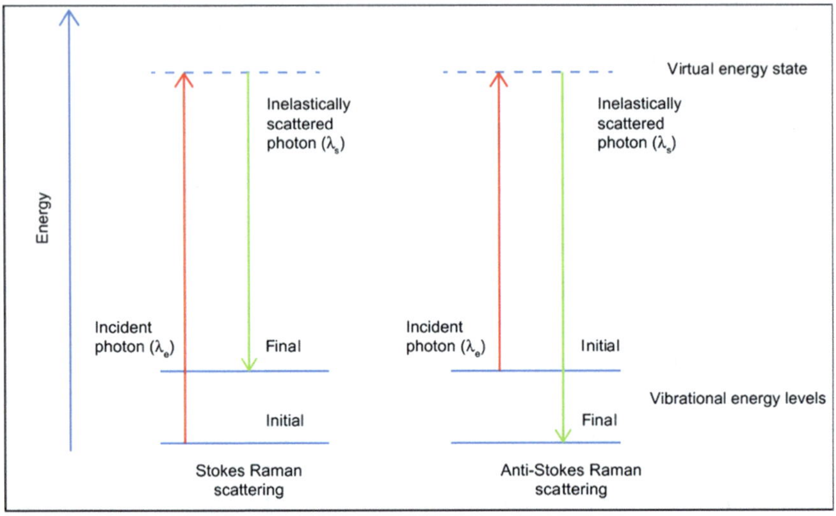

Figure 8.3. *Schematic showing the difference between stokes and anti-stokes Raman spectroscopy.*

Source: https://www.researchgate.net/publication/304341022/figure/fig3/AS:669038971744263@1536522729466/Schematic-showing-the-difference-between-Stokes-and-anti-Stokes-Raman-scattering-Notes.png.

8.1.3. Role of Molecular Vibrations in Raman Effect

Understanding and using Raman spectroscopy requires a basic understanding of how molecular vibrations affect the Raman effect. Raman spectroscopy is based on Raman scattering, which happens when light of one color interacts with molecular vibrations. These vibrations are the same as the movement of atoms within a molecule, and they show up as different vibrational modes, like stretching and bending. When light hits a sample, most of the photons bounce off of it elastically (Rayleigh scattering). However, a small number of them bounce off of it inelastically, which causes energy shifts that are unique to certain molecular vibrations. This change gives important information about the molecular structure, dynamics, and interactions. This makes Raman spectroscopy a very useful tool for chemists and materials scientists. The molecular environment has a big effect on molecular vibrations, which in turn affects the Raman scattering process. Different types of molecules have different vibrational frequencies because of how their atoms are arranged and how they bond with each other. You can measure these frequencies as Raman shifts, which can help you figure out what molecules are and how they interact with each other. Recent research has also shown that new developments in surface-enhanced Raman scattering (SERS) have made traditional Raman spectroscopy even better. For instance, biocompatible Raman probes that use anisotropic silver nanoparticles can greatly improve Raman signals, giving them high analytical enhancement factors. This improvement is due to both electromagnetic and chemical processes that happen when molecular vibrations interact with nanostructures. This makes it possible to detect vibrational information even in complicated biological systems (Salazar *et al.,* 2019). Improvements in time-resolved vibrational spectroscopy help us better understand how molecular vibrations affect our understanding of dynamic processes. Recent conferences have shown how important ultrafast Raman spectroscopy is for studying temporary molecular states and reaction dynamics that happen on a picometer length scale. Time-resolved experiments give us a lot of information about structural and kinetic changes that happen in real time. This information is useful in many areas of chemistry, biology, and material science. Researchers can figure out complicated chemical mechanisms and come up with new technical uses by studying how molecular vibrations work together and change over time. As vibrational spectroscopy research progresses, the examination of molecular vibrations persists as a pivotal focus in fostering scientific advancements within this field (Laubereau *et al.,* 2007).

8.2. INSTRUMENTATION AND TECHNIQUES

Raman spectroscopy is based on many different types of instruments and methods that make it better at analyzing things. The main idea behind the technique is that monochromatic light, usually from a laser, scatters inelastically and interacts with molecular vibrations in a sample. The resulting spectrum gives information about the molecular makeup and structure of the sample. To make Raman spectroscopy work better, new technologies like Fourier transform Raman (FT-Raman) spectroscopy and surface-enhanced Raman spectroscopy (SERS) have been developed. These methods make traditional Raman techniques more sensitive and precise, which makes it possible to find very small amounts of analytes. Moreover, novel applications that integrate Raman spectroscopy with imaging techniques, such as mapping, enhance its effectiveness in materials characterization, especially in the analysis of complex matrices like painted textiles, which frequently require multifaceted analytical strategies (Lennard *et al.,* 2017). Another important part of instrumentation is the use of advanced sample environments, like diamond anvil cells (DACs), which let scientists study materials at different temperatures and pressures.

These devices make it easier to study metastable states of materials by giving researchers precise control over the conditions of their experiments. In this case, Raman spectroscopy is a very important tool because it can show how structures change across phase boundaries when conditions are very extreme. For example, studies have demonstrated the effective mapping of phase stability in metallic systems, showcasing the robustness and adaptability of the techniques (Hinton *et al.,* 2019). This capacity to observe complex molecular interactions in real time enhances the comprehension of material properties and guides the creation of novel materials and applications across various domains, including condensed matter physics and materials science. Also, as instruments continue to improve, user-friendly software and data acquisition systems have become essential for getting the most out of Raman spectroscopy. Advanced algorithms for data analysis and interpretation make it easier to get useful information from complicated spectra, which makes the technique more useful. Collaboration between scientists and practitioners in related fields, such as conservation, has been shown to be necessary to get the most out of these tools. Recent literature emphasizes that the amalgamation of Raman spectroscopy with various analytical techniques, such as mass spectrometry and chromatographic methods, facilitates a thorough characterization of materials and their degradation mechanisms. This multidisciplinary approach highlights the potential of Raman spectroscopy to uncover insights essential for both research and practical applications, facilitating progress in material conservation and analysis (Lennard *et al.,* 2017) (Figure 8.4).

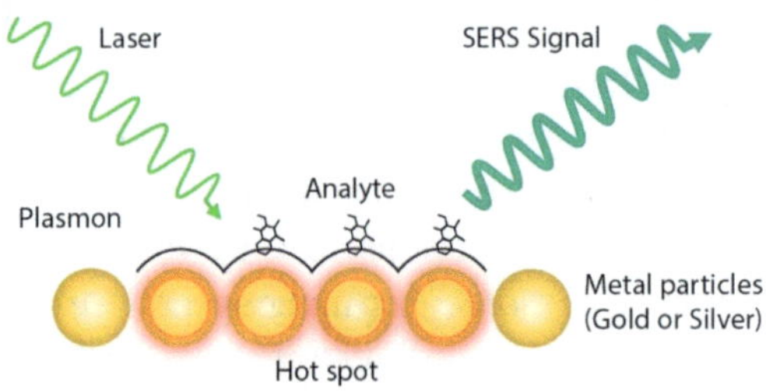

Figure 8.4. *Surface-enhanced Raman scattering (SERS).*

Source: https://static.horiba.com/fileadmin/Horiba/_processed_/7/5/csm_Raman_FAQ_SERS_principle_a4d2533210.jpg.

8.2.1. Components of a Raman Spectrometer

A Raman spectrometer has several important parts that work together to look at molecular vibrations and give information about the makeup of a material. The laser source is the main part of the spectrometer and is what excites the light. The laser makes it possible for photons to scatter by sending out light of only one color. This scattering process is very important because it causes energy shifts that show the molecular structure. Solid-state lasers or diode lasers are used in many modern applications because they are stable and accurate. After the laser, a series of optical components, such as lenses and filters, focus the light on the sample and improve the collection of scattered light for later analysis. These parts work together to make Raman spectroscopy work well. This affects how it can be used in many fields, such as materials science and biomedicine (Singha *et al.,* 2004). The spectrometer uses a spectrograph, which is a very important part that looks at the scattered light to make a spectrum. It is possible to find different vibrational modes in the sample because the spectrograph separates the incoming light by wavelength. A diffraction grating is usually used to separate the light into its different wavelengths. You can look at the spectral data to learn important chemical information about the sample, such as its molecular structure and composition. Some advanced spectrometers may also use charge-coupled devices (CCDs) as detectors. These devices are very sensitive and can quickly collect data. The design and quality of these parts have a big impact on the spectrometer's resolution and sensitivity. This makes them important for a wide range of uses, from nanotechnology research to medical diagnostics, as shown in new studies that use Raman spectroscopy (Darvin *et al.,* 2018). A Raman spectrometer may come with extra parts and settings that are made for certain uses, in addition to the main parts. For example, imaging systems can be combined to make spatially resolved Raman spectroscopy possible, which makes it easier to analyze a lot of samples at once. This integration makes it possible to quickly record full Raman images, which makes it easier to tell the difference between healthy and diseased tissues. This is very important for cancer diagnosis. Additionally, technological advancements have resulted in the creation of portable Raman spectrometers, enhancing the accessibility of Raman spectroscopy for field applications. These advancements highlight the adaptability of Raman spectrometers, as they conform to the requirements of diverse research and industrial settings, thus enhancing their significance in nascent scientific fields, as indicated by Singha et al. (2004). The interaction of these elements and innovations underscores the importance of Raman spectrometry in modern scientific research.

8.2.2. Types of Raman Spectroscopy (e.g., Dispersive, FT-Raman)

Raman spectroscopy has developed into various distinct methodologies, each catering to specific applications and providing particular advantages. Dispersive

Raman spectroscopy is the most notable of these because it has always used diffraction gratings to separate spectra. This method uses a monochromator to look at the light that has been scattered by a sample. By measuring changes in energy, scientists can find molecular vibrations. Dispersive systems may have trouble with low signal intensity, especially when the sample shows weak Raman scattering, even though they have some benefits, like high sensitivity and resolution. Because of this, technology is always trying to make dispersive systems more efficient and sensitive so that they can give more accurate analytical results.

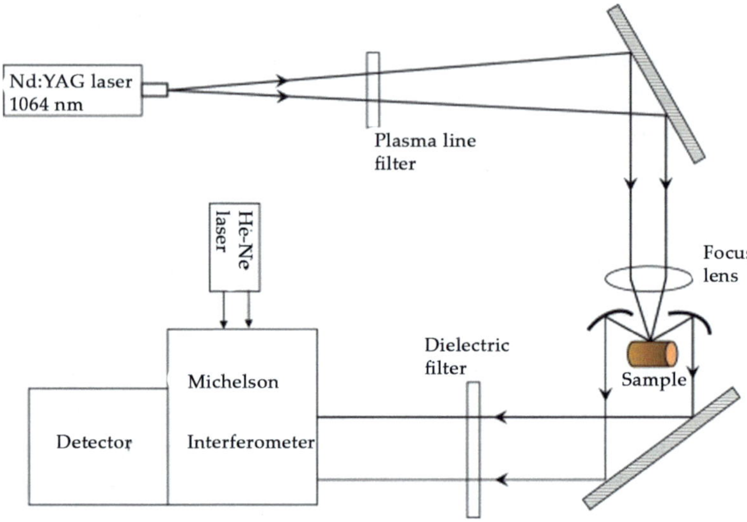

Figure 8.5. *Fourier transform Raman spectroscopy.*

Source: https://ars.els-cdn.com/content/image/3-s2.0-B9780128142646000062-f06-01-9780128142646.jpg.

Fourier-transform Raman spectroscopy (FT-Raman), on the other hand, has gotten a lot of attention because it has high spectral resolution and throughput. This method uses Fourier-transform techniques to process the scattered light in a new way that speeds up data acquisition and cuts down on the time needed for spectral analysis. Recent studies have shown that improvements in high-performance interference filters have closed the performance gap that was thought to be caused by the limitations of FT-Raman systems in the visible wavelength range. Some new ideas even suggest that FT-Raman can get results that are just as good as or better than those of standard dispersive spectrometers, especially when working with samples that don't give off light. These results prompt a reevaluation of the situational applicability of various

Raman techniques, highlighting FT-Raman's potential in diverse analytical settings (Bernáth *et al.,* 2014). The combination of Raman spectroscopy and multivariate data analysis is another step toward getting more accurate measurements. For example, research using vibrational spectroscopy has shown a lot of promise for things like using Raman spectral data to measure the amount of trans fatty acids in dairy products. Researchers can use advanced analytical frameworks to create multivariate models that accurately predict the presence of trans fatty acids in dairy products that come from nature and from industry. This demonstrates the applicability of Raman techniques beyond simple qualitative analysis, facilitating comprehensive quantitative evaluations in food science and other fields. The ongoing enhancement of dispersive and FT-Raman methodologies underscores their increasing significance in intricate scientific investigations, highlighting the necessity for continuous research in this domain (Beattie *et al.,* 2015) (Figure 8.5).

8.2.3. Sample Preparation and Handling Techniques

For Raman spectroscopy to work well, it is very important to use good sample preparation and handling methods, because they have a direct effect on the quality of the spectral data collected. The first step in this process is to carefully choose samples that are representative of the material being studied. It is important to keep contamination to a minimum and make sure that the samples are the right size for Raman analysis. For example, samples should be finely ground and mixed evenly so that the laser source can work best. Also, the choice of substrate can have a big impact on the spectra, so materials that don't interfere too much but still allow for reproducibility need to be used. Getting high-resolution Raman spectra that accurately show the molecular characteristics of the samples starts with properly preparing the samples. After choosing and preparing the samples, the way they are handled before analysis can make Raman spectroscopy even more useful. It's very important to pay close attention to the environment. Samples should be kept dry and free of contaminants, which can mess up the spectral data. Using special transfer tools and micro-manipulators might also help with the accuracy of sample placement during spectroscopic analysis. The literature indicates that implementing protective measures, such as nitrogen purging, can alleviate the impact of atmospheric interference on sensitive samples. Additionally, keeping samples in controlled conditions until they are analyzed can help keep them intact by reducing the risk of degradation or change. These careful handling practices not only keep the results reliable, but they also help the findings be reproduced in different analytical situations. Advanced methods for extracting and manipulating samples are also very important for the effectiveness of Raman spectroscopy, especially when dealing with complicated samples like those found in astrophysics and materials science. For example, new techniques using keystones have been created to get specific particle impacts from aerogel collectors. These methods cause less damage to the areas around them, which keeps the important features needed for analysis. Using custom-made fixtures, samples can be made to stand on their own, which is helpful because it keeps supporting substrates from getting in the way during spectroscopic acquisition. The capacity to uniformly crush samples while preserving their intrinsic molecular structure underscores the significance of meticulous sample preparation in acquiring high-quality spectral data, as evidenced in multiple spectroscopic investigations (Adya *et al.,* 2009; Brennan *et al.,* 2003). Such advancements are essential for comprehending complex molecular interactions and for broadening the applicability of Raman techniques across various scientific disciplines.

8.3. DATA ACQUISITION AND ANALYSIS

Raman spectroscopy depends a lot on good methods for collecting and analyzing data that can show important information about the molecular makeup of samples. In the past, Raman images were made by scanning specific areas of interest in a labor-intensive way. But these methods can be very slow, which is a problem for applications that need quick evaluations, like clinical diagnostics. Recent improvements, especially those that combine integral field spectroscopy (IFS) with Raman imaging, have changed this process completely. Researchers can greatly speed up the process of collecting data without losing the quality of the spectral information by allowing them to quickly capture whole Raman images in one exposure. This advancement facilitates prompt and precise assessments of tissues, particularly in the detection of conditions like cancer in *ex vivo* or *in vivo* contexts, as demonstrated by the effective utilization of this technology in skin biopsy analysis (Darvin *et al.,* 2018). Aside from faster imaging, new developments in spatially offset Raman spectroscopy (SORS) have made it even easier to collect data in Raman spectroscopy. Researchers can precisely control the collection of Raman signals from subsurface layers of optically turbid samples by using technologies like digital micro-mirror devices (DMD). This method can use different detection geometries, like annular and multi-offset modes, which can be changed to fit different types of samples. This makes it good for both macro- and micro-scale measurements. The flexibility of DMD-based SORS improves data collection without making the experiment too complicated. As a result, researchers can more accurately analyze the underlying substance of samples, which could lead to new uses in many fields, such as pharmaceuticals, material science, and biomedical research (Gibson *et al.,* 2016). Analyzing the collected Raman data is very important for getting useful information that can be used in science and real life. Researchers can now use advanced algorithms and software systems to make sense of spectral data. These tools help them find specific molecular features and measure the amounts of different compounds in a sample. These analytical methods are very important for telling the difference between healthy and sick tissues. This is especially true in cancer diagnostics, where small changes in the spectrum can show pathology. Moreover, combining data collection methods with machine learning techniques can improve the accuracy of predictions and make it easier to find complex patterns in the data. Raman spectroscopy is a powerful tool for both research and clinical use because it uses advanced data collection methods and strong analytical frameworks. This opens the door to better health outcomes and new ways to characterize materials (Figure 8.6).

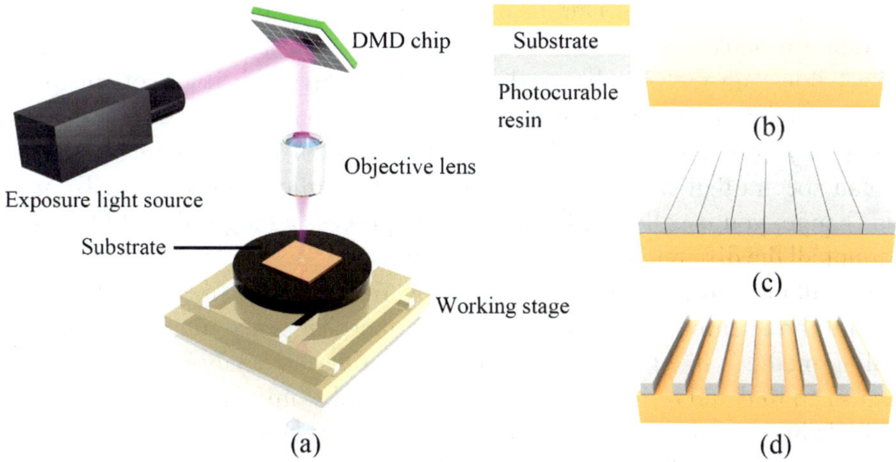

Figure 8.6. *Digital micromirror device (DMD).*

Source: https://www.researchgate.net/publication/336883905/figure/fig1/AS:830637099458561@1575050725375/a-Schematic-view-of-the-digital-micromirror-device-DMD-based-maskless.jpg.

8.3.1. Process of Data Collection in Raman Spectroscopy

When you use Raman spectroscopy to collect data, the first step is to excite a sample with a single-color light source, usually a laser. This first step is very important because the wavelength of the light that hits the sample determines which molecular vibrations can be studied. When the light hits the sample, a small number of the photons scatter inelastically, which causes changes in energy that correspond to molecular vibrations. The Raman scattered light, which is usually much weaker than the beam that hits it, has a lot of information about the sample's molecular structure and composition. To improve signal intensity and make sure that data is captured accurately, it is important to use techniques like optimizing laser power and focusing the beam. These techniques have a big effect on the overall reliability and quality of the collected spectra. After excitation, the scattered light needs to be collected and filtered so that the Raman signal can be separated from other sources of interference, like Rayleigh scattering and fluorescence. This usually means using optical parts like filters and lenses to separate the Raman signal from the light that is scattered elastically. The refined Raman signal is then sent to a spectrometer, which breaks the light down into its different wavelengths. This dispersion makes it possible to find specific spectral lines that are linked to certain vibrational modes of the molecules in the sample. During this stage, high-resolution detection is very important to make sure the resulting spectra are clear and correct. In order to get reliable data that can be used for further analysis, it is important to use techniques that improve signal-to-noise ratios, such as using the right detector technology or averaging signals. The last step in collecting data is to turn the Raman spectra that were collected into useful chemical information. To make this change, you usually need complex data processing algorithms to take into account all the different things that could affect the spectra you get. These kinds of algorithms can make the data better by fixing things like baseline shifts, spectral broadening, and other problems that happened during collection. Adding methods like machine learning or multivariate analysis can also make it easier to find and measure compounds in complex mixtures. The progressive methods in data processing not only make Raman spectroscopic analyses more accurate, but they also make them useful in more fields, such as biomedical diagnostics. For example, quantum cascade lasers can be used for higher fidelity measurements (McCrow *et al.,* 2011). Such improved methods could have a big effect on the translational side of Raman spectroscopy in clinical settings, as shown by current research (Fletcher *et al.,* 2025).

8.3.2. Interpretation of Raman Spectra

Understanding Raman spectra is an important part of using Raman spectroscopy in many areas, such as materials science and art conservation. The Raman effect, which is the inelastic scattering of monochromatic light, gives us important information about how molecules vibrate and how compounds are made up. However, interpreting the spectra can be very hard because different materials' spectral features may overlap or be similar. Each spectrum can show a different fingerprint, but figuring out what makes them different, especially when it comes to polymorphs or synthetic versus natural pigments, requires careful study (Lennard *et al.,* 2017). This complexity highlights the necessity for robust analytical methodologies that enable precise interpretation, guaranteeing that the insights obtained from the spectra are dependable and enlightening. Also, new developments in spectroscopic methods have made Raman spectroscopy much better at getting high-resolution spectra. New technologies like surface-enhanced Raman spectroscopy (SERS) have made weak Raman signals stronger, making it easier to find trace compounds. These improvements make it possible to find pigments and materials that were hard to find with older methods. This makes them very useful for art conservation and historical research (Lennard *et al.,* 2017).

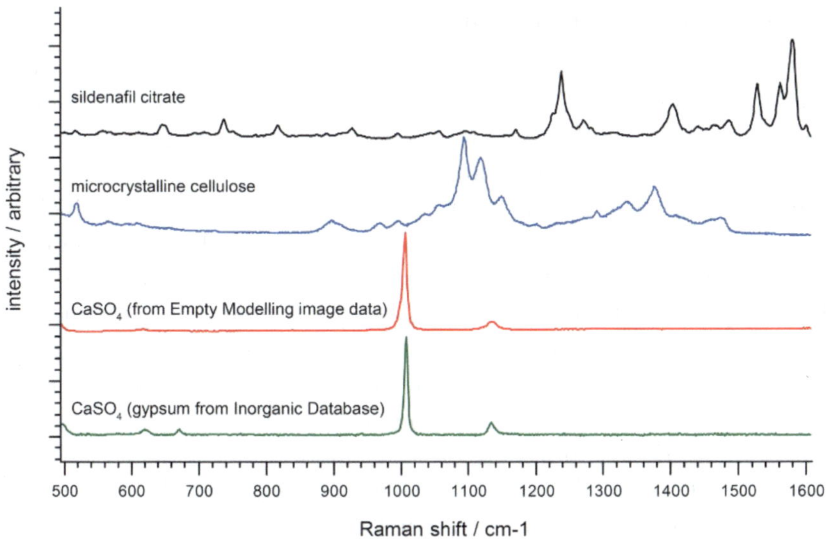

Figure 8.7. *Interpretation of Raman spectra.*

Source: https://www.renishaw.com/media/img/en/12dd10113c744d81b498da73b9ba56f4.jpg.

Nonetheless, notwithstanding these technological advancements, the manual interpretation of spectra persists, chiefly owing to the subjective characteristics of spectral analysis. This human factor can sometimes cause mistakes or inconsistencies in the classifications, which means that a systematic approach to data interpretation that uses chemometric methods to make things more accurate and efficient is needed. Researchers are increasingly using automation and machine learning to classify Raman spectra because manual interpretation has its limits. The development of algorithms that can automatically sort spectral data into pre-defined reference classes is a big step forward in analytical methods. These kinds of systems use chemometric methods to improve spectral separability, which makes it possible to classify closely related spectral profiles with little help from the user. This automated classification not only makes the analysis process faster, but it also reduces the subjective biases that can come from how people interpret things (González *et al.,* 2016). The successful validation of these systems underscores their potential utility in the interpretation of Raman spectra, thereby facilitating the convergence of scientific inquiry and practical application in domains such as art conservation and materials analysis (Figure 8.7).

8.3.3. Software and Tools for Spectral Analysis

In the field of Raman spectroscopy, software and tools for spectral analysis are very important for understanding and analyzing the data that is collected during experiments. Modern spectroscopic tools come with a lot of software that can quickly process large amounts of spectral data. These programs make it easier to get useful information from complex spectra, which helps researchers understand molecular structures, chemical compositions, and interactions on a basic level. For example, advanced algorithms can separate overlapping peaks and figure out the ratios of peak intensities, which are very important for quantitative analysis. Also, software solutions usually come with calibration protocols to make sure that results are accurate and can be repeated in different experimental settings. Raman spectroscopy is becoming more popular in many fields, so it's important to improve these analytical tools so they can be used effectively in areas like materials science and pharmaceuticals. The incorporation of process analytical techniques, like those used in hot-melt extrusion (HME), illustrates the significance of software tools for real-time spectral analysis. Recent advancements have shown that spectroscopic techniques can be employed to oversee and regulate essential product parameters during HME. This is especially important in the pharmaceutical industry, where the quality of extrudates must be carefully checked at every step of the manufacturing process. Researchers can better understand the extrusion process by using advanced software platforms to see how materials behave during the process. This combination of Raman spectroscopy and process analytics helps create a more reliable manufacturing framework that ensures product quality and improves production efficiency at the same time. These kinds of new ideas show how important it is to have advanced software tools that make it easier to do full spectral analysis in real-world situations (Beer *et al.,* 2013). Furthermore, advanced applications of Raman spectroscopy, especially in micromanipulation, demonstrate the versatility and complexity of contemporary analytical instruments. Using optical micromanipulation and

Raman microspectroscopy together makes it possible to get a detailed picture of individual cells, which can then be sorted and studied based on their chemical makeup. This method makes spectral analysis more accurate by letting scientists focus on certain microorganisms and see how they react to different things. In order to analyze stress responses and improve experimental conditions, it is important to combine software tools that manage the data generated by these complex techniques (Pilát *et al.,* 2015). Researchers are using Raman spectroscopy to look into the biochemical pathways of living cells. The tools and software made for spectral analysis will be very important for helping us learn more about how cells behave in different situations. This will open up new possibilities for biotechnology and medicine.

8.4. APPLICATIONS IN CHEMISTRY

Raman spectroscopy has many uses in chemistry and is very useful in many areas, such as material science and pharmaceutical analysis. One of the most important uses is to improve Raman signals by using nanostructured surfaces in a smart way, like polymer nanofibers. When metal nanostructures and particles are added to these nanofibers, they become strong substrates for surface-enhanced Raman spectroscopy (SERS). Because of their complicated three-dimensional structure, these substrates can produce Raman signals that are many times stronger than those produced by traditional methods. The complex relationship between a larger surface area and the way light scatters in these nanofibrous systems is very important for making the analytical sensitivity better so that small amounts of substances can be found in different chemical environments (Camposeo *et al.,* 2015). These improvements show how promising nanofiber-based platforms are for moving analytical technologies in chemistry forward. Another important use of Raman spectroscopy is the addition of two-dimensional (2D) materials, which makes signal enhancement easier. Materials like graphene and molybdenum disulfide (MoS_2) have shown a lot of promise in surface-enhanced Raman spectroscopy (SERS). 2D materials can be used as both a substrate and a signal enhancer in Raman experiments because they have a large surface area and electronic properties that can be changed. Recent research underscores the efficacy of employing uncoated 2D materials and hybrid structures that integrate metals with 2D substrates, resulting in significant Raman enhancement attributed to localized electromagnetic fields. This integration greatly increases the usefulness of Raman spectroscopy in chemical analysis, making it possible to find low-concentration species and describe complex materials (Xia *et al.,* 2017). The ongoing investigation and advancement in this domain signify a burgeoning field poised for innovative applications. The integration of advanced materials like nanofibers and 2D materials into Raman spectroscopy shows how important it is becoming in chemistry. Researchers have powerful tools for a wide range of uses, from environmental monitoring to biomedical diagnostics, thanks to the ability to fine-tune signal strength and sensitivity. Chemists can learn more about complicated chemical interactions and molecular behaviors that were hard to study before by using the benefits of custom nanostructures and better materials. This shows how important Raman spectroscopy is not only for academic research but also for real-world uses in many fields. It shows a very important link between advanced material science and analytical chemistry. As researchers keep coming up with new ideas, the future of Raman spectroscopy looks bright for finding new ways to solve difficult chemical problems (Figure 8.8).

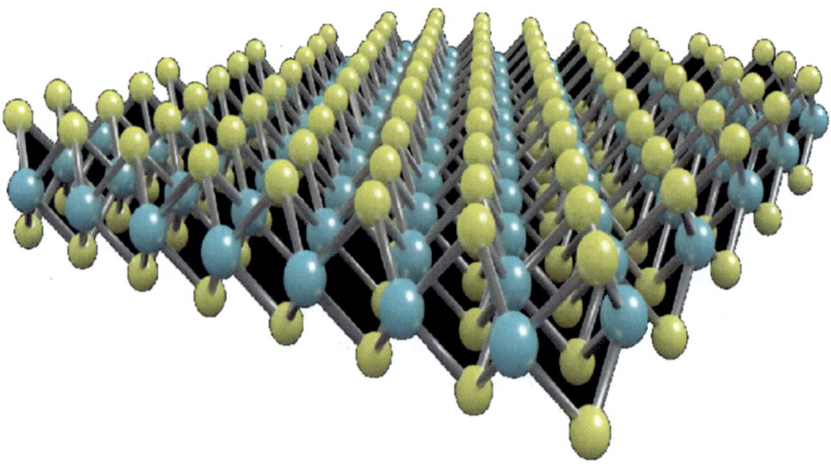

Figure 8.8. *Molybdenum disulfide.*

Source: https://www.acs.org/content/dam/acsorg/molecule/archive-new/m/molybdenum-disulfide-3d.png.

8.4.1. Identification of Chemical Compounds

The identification of chemical compounds is an important part of many scientific fields, including materials science and biomedical research. Raman spectroscopy is a powerful vibrational spectroscopic method that is great at giving information about molecular vibrations that can be linked to molecular structure. This method is especially useful for characterizing organic matter (OM), which is hard to do because it is so complex. Recent advancements have underscored the constraints of conventional peak-fitting algorithms in Raman spectroscopy, especially concerning the diffraction-limited spatial resolution that obstructs precise chemical identification. Recent advancements utilizing genome-inspired collective molecular structure fingerprinting are facilitating enhanced molecular characterization, allowing researchers to obtain more intricate chemical data from Raman spectra, thus deepening our comprehension of organic matter composition and reactivity (Bryndzia *et al.,* 2016). The combination of Raman spectroscopy with cutting-edge computer methods is a big step forward in the non-destructive identification of chemical compounds. Researchers can now use ab initio calculations and data mining techniques to get molecular fingerprints that are directly related to chemical properties like the size distribution of aromatic clusters and the ratio of hydrogen to carbon. This method is a paradigm shift that makes it possible to measure chemical signatures that could greatly improve environmental monitoring and biomolecular research. Being able to find and measure these chemical properties without damaging them is important for finding contamination in different places and for studying organic matter in fields like biogeochemistry and astrobiology (Bryndzia *et al.,* 2016). Furthermore, Raman spectroscopy has been successfully utilized in the examination of painted textiles, demonstrating its adaptability in material characterization. The application of vibrational spectroscopic techniques, especially in the realm of cultural heritage preservation, illustrates how Raman and associated methods augment our comprehension of materials, their deterioration, and their conservation requirements. Recent literature underscores the significance of collaborative endeavors between conservators and scientists in the interpretation of Raman findings, which can enhance both the identification of chemical compounds and the methodologies utilized in conservation practices. This collaborative approach makes sure that artifacts are kept safe by having a deep understanding of what they're made of and the chemical interactions that affect their stability and integrity. Consequently, the identification of chemical compounds via Raman spectroscopy is essential for the progression of scientific inquiry and the preservation of cultural heritage.

8.4.2. Monitoring Chemical Reactions

Monitoring chemical reactions is essential in both academic and industrial settings, as it enables researchers to comprehend reaction mechanisms and enhance conditions for optimal results. Raman spectroscopy, which can give molecular fingerprints based on vibrational modes, has become a powerful way to see chemical changes as they happen. This method makes it easier to see small changes in molecular vibrations, which can show how a reaction is going. Scientists can learn about intermediate species that form during a reaction by using Raman spectroscopy. This helps them understand how catalysts work and how well they work. Real-time monitoring not only helps us understand chemical pathways better, but it also helps us make better catalysts and find the best reaction conditions. This has led to progress in many chemical fields, such as pharmaceuticals and materials science. Recent studies have shown that combining Raman spectroscopy with advanced catalytic research can give us new information about how reactions work. For instance, ethylene epoxidation has historically necessitated elevated pressures and temperatures; however, recent research has shown that plasmonic photoexcitation of silver catalysts facilitates these transformations under conditions close to ambient levels. Using Raman spectroscopy techniques, scientists have been able to directly observe this plasmon excitation-assisted reaction on the surfaces of silver nanoparticles. This has led to the discovery of new ways that catalysis works. As noted by Heo et al. (2018), it was found that the photogenerated graphene on the catalyst's surface is very important for epoxidation.

Figure 8.9. *Monitoring reactions.*

Source: https://d1ymz67w5raq8g.cloudfront.net/Pictures/1024x536/0/9/2/140092_GettyImages-185588711.jpg.

This shows how useful Raman spectroscopy can be for understanding complicated catalytic behaviors and making new materials for chemical changes. Operando studies have elucidated catalyst behavior under reaction conditions, further enhancing the capabilities of Raman spectroscopy. For example, keeping an eye on supported CrO_3/Fe_2O_3 catalysts during high-temperature water-gas shift reactions has led to important discoveries about how catalysts change. Using *in situ* Raman techniques, researchers found structural changes in the catalysts, such as the decrease of Cr^{6+} species and the change of Fe_2O_3 to Fe_3O_4 when the catalysts were in use. The insights obtained from these studies, as illustrated by Daturi et al. (2016), elucidate the essential surface and bulk transformations of the catalysts, which directly influence catalytic activity. Thus, incorporating Raman spectroscopy into catalytic research not only deepens the comprehension of reaction dynamics but also aids in the development of more efficient and resilient catalytic systems, fostering future advancements in chemical engineering (Figure 8.9).

8.4.3. Analysis of Complex Mixtures

The examination of intricate mixtures is a pivotal application of Raman spectroscopy, utilizing its capacity to deliver comprehensive molecular data in heterogeneous samples. These kinds of complicated mixtures are common in many areas, such as food science, pharmaceuticals, and environmental monitoring. In these situations, Raman spectroscopy is very useful because it can identify chemical components without needing a lot of sample preparation. Its non-destructive nature is especially useful because it lets you study sensitive compounds without changing them. The technique's ability to provide real-time analysis also helps with process control in the making of pharmaceuticals. For instance, in hot-melt extrusion (HME), it is important to keep an eye on the quality of the extrudates. Recently, process analytical technology has focused on using spectroscopic techniques to better understand and control how materials behave during production (Beer *et al.,* 2013). Raman spectroscopy is unique in that it uses a multiscale approach to analyze different parts of complex mixtures at the same time. This multiscale methodology is crucial for polymorphic substances, as changes in the crystalline forms can have a big effect on their physical properties. Research has shown that crystallizing compounds can produce different polymorphs depending on the conditions. Raman spectroscopy can accurately describe these polymorphs by finding small changes in vibrational modes. Such insights are essential for refining crystallization processes in pharmaceuticals to guarantee product uniformity and quality. By clarifying the link between molecular structure and physical behavior, analysts can make predominance diagrams that help control crystallization better, which speeds up the production process (Biscans *et al.,* 2011). The practical incorporation of Raman spectroscopy into the analysis of intricate mixtures demonstrates its adaptability in practical applications. For example, in food science, the method has been used to find adulterants and figure out how good food is. It can tell the difference between compounds that are very similar and often found together in food mixtures because it is sensitive to molecular vibrations. Raman spectroscopy is a very useful tool for making sure that safety standards are met and building consumer trust because it can analyze samples *in situ*. In environmental monitoring, the method can find and describe pollutants in complicated mixtures, which helps us understand where they came from and how they got there. These applications show how important Raman spectroscopy is as a powerful tool for research and industry to use when evaluating and managing complex mixtures.

8.5. APPLICATIONS IN BIOLOGY AND MEDICINE

Raman spectroscopy has become an essential analytical technique in biology and medicine, primarily because it offers comprehensive information about the molecular structure of biological specimens. This method works by using Raman scattering, which happens when incoming photons interact with the vibrational modes of molecules. This causes energy shifts that reveal important information about the molecules' structural properties. Life science applications include a wide range of methods, from counting certain biomolecules to making medical diagnostic processes better. Raman spectroscopy provides a non-invasive method for *in situ* sample analysis, safeguarding the integrity of fragile biological materials while delivering high-resolution imaging that facilitates accurate identification of cellular structures and functions (Notingher *et al.,* 2017). Raman spectroscopy is especially useful in the field of medical diagnostics. For example, the method can be used to find diseases like cancer early by looking at tissue biopsies. Raman spectra give us molecular signatures that help us tell the difference between cancerous and healthy tissues. This lets doctors do more accurate and timely interventions. Additionally, the ability to visualize and characterize cellular interactions and the biochemical environment provides significant insights that can improve our comprehension of pathophysiological processes. Hyperspectral imaging, which uses Raman spectroscopy, is another useful technique because it lets you look at more than one spectral band at a time, which makes diagnosis more reliable and efficient (Notingher *et al.,* 2017). Raman spectroscopy is also being used in new ways outside of the lab, such as for remote monitoring and environmental studies of biological samples. One big step forward is the creation of a simple Raman spectrometer that can be used to monitor the temperature of water from a distance. This technology can also be used for other liquids and live cells. This setup shows how flexible Raman spectroscopy can be by using common parts like a cheap diode pump laser and standard optical equipment to get useful data. This method not only lowers costs by allowing visible inspection and Raman analysis to happen at the same time through the same optical path, but it also makes the instrument more useful for different types of research. These kinds of new ideas show how Raman spectroscopy can connect complicated biological research with real-world environmental assessments, making it useful in both basic and applied sciences.

8.5.1. Use in Cellular and Tissue Analysis

Raman spectroscopy has become a very useful tool for studying cells and tissues, giving scientists a very accurate way to look at complicated biological systems. Its natural sensitivity to molecular vibrations gives us information about the biochemical makeup of tissues, which helps us find and describe different parts of cells. This is especially useful in research on diseases like myocardial infarction, where inflammation is a key part of the process. Researchers can use Raman spectroscopy to see changes in the structure and composition of tissues. This makes it easier to keep an eye on the inflammatory phases that happen after an injury. This non-invasive technique diminishes the necessity for extensive sample preparation and facilitates real-time analysis, thereby augmenting the comprehension of cellular dynamics during essential physiological processes. Raman spectroscopy enables the meticulous examination of protein distributions linked to inflammation within cardiac tissue analysis. Confocal Raman microscopy is one of the techniques that combines the high spatial resolution of confocal imaging with the molecular specificity of Raman scattering to make three-dimensional images of tissue samples. This makes it possible to find and do semi-quantitative analyses of proteins that are part of the inflammatory cascade, such as cytokines and chemokines. These abilities are very important when figuring out how bad the inflammation and tissue damage are after a heart attack. Combining Raman spectroscopy with other imaging techniques like Western blotting and confocal laser scanning microscopy gives a more complete picture of inflammatory processes at the cellular level (Cantatore *et al.,* 2015). Furthermore, improvements in Raman spectroscopic methods, such as remote monitoring setups, have expanded the range of cellular analysis beyond conventional uses. Recent advancements enable the examination of live cells in real-time, thereby facilitating the evaluation of physiological alterations in response to diverse stimuli. For example, monitoring certain Raman shifts linked to water molecules can help us find temperature changes in biological systems. This method, which uses cheap parts like diode lasers and commercial spectrometers, shows how easy and adaptable Raman techniques are for cellular research. These advancements not only facilitate economical implementations but also enable extensive applications across various domains, such as drug delivery, cancer research, and regenerative medicine, thereby solidifying Raman spectroscopy as a fundamental component of contemporary biological analysis (Pikov *et al.,* 2010) (Figure 8.10).

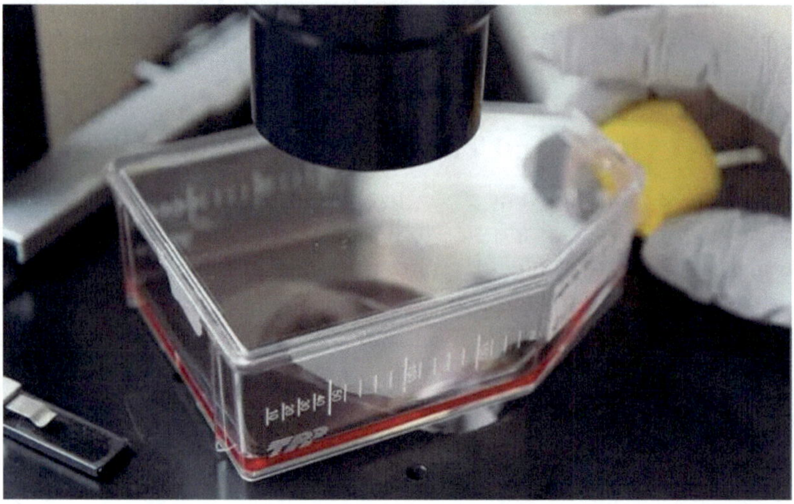

Figure 8.10. *Cell and tissue analysis.*

Source: https://institut-curie.org/_ipx/fit_cover&f_webp&q_80&s_650x415/ https://cdn.curie.fr/media/paragraph_header/L%2527analyse-des-tissus.jpg.

8.5.2. Role in Drug Development and Testing

Raman spectroscopy has become an important tool in drug development and testing because it doesn't hurt anyone and can give information at the molecular level. This spectroscopic method helps to describe pharmaceutical compounds by finding molecular vibrations, which gives us information about chemical structures and interactions. Researchers can keep an eye on the quality of drug formulations with such detailed spectral fingerprints, which make sure that active pharmaceutical ingredients (APIs) are always present in the same amounts. Raman spectroscopy not only makes the drug manufacturing process more efficient by allowing for real-time analysis, but it also cuts down on expensive recalls caused by purity problems. The Boston University Photonics Center says that progress in this field has led to new uses that help meet ongoing regulatory requirements and make drug development safer in the pharmaceutical industry (ANONYMOUS, 2015). Additionally, the use of Raman spectroscopy in pharmacokinetics has changed the way scientists look at how drugs are absorbed, distributed, metabolized, and excreted (ADME). Scientists can use this method to keep an eye on how drugs work in living organisms, giving them real-time feedback on how they work in biological systems. Researchers can use Raman spectroscopy to get a better idea of how effective and safe a drug is, which helps them figure out the best way to give it to people. This method can also help us understand how drugs and biological molecules interact, which is important for figuring out what side effects might happen. The Boston University Photonics Center is dedicated to interdisciplinary research that seeks to incorporate these analytical methods into drug testing protocols, facilitating personalized medicine approaches tailored to individual patient requirements (ANONYMOUS, 2013). Raman spectroscopy is also very important for making sure that drugs meet safety and quality standards. Regulatory bodies need strong analytical methods to make sure that drugs on the market are safe and work as they should. Raman spectroscopy is a good option because it gives quick and accurate results. It can analyze complicated mixtures without needing a lot of sample preparation, which makes it easier to follow Good Manufacturing Practices (GMP) rules. It is necessary to use such advanced optics not only to meet strict regulatory standards but also to build public trust in new drugs. The ongoing development of Raman technologies continues to shape the future of pharmaceuticals, making drug approval processes stronger and ultimately improving patient outcomes (ANONYMOUS, 2015). This is shown by the work of institutions like the Boston University Photonics Center.

8.5.3. Applications in Disease Diagnosis

Raman spectroscopy is one of the most cutting-edge ways to diagnose diseases because it can give molecular information through vibrational spectra. In the field of heart care, finding cardiac troponin is very important for diagnosing acute myocardial infarction (AMI), which is one of the most common causes of death around the world. Raman techniques are very sensitive and can find very small amounts of cardiac troponin, which quickly rises in the blood after an AMI event. This is a great chance for diagnosis. Researchers can make very sensitive immunosensors that make AMI diagnoses more accurate and timelier by using optical methods like Raman spectroscopy. This application highlights the capacity of Raman spectroscopy to improve clinical practices and patient outcomes by delivering non-invasive and expedited evaluations essential for informed treatment decisions (Abdolrahim *et al.,* 2015). Raman spectroscopy shows a lot of promise for finding plant diseases by looking at volatile organic compounds (VOCs). It can also be used to diagnose heart problems. For example, keeping an eye on certain VOCs that are linked to plant stress makes it much easier to find Huanglongbing (HLB), a terrible disease that affects citrus plants. The creation of biosensor arrays, especially those made from single-walled carbon nanotubes (SWNTs) altered with single-stranded DNA, offers a sensitive means of identifying these biomarkers. Raman spectroscopy is very important for figuring out what these SWNT-ssDNA hybrids are like, which helps us learn more about how they interact with VOCs. By concentrating on the asymptomatic phases of infection, Raman-based techniques could transform disease detection, facilitating earlier intervention and containment strategies crucial for safeguarding the agricultural economy from losses linked to HLB (Liu *et al.,* 2019). In the end, adding Raman spectroscopy to diagnostic practices could change a lot of different fields, from cardiology to agriculture. Because it can make molecular structures and interactions clearer, it opens the door to the creation of non-invasive diagnostic tools that could change the way diseases are found and tracked. Raman techniques not only improve the accuracy and sensitivity of current methods, but they also help the field of personalized medicine, where tests can be made to fit each patient's needs. As research advances, the integration of Raman spectroscopy with emerging technologies, including machine learning and biosensors, is expected to yield breakthroughs that tackle enduring challenges in disease diagnosis and management, signaling a new era of precision in healthcare (Abdolrahim *et al.,* 2015; Liu *et al.,* 2019) (Figure 8.11).

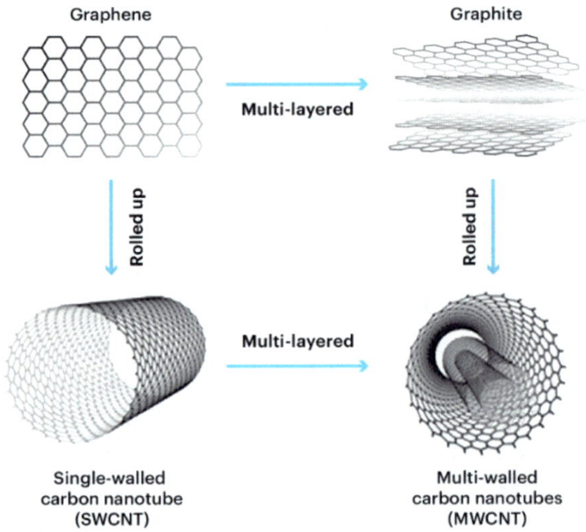

Figure 8.11. *Single-walled carbon nanotubes.*

Source: https://tuball.com/media/imperavi_redactor_content/e7/b1/e7b13bed-6a23-4c89-acde-67c53fd3c775/%D0%BA%D0%B0%D1%80%D1%82%D0%B8%D0%BD%D0%BA%D0%B8-06%20(1).png.

8.6. ADVANTAGES AND LIMITATIONS

Raman spectroscopy is well-known for its unique benefits, especially when it comes to characterizing materials. One of the best things about it is that it doesn't hurt the samples, so you can look at them without changing their physical or chemical state. This quality is very important in fields like art conservation, where it is important to keep historical artifacts safe. Raman spectroscopy also has high spatial resolution, which makes it possible to make detailed maps of the chemical compositions of different materials. Recent improvements in methods, like surface-enhanced Raman spectroscopy (SERS), have made it even more sensitive, allowing it to find very small amounts of substances. Raman spectroscopy is an important tool in both basic and applied science because it gives us information about molecular structures and interactions without needing complicated sample preparation (Lennard *et al.,* 2017). Even though Raman spectroscopy has its pros, it also has some cons that need to be taken into account in a full analysis. One major problem is that it has a low signal intensity by nature, which is especially true for samples with weak Raman scattering. This may require longer acquisition times and could restrict the technique's applicability in rapidly changing environments or for kinetic studies. Fluorescence can also mess with the Raman signal, making it harder to understand the data, especially when working with biological samples. Raman spectroscopy does not necessitate extensive sample preparation; however, specific sample types, particularly those exhibiting low vibrational energy levels, may present challenges. The advancement of complementary techniques is essential, underscoring the significance of integrated analytical methodologies that amalgamate various spectroscopic techniques to alleviate these constraints (Denschlag *et al.,* 2004). Another thing to think about when judging Raman spectroscopy is that it needs special equipment and trained staff. The necessary equipment can be expensive, which may make it hard for smaller labs or groups to get to. Also, to understand Raman spectra, you need to know a lot about molecular vibrations and chemical behaviors. This means that people who work with them need to have a technical background. This complexity can make it hard for Raman techniques to be used widely in many fields. Still, improvements in technology and technique optimization make things easier to get to and use. Researchers are looking into the full potential of Raman spectroscopy. To make more progress in analytic applications, they will need to find ways to get around these problems while making the most of its unique benefits (Lennard *et al.,* 2017).

8.6.1. Benefits of Using Raman Spectroscopy

Raman spectroscopy is a strong analytical method that has many uses in chemistry, biology, materials science, and other fields. One of the main benefits is that it doesn't destroy samples, which lets researchers look at them without changing or harming them. This feature is especially useful in fields like art conservation, where keeping the original work's integrity is very important. Raman spectroscopy also gives information about molecular vibrations, which makes it possible to find chemical compounds even when they are very small amounts. This sensitivity is important for many uses, such as biomedical diagnostics and environmental monitoring, where finding small amounts of substances can lead to big breakthroughs and improvements in public health and safety. Raman spectroscopy is also very flexible because it works with a wide range of sampling methods.

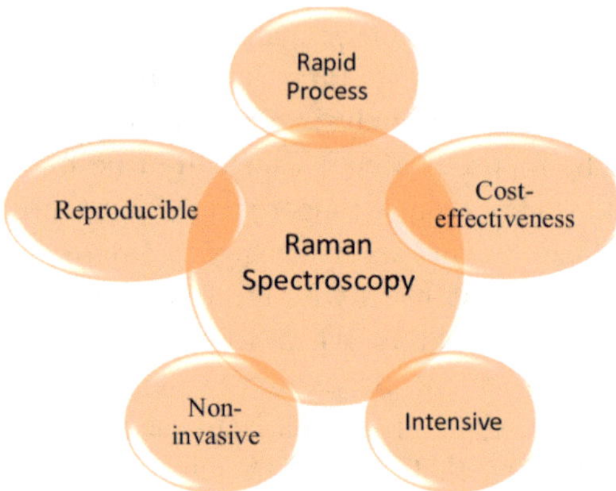

Figure 8.12. *Advantages of Raman spectroscopy.*

Source: https://www.researchgate.net/profile/Yu-Kiat-Lin/publication/348043494/figure/fig1/AS:1056364512153600@1628868334343/Advantages-of-Raman-spectroscopy.png.

For example, surface-enhanced Raman spectroscopy (SERS) can greatly increase the Raman signal. Researchers can increase the intensity of Raman scattering by up to 10^6 times by using methods that use nanostructured sensors, like those that have been developed recently. This increase in sensitivity makes it possible to take measurements of multiple parameters at once, which can give you a better dataset than traditional methods (Charlton *et al.,* 2014). Raman spectroscopy can also be used in real-time monitoring systems, which makes it easier to keep track of quality in manufacturing processes like hot-melt extrusion (HME). This flexibility makes Raman spectroscopy an essential tool for ensuring product quality and meeting regulatory standards in the production of pharmaceuticals (Beer *et al.,* 2013). Another great thing about Raman spectroscopy is that it can give quick results without needing a lot of sample preparation. This efficiency makes it possible to analyze a lot of data quickly, which is especially useful in situations where time is very important. For example, in pharmaceutical analysis, the ability to continuously monitor important parameters during manufacturing processes improves understanding and control of the process (Beer *et al.,* 2013). Raman spectroscopy's ability to give chemical information from just one measurement makes workflows easier and cuts down on the number of analyses needed. As this powerful spectroscopic technique finds more uses, its role in advancing scientific knowledge and practical applications keeps growing. This shows how useful and relevant it is in many different fields (Figure 8.12).

8.6.2. Limitations and Challenges Faced

Raman spectroscopy is a powerful way to analyze things, but it has its limits and problems. One big problem is that the method is very sensitive to how the sample is prepared and the environment. To get accurate results, samples must be carefully prepared so that fluorescence doesn't get in the way of Raman signals. Because of this sensitivity, it is important to handle and optimize experimental conditions very carefully, especially when looking at complicated biological materials or analytes that are not very concentrated. Conventional characterization techniques frequently lack the necessary spatial resolution and efficiency for comprehensive analysis of such samples. As mentioned, traditional methods have problems that make them less useful, so Raman spectroscopy needs to change so that it can be used as both a main and secondary analytical tool in many different situations (Chougule *et al.,* 2017). Consequently, comprehending and tackling these challenges is essential for enhancing the applicability of Raman spectroscopy in various domains. Raman spectroscopy also has a big problem with its quantitative analysis, which is hard to do. It is often hard to use Raman signals to figure out how much of an analyte is present because samples are often not uniform and the intensity of the scattering can change. Because the method depends so much on the inelastic scattering of monochromatic light, changes in the structure of the sample can cause results to be inconsistent. The presence of structural defects or irregularities may further complicate the interpretation of spectral data, thereby affecting the overall reliability of the results obtained. This issue is especially important in the pharmaceutical industry, where it is very important to accurately describe the physical-mechanical quality attributes of tablet formulations (Chougule *et al.,* 2017). As a result, Raman spectroscopy needs to keep getting better in terms of methods and instruments so that it can be used in more routine analyses. The high cost of equipment and maintenance is also a major reason why Raman spectroscopy is not used more widely in many fields. High-quality Raman systems can be expensive to buy at first, and because they are so complicated to use, you will need ongoing training and support. Also, even though Raman instruments have come a long way, adding them to current workflows often requires more money and planning. For example, when looking at manufacturing methods like particulate injection molding, it may be clear that making custom inserts with complicated shapes is difficult. These financial and operational problems can stop companies from fully using Raman spectroscopy, even though it has the potential to give them accurate analytical information. To make Raman spectroscopy more widely used in both academic and industrial settings, it will be important to deal with these barriers.

8.6.3. Comparison with Other Spectroscopic Techniques

People often compare Raman spectroscopy to other types of spectroscopies, like infrared (IR) spectroscopy and X-ray fluorescence (XRF), especially when it comes to how well they work together. IR spectroscopy is great at finding molecular vibrations by looking at changes in dipole moments. Raman spectroscopy, on the other hand, looks at changes in polarizability to give us information about non-polar and symmetric molecules that IR might miss. Using both of these methods together gives a more complete picture of materials because IR spectroscopy can look at functional groups and Raman can look at molecular structure and bonding. This synergy can be especially useful in fields like art conservation, where it's important to know how materials are made up and how strong they are. Raman and IR methods work better together, which makes analysis deeper and helps conservation decisions (Lennard *et al.,* 2017). Along with IR and XRF, improvements in multimodal imaging methods have created a new standard for characterizing materials, especially when it comes to studying painted textiles and works of art. Researchers can get a full picture of the pigments and materials used in artworks by using a mix of methods, such as micro-Raman spectroscopy, VIS-NIR reflectance spectroscopy, and optical coherence tomography (OCT). Micro-Raman spectroscopy is very good at finding pigments in complicated mixtures. XRF, on the other hand, gives information about the elemental composition, which helps conservators figure out the historical context and production methods of artworks. These methods are even more useful because they don't hurt the artworks and still get important information. This approach shows how important multimodal techniques are in art conservation settings by showing how they all work together to help us understand material properties better (Bailey *et al.,* 2015). The comparative analysis of spectroscopic techniques encompasses their limitations and applicability across diverse contexts. For example, Raman spectroscopy is useful because it has a high spatial resolution and doesn't require much sample preparation. However, it might not work as well for samples that are very fluorescent because the fluorescence can hide the Raman signals. To get around these problems, you can use techniques like photoluminescence spectroscopy or surface-enhanced Raman spectroscopy. In the end, the success of any spectroscopic method depends on the material and application being studied. As new methods keep coming out, the way that spectroscopic techniques work together will affect the direction of future research. This is especially true in fields that cross disciplines, like heritage science, where the goal is to understand and protect cultural artifacts using advanced analytical methods (Lennard *et al.*, 2017) (Figure 8.13).

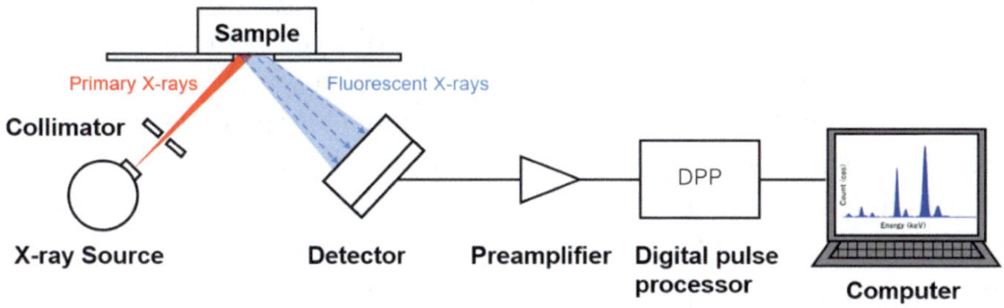

Figure 8.13. *X-ray fluorescence (XRF).*

Source: https://static.horiba.com/fileadmin/Horiba/_processed_/a/6/csm_Schematic_diagram_of_a_basic_benchtop_EDXRF_analyzer_0fc48a8be6.png.

SUMMARY

When you look into Raman Spectroscopy, it's clear that this powerful analytical tool has changed many areas, such as chemistry, biology, and materials science. Raman Spectroscopy has many advantages over traditional methods because it gives detailed information about molecular vibrations and structural properties without needing a lot of sample preparation. Because it doesn't damage samples, it lets you look at them in their natural state, which is especially helpful for complicated biological systems and delicate materials. This built-in ability not only speeds up research but also makes the data that is collected more reliable. So, the ongoing development of Raman Spectroscopy is essential for pushing the limits of scientific research in many fields. Additionally, combining Raman Spectroscopy with other analytical methods, like imaging techniques and machine learning algorithms, helps us understand complex phenomena better. Researchers are always looking into the complicated links between molecular structure and function. Using these methods together could lead to big breakthroughs. For example, combining Raman data with spectroscopic imaging has made it easier to look at samples that are not uniform, giving us more information about how molecular species are spread out in space.

These improvements mark the start of a new era in research and diagnostics, showing how important it is to use different fields to solve scientific problems. As the field advances, the ramifications of Raman Spectroscopy transcend academic research, impacting industrial applications, clinical diagnostics, and environmental monitoring. The ability to analyze data in real time in many different situations shows how useful and flexible the technique is. Nonetheless, it is imperative to remain aware of the difficulties inherent in interpreting Raman spectra and to persist in enhancing methodologies for data analysis. The conversations about Raman Spectroscopy, like those in recent research on digital humanities and authenticity evaluation, stress how important it is to be skeptical and do thorough scientific research. Consequently, as scientific progress persists, prioritizing methodological integrity and innovation will be essential to fully harnessing the potential of Raman Spectroscopy across various applications (Alastair *et al.,* 2018; McGrath *et al.,* 2019).

Raman spectroscopy is an important analytical method that gives scientists a lot of information about molecular vibrations. It is very useful in many areas of science. The method uses inelastic scattering of monochromatic light, usually from lasers, to create a molecular signature based on the molecules' vibrational states in a sample. One of the best things about Raman spectroscopy is that it doesn't destroy samples, which means you can analyze them without changing or damaging them. This method can also give researchers detailed information about the structure and dynamics of molecules, which lets them figure out not only what chemicals are present but also how much of each one is present and how they interact with each other. This ability shows how important it is in fields like chemistry, biology, and material science, where knowing how molecules behave is important for making progress and using these fields. Raman spectroscopy is very important for figuring out which biomolecules are responsible for the bioactivity of natural products like honey, which have unique compositions. For example, studies have shown that Fourier Transform Raman spectroscopy (FT-Raman) works well for looking at nectar from Mānuka (*Leptospermum scoparium*), which is known for having high levels of dihydroxyacetone (DHA) that are linked to its health benefits. The capacity to differentiate among various cultivars based on DHA content underscores the methodology's relevance in agronomy and food science, supplying essential data that guide breeding initiatives designed to improve the medicinal properties of honey. So, Raman spectroscopy is not only a way to analyze things, but it also connects molecular analysis to real-world uses in the honey industry for product development and quality control (Nickless *et al.,* 2015). Additionally, the versatility of Raman spectroscopy is underscored by its comparative effectiveness with alternative analytical methods. In materials science, the investigation of hydrogenated diamond-like carbon films demonstrates the capabilities of Raman spectroscopy in assessing essential film characteristics, such as *sp*^{3} content and hardness. Raman measurements often give us information that is useful in addition to what we get from X-ray photoelectron spectroscopy and ellipsometry. This shows how this method can make accurate predictions about the properties of materials. The ability to link spectroscopic data to physical properties makes Raman methods even more important for research and industry, since they help people understand materials better and keep quality control in mind when making them. Raman spectroscopy is an essential tool for both scientific research and industrial applications because it can be used in so many different ways. It can help us understand complex molecular interactions and material properties (Singha *et al.,* 2006).

REFERENCES

1. Abdolrahim, M., Alhosseini, S. N., Rabiee, M., Tahriri, M., et al. (2015). Development of optical biosensor technologies for cardiac troponin recognition. https://core.ac.uk/download/213059311.pdf (accessed on 29th October 2025).
2. Adya, A. K., Akatsuka, H., Matsuura, H., Okamoto, Y., et al. (2009). XAFS analyses of molten metal fluorides. https://core.ac.uk/download/228176816.pdf (accessed on 29th October 2025).
3. Baianu, I. C. (2011). Food Physical Chemistry and Biophysical Chemistry. https://core.ac.uk/download/pdf/290434.pdf (accessed on 29th October 2025).
4. Bailey, K., Bellesia, S., Brooks, C., Burgio, L., et al. (2015). A holistic multimodal approach to the non-invasive analysis of watercolor paintings. https://core.ac.uk/download/30637216.pdf (accessed on 29th October 2025).
5. Beattie, R. J., Downey, G., Fearon, A. M., O'Donnell, C., et al. (2015). Prediction of naturally-occurring, industrially-induced and total trans fatty acids in butter, dairy spreads and Cheddar cheese using vibrational spectroscopy and multivariate data analysis. https://core.ac.uk/download/45656662.pdf (accessed on 29th October 2025).
6. Bengtsson, P. E., Marrocco, M., & Nordström, E. (2014). Sensitivity of coherent anti-Stokes Raman lineshape to time asymmetry of laser pulses. http://arxiv.org/abs/1407.1719 (accessed on 29th October 2025).
7. Bernáth, B., Dzsaber, S., Fehér, T., Gyüre, B., et al. (2014). A Fourier transform Raman spectrometer with visible laser excitation. http://arxiv.org/abs/1406.4027 (accessed on 29th October 2025).
8. Bianco, A., Bruno, M., Ciccola, A., Gagliano Candela, R., et al. (2019). Deepening inside the pictorial layers of Etruscan sarcophagus of Hasti Afunei: An innovative micro-sampling technique for Raman/SERS analyses. https://core.ac.uk/download/231748462.pdf (accessed on 29th October 2025).
9. Biscans, B., & Teychené, S. (2011). A multiscale approach for the characterization and crystallization of Eflucimibe polymorphs: From molecules to particles. https://core.ac.uk/download/12043420.pdf (accessed on 29th October 2025).
10. Brennan, F., Graham, G., Heaney, P., Hörz, F., Kearsley, A., Kilcoyne, A. L. D., et al. (2003). Aerogel keystones: Extraction of complete hypervelocity impact events from aerogel collectors. http://arxiv.org/abs/astro-ph/0312460 (accessed on 29th October 2025).

11. Bryndzia, L. T., Ferralis, N., Grossman, J. C., Liu, Z., et al. (2016). Genome-inspired molecular identification in organic matter via Raman spectroscopy. http://arxiv.org/abs/1602.02083 (accessed on 29th October 2025).
12. Camposeo, A., Gucciardi, P. G., Magrì, D., Maragò, A. M., et al. (2015). Surface-enhanced Raman spectroscopy in 3D electro spun nanofiber mats coated with gold nanorods. http://arxiv.org/abs/1507.07467 (accessed on 29th October 2025).
13. Cantatore, S., De Carlo, D., Fineschi, V., Fiore, C., et al. (2015). Confocal laser scanning microscope, Raman microscopy and Western blotting to evaluate inflammatory response after myocardial infarction. https://core.ac.uk/download/54248088.pdf (accessed on 29th October 2025).
14. Caro Salazar, C., Franco, J., García Martín, M. L., Oliva Montero, M., et al. (2019). Synthesis and characterization of elongated-shaped silver nanoparticles as a biocompatible anisotropic SERS probe for intracellular imaging: Theoretical modeling and experimental verification. https://core.ac.uk/download/190375878.pdf (accessed on 29th October 2025).
15. Charlton, M. D. B., Chen, R., Kalsi, S., Oo, Z., et al. (2014). Integrated waveguide and nanostructured sensor platform for surface-enhanced Raman spectroscopy.
16. Chougule, M. B., Dave, V. S., Haware, R. V., Shahin, H., et al. (2017). Emerging technologies for the non-invasive characterization of physical-mechanical properties of tablets. https://core.ac.uk/download/129153450.pdf (accessed on 29th October 2025).
17. Darvin, M. E., Gersonde, I., Moralejo, B., Roth, N., et al. (2018). Non scanning large-area Raman imaging for *ex vivo/in vivo* skin cancer discrimination. http://arxiv.org/abs/1810.03363 (accessed on 29th October 2025).
18. Daturi, M., Frenkel, A. I., Gibson, E. K., Keturakis, C., et al. (2016). Dynamics of CrO_3–Fe_2O_3 catalysts during the high-temperature water-gas shift reaction: Molecular structures and reactivity. https://core.ac.uk/download/96882578.pdf (accessed on 29th October 2025).
19. De Beer, T., Remon, J. P., Saerens, L., Vervaet, C., et al. (2013). Process monitoring and visualization solutions for hot-melt extrusion: A review. https://core.ac.uk/download/55761837.pdf (accessed on 29th October 2025).
20. Denschlag, J. H., Grimm, R., Thalhammer, G., Theis, L. S., et al. (2004). Inducing an optical Feshbach resonance via stimulated Raman coupling. http://arxiv.org/abs/cond-mat/0409552 (accessed on 29th October 2025).

21. Draxl, C., Giantomassi, M., Gillet, Y., Gonze, X., et al. (2017). Ab initio approach to second-order resonant Raman scattering including exciton-phonon interaction. http://arxiv.org/abs/1708.05874 (accessed on 29th October 2025).

22. Ewart, P. (2019). A comparison of processing techniques for producing prototype injection molding inserts. https://core.ac.uk/download/322841895.pdf (accessed on 29th October 2025).

23. Fletcher, D. G., & McKenzie, R. L. (2025). Laser-spectroscopic measurement techniques for hypersonic, turbulent wind tunnel flows. https://core.ac.uk/download/pdf/42813984.pdf (accessed on 29th October 2025).

24. Gardiner, A. T., Britton, G., Britton, G., Britton, G., Uragami, C., Frank, H. A., Hashimoto, H., et al. (2018). Understanding/unravelling carotenoid excited singlet states. https://core.ac.uk/download/157582964.pdf (accessed on 29th October 2025).

25. Ghatak, A., & Pathak, A. (2017). Classical light vs. nonclassical light: Characterizations and interesting applications. http://arxiv.org/abs/1705.00650 (accessed on 29th October 2025).

26. Gibson, G., Liao, Z., Notingher, I., Padgett, M. J., et al. (2016). DMD-based software-configurable spatially-offset Raman spectroscopy for spectral depth-profiling of optically turbid samples. https://core.ac.uk/download/42372091.pdf (accessed on 29th October 2025).

27. Glazov, M. M., & Zapasskii, V. S. (2015). Linear optics, Raman scattering, and spin noise spectroscopy. http://arxiv.org/abs/1502.01155 (accessed on 29th October 2025).

28. González Vidal, J. J., Pérez Pueyo, R., & Soneira Ferrando, M. J. (2016). Automatic classification system of Raman spectra applied to pigments analysis. https://core.ac.uk/download/81581746.pdf (accessed on 29th October 2025).

29. Heo, J., Jain, P. K., Kumari, G., Zhang, W., et al. (2018). *In situ* formation of catalytically active graphene in ethylene photo-epoxidation. https://core.ac.uk/download/323308648.pdf (accessed on 29th October 2025).

30. Hinton, J. K. (2019). High-pressure high-temperature exploration of phase boundaries using Raman spectroscopy. https://digitalscholarship.unlv.edu/cgi/viewcontent.cgi?article=4616&context=thesesdissertations (accessed on 29th October 2025).

31. Laubereau, A., Mantel, K. H., & Zinth, W. (2007). Proceedings of the Thirteenth International Conference on Time-Resolved Vibrational Spectroscopy. https://core.ac.uk/download/12162307.pdf (accessed on 29th October 2025).

32. Lennard, F., Smith, M., & Thompson, K. (2017). A literature review of analytical techniques for materials characterization of painted textiles–Part 2: Spectroscopic and chromatographic analytical instrumentation. https://core.ac.uk/download/84591645.pdf (accessed on 29th October 2025).
33. Liu, G., Mulchandani, A., Pham, T., Ramnani, P., et al. (2019). Gas biosensor arrays based on single-stranded DNA-functionalized single-walled carbon nanotubes for the detection of volatile organic compound biomarkers released by Huanglongbing-disease infected citrus trees. https://core.ac.uk/download/323067683.pdf (accessed on 29th October 2025).
34. McCrow, A., & McCrow, A. (2011). Laser-based mid-infrared spectroscopic imaging–Exploring a novel method for application in cancer diagnosis. https://core.ac.uk/download/1592070.pdf (accessed on 29th October 2025).
35. McGrath, J. F. (2019). Learning from Jesus' Wife: What does forgery have to do with the digital humanities? https://core.ac.uk/download/212825834.pdf (accessed on 29th October 2025).
36. ANONYMOUS. (2013). The Boston University Photonics Center annual report 2013–2014. https://open.bu.edu/bitstream/2144/29752/1/Photonics-Center-Final-Annual-Report-2013-2014-v2.pdf (accessed on 29th October 2025).
37. Nickless, E. M. (2015). Influential factors in nectar composition and yield in Leptospermum scoparium: A thesis presented in partial fulfilment of the requirements for the degree of Doctor of Philosophy in Plant Science, Institute of Agriculture and the Environment, College of Sciences, Massey University, Palmerston North, New Zealand. https://core.ac.uk/download/159449396.pdf (accessed on 29th October 2025).
38. Notingher, I., Shipp, D., & Sinjab, F. (2017). Raman spectroscopy: Techniques and applications in the life sciences. https://core.ac.uk/download/80838463.pdf (accessed on 29th October 2025).
39. Pikov, V., & Siegel, P. H. (2010). Thermal monitoring: Raman spectrometer system for remote measurement of cellular temperature on a microscopic scale. https://core.ac.uk/download/4884179.pdf (accessed on 29th October 2025).
40. Pilát, Z. (2015). Optical micromanipulation techniques combined with micro spectroscopic methods. https://core.ac.uk/download/30311630.pdf (accessed on 29th October 2025).

41. Singha, A., Ghosh, A., Roy, A., Ray, N. R., & Wazumi, K. (2006). Quantitative analysis of hydrogenated DLC films by visible Raman spectroscopy. http://arxiv.org/abs/cond-mat/0601030 (accessed on 29th October 2025).

42. Singha, A., Roy, A., Grethlein, C. E., Lamb, H., Dhar, P., & Rytov, S. M. (2004). A non-destructive analytic tool for nanostructured materials: Raman and photoluminescence spectroscopy. http://arxiv.org/abs/cond-mat/0406096 (accessed on 29th October 2025).

43. Weiner, A. M., Zheltikov, A. M., von Vacano, B., Ooi, C. H. R., Meshulach, D., Oron, D., Oron, D., et al. (2007). Noise auto-correlation spectroscopy with coherent Raman scattering. http://arxiv.org/abs/0710.3173 (accessed on 29th October 2025).

44. Xia, M. (2017). A review on applications of two-dimensional materials in surface enhanced Raman spectroscopy. https://core.ac.uk/download/190749177.pdf (accessed on 29th October 2025).

CHAPTER 9

MOLECULAR LUMINESCENCE SPECTROSCOPY

Contents

INTRODUCTION

Molecular luminescence spectroscopy is a flexible and powerful way to analyze things that gives you information about molecular structure, dynamics, and interactions. This technique takes advantage of the light that comes from a substance that has absorbed photons, which lets scientists learn important things about how different molecules change their electronic states. This method is important for more than just basic research; it is also very useful in a wide range of fields, including environmental monitoring, biomedical diagnostics, and materials science. Scientists can find small amounts of things by using luminescence, which is very useful when other methods don't work. As the field progresses, the creation of new luminescent materials and methods is expanding the range of molecular luminescence spectroscopy, emphasizing its essential role in modern scientific research and application. Recent technological improvements in molecular luminescence spectroscopy have made it much more sensitive and specific, opening up new possibilities for its use. One notable example is the use of lanthanide-based compounds, which are known for having sharp emission lines and long luminescence lifetimes. Recent research has demonstrated techniques to enhance the stability and efficacy of these materials, including the synthesis of doped UiO-66, a zirconium-based metal-organic framework that maintains exceptional luminescent characteristics under diverse conditions.

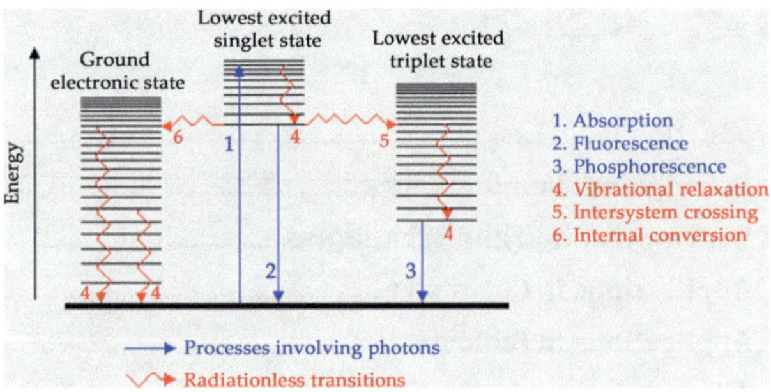

Figure 9.1. *UV-Vis luminescence spectroscopy.*

Source: https://teaching.shu.ac.uk/hwb/chemistry/tutorials/molspec/photlum1.gif.

When these new materials are combined with older spectroscopic methods, it makes analyses more reliable and helps us understand how luminescent behavior works in complicated settings. As a result, luminescent compounds create rich energy landscapes that can be understood both through empirical measures and mathematical models. This gives us a complete picture of how molecules behave that is useful in many scientific fields (Vos *et al.,* 2018). As molecular luminescence spectroscopy progresses, the convergence of chemistry, physics, and engineering becomes more evident. For example, using nuclear physics to study cultural heritage shows how luminescence can be used in many different fields, connecting science and the humanities. The transfer of knowledge from nuclear physics to fields like archaeology and history shows how useful the techniques are for finding historical artifacts and improving preservation efforts. Projects throughout Europe demonstrate a dedication to utilizing molecular luminescence spectroscopy for the preservation of cultural heritage through innovative research and collaboration among diverse scientific communities (ANONYMOUS, 2016). This interdisciplinary approach not only enhances our comprehension of molecular luminescence but also demonstrates its potential influence across various domains, enriching both scientific exploration and cultural appreciation (Figure 9.1).

Definition of Molecular Luminescence Spectroscopy

Molecular luminescence spectroscopy is a method that looks at the light that molecules give off when they move from an excited state to a ground state. At its core, this method helps us understand how different molecules are structured and how they behave, especially when it comes to their electronic configurations and surroundings. The basis of luminescence spectroscopy is the way photons interact with molecules, which causes them to absorb light and then emit it as fluorescence or phosphorescence. This special way of giving off light means that the energy state has changed and gives us a lot of information about molecular properties, such as energy levels, concentration, and how molecules interact with the media around them. As a result, it is a very useful tool for researchers in many fields, from chemistry to biology, because it helps them understand how molecules move and makes it easier to study complicated systems. Molecular luminescence spectroscopy is important for more than just basic molecular characterization; it is also a quantitative analytical method that can be used in many different ways. The method is especially useful in biochemistry and environmental science for finding certain compounds or keeping an eye on biochemical processes. For instance, fluorescence microscopy uses this method to see parts of cells, giving researchers important spatial information for biological studies. Moreover, improvements in instruments, like those discussed by Ebert et al. (2005), show how important it is to use fluorescence standards to ensure that measurements are accurate. These standards help reduce instrumental errors and make fluorescence data more reliable. This shows how important it is to be very careful about quality control in experimental work. Furthermore, recent research has investigated the theoretical foundations of luminescence spectroscopy utilizing computational chemistry methodologies. Methods like density functional theory (DFT) and time-dependent DFT (TD-DFT) are being used more and more to predict how different chemical systems will glow. The study by Amolo et al. (2017) demonstrates how these computational models successfully link theoretical predictions with experimental outcomes, providing further insight into the connection between molecular structure and luminescent properties. These improvements not only help us better understand how luminescence works, but they also make it easier to make new luminescent materials that can be used in sensor technology, light-emitting devices, and other things. Consequently, the amalgamation of experimental and theoretical insights advances the domain of molecular luminescence spectroscopy into novel areas of scientific inquiry and application (Figure 9.2).

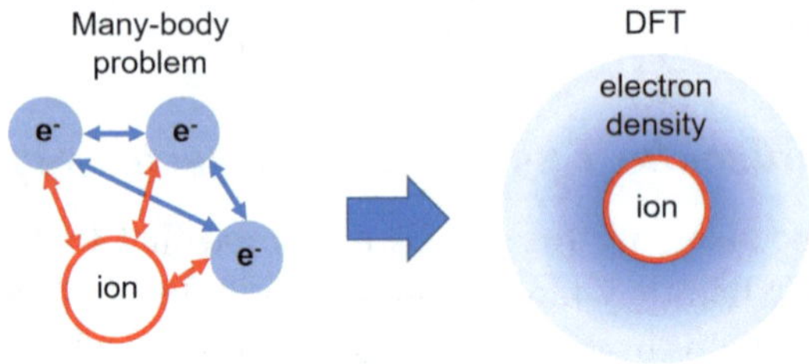

Figure 9.2. *Density functional theory (DFT).*

Source: https://figures.semanticscholar.org/3262f9f1418d1bcfd6be316fef39103b8b5fbc53/22-Figure3.1-1.png.

9.1. PRINCIPLES OF LUMINESCENCE

Understanding molecular luminescence spectroscopy requires knowing the basics of luminescence, which helps us understand how excited electronic states behave in molecules. When an electron moves from a higher energy state to a lower one, it releases energy in the form of photons, which is what luminescence is. There are two main types of this process: fluorescence and phosphorescence. The main difference between them is how long they last and how they work. Fluorescence happens almost right away, but phosphorescence takes longer because it involves triplet states, which means the emissions last longer. It is important to understand these interactions because they have a big effect on how well luminescent materials work and how efficient they are, especially in displays, sensors, and biological imaging. Environmental factors and molecular interactions are very important to luminescence because they can change the emission properties a lot. The examination of *solvatochromism*, which refers to the alteration in absorption and emission spectra in relation to solvent polarity, provides significant insight into how solvent interactions can influence luminescent characteristics. For example, some dyes that are mixed into liquid crystals show significant bathochromic shifts in their absorption maxima when compared to isotropic solvents. This has been shown in studies with liquid crystal hosts. These kinds of things make it seem like the energy levels of the absorbing species change because of the unique orientation and interaction dynamics in an anisotropic medium, which affects luminescence.

These results show how important the environment is for luminescence behavior and are key to making new luminescent materials for specific uses, as shown by research on dye molecules in liquid crystals (Bloor *et al.,* 2003). The capacity to alter luminescent characteristics via structural modifications, including surface passivation and nanoscale engineering, has surfaced as a promising pathway for the enhancement of luminescent materials. Studies show that silicon clusters, for example, can fluoresce under certain conditions, especially when they are passivated or put in the right host. The precise mechanism underlying this fluorescence is frequently ascribed to quantum confinement effects and localized transitions at the nanoparticle surface, which are essential for facilitating efficient light emission. The fabrication of luminescent silicon structures at the nanoscale has significant ramifications for electronics and biomedicine, providing a means for the advancement of sophisticated light-emitting devices. Grasping the fundamental principles of luminescence enables the investigation of novel materials and techniques in molecular luminescence spectroscopy, thereby promoting advancements in technology and applications (Dasog *et al.,* 2017) (Figure 9.3).

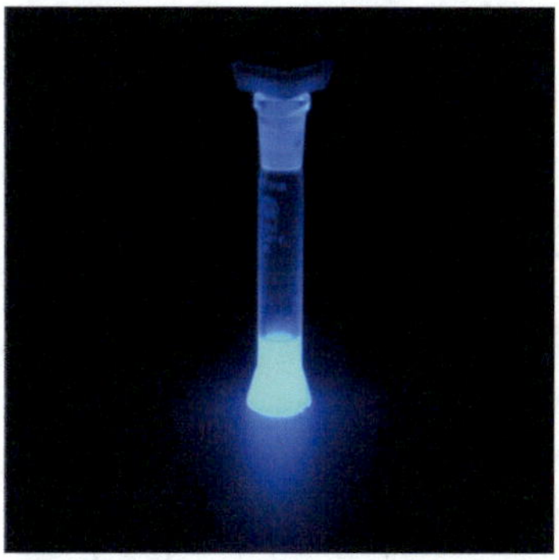

Figure 9.3. *Basic principles of luminescence.*

Source: https://www.leica-microsystems.com/fileadmin/_migrated/media/Luminol2006_06.jpg.

9.1.1. Mechanisms of Luminescence: Fluorescence and Phosphorescence

Fluorescence and phosphorescence are two important types of luminescence that are important in molecular luminescence spectroscopy. When a molecule absorbs photons and then re-emits light within a short period of time, usually in the nanosecond range, it fluoresces. The Stokes shift makes this quick emission happen because the excited singlet state goes back to the ground state, which gives off light at a longer wavelength. Fluorescence works better in places that allow electronic and vibronic transitions, which makes the light more pronounced. As previously mentioned, individual C_{60} molecules exhibit unique fluorescence properties when analyzed through methods like scanning tunneling microscopy, providing essential understanding of the electronic states associated with these transitions (Blüm *et al.,* 2005). Phosphorescence, on the other hand, has a more complicated transition mechanism that can last for much longer periods of time, from milliseconds to hours. This longer life is because the system goes from a singlet to a triplet state, which is spin-forbidden. This makes it less likely to happen and leads to excited states that last longer. In this context, the crystal structure can have a big effect on vibrational coupling and the dynamics of the transition.

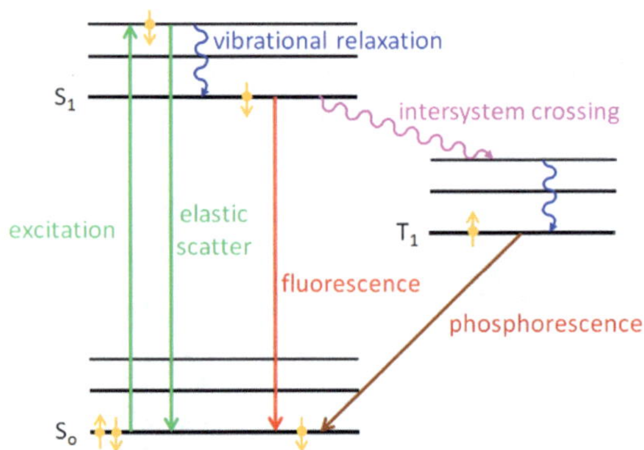

Figure 9.4. *Theory of fluorescence and phosphorescence.*

Source: https://www.oceanopticsbook.info/packages/iws_l2h/conversion/files/Jablonski-fluor-phos-600.png.

This has been shown in studies of crystalline materials like bibenzyl. The geometry of this type of crystal makes exciton coupling more efficient, which improves the information we get from phosphorescence spectra and tells us about the lifetime and decay dynamics of the excited states that form (Ake *et al.,* 2025). Understanding how triplet states work is very important for many fields, including materials science and bioimaging, where the light's temporal resolution can help us understand how molecules interact with each other. To make molecular luminescence spectroscopy more useful in different fields, we need to understand how these mechanisms work. Researchers can study molecular environments, keep an eye on dynamic processes, and make sensors and imaging technologies by using both fluorescence and phosphorescence. Each mechanism provides different information about the molecular dynamics and environmental conditions occurring at the molecular level, such as how temperature affects behavior and how competing radiationless processes operate. This can lead to a wide range of uses. For example, using luminescence as a probing method in biochemical systems can give you real-time information about how molecules are shaped and how they react. So, understanding fluorescence and phosphorescence not only helps us understand the theory better, but it also opens up new possibilities for scientific research and technology development (Figure 9.4).

9.1.2. Energy Transitions and Electronic States

Understanding molecular luminescence spectroscopy requires a basic understanding of energy transitions and electronic states. In this case, energy transitions are the movement of electrons between different energy levels in a molecule, which usually happens when photons are absorbed. During these transitions, electronic states can be divided into ground and excited states. The excited states are arrangements of electrons that have more energy. These transitions are very important because they determine the spectral properties that we see during luminescence events like fluorescence and phosphorescence. Researchers can learn a lot about molecular structures and dynamics by looking at the specific wavelengths that come out during these processes. The connection between energy levels and electronic states is what makes molecular luminescence spectroscopy work. It shows how energy is taken in and given off at the molecular level. A significant application of comprehending electronic states in molecular luminescence is demonstrated through sophisticated methodologies like Resonant X-ray Emission Spectroscopy (RXES). This method makes it possible to characterize ligand field states in complex molecules, such as the square-planar complex [Pt(tpy)Cl]Cl, where *tpy* stands for 2,2′:6′,2″-terpyridine. RXES not only shows how weak metal-metal interactions change the energy levels of electronic states, but it also gives us a lot of information about how these changes show up in luminescence phenomena. In fields like material design, catalysis, and photonics, where knowing exactly what electronic states are can lead to new ideas, it's important to be able to tell these small differences apart. By linking electronic characterization and luminescence, RXES is a good example of how theoretical ideas and real-world uses come together in the study of energy transitions. Another important advancement is the use of Scanning Tunneling Microscopy (STM) to study single molecules. This makes it possible to excite luminescence from individual C_{60} molecules that are part of fullerene nanocrystals. When electrons tunnel through a double barrier STM junction, they excite different electronic states, which can be seen as luminescence. The resulting spectra display both fluorescence and phosphorescence, indicating intricate interactions between electronic transitions and vibrational modes of the C_{60} molecule. This technique demonstrates the ability for precise chemical identification at the single-molecule scale and elucidates the relationship between energy transitions and the resulting luminescent characteristics. By combining STM with molecular luminescence spectroscopy, scientists learn a lot about how energy states affect luminescence. This opens up new possibilities for nanotechnology and molecular engineering.

9.1.3. Role of Molecular Structure in Luminescence

To understand how luminescent materials emit light, you need to know about their molecular structure. The electronic transitions that happen within their structural framework are very important to this process. Different molecular configurations can lead to different electronic states, energy gaps, and radiative properties, which affect how these materials interact with light. In luminescence spectroscopy, the presence of conjugated pi systems, functional groups, and molecular symmetry are important factors that affect the wavelengths and intensities of emissions. For example, changes in these structural parameters can cause changes in fluorescence properties. This is seen in some organic dyes that show *solvatochromism*, which is a reaction to changes in solvent polarity and the environment. This shows how molecular architecture and luminescent behavior are always changing, which shows how important it is to accurately describe molecular structures in molecular luminescence spectroscopy. When it comes to solid-state materials like hexagonal boron nitride (h-BN), the molecular structure plays an even bigger role. Defects like nitrogen vacancies (VN) have a big effect on the electronic properties of h-BN, and they can also have a big effect on the luminescent response. Research demonstrating that UV irradiation of h-BN produces variable fluorescence contingent upon environmental conditions exemplifies the impact of molecular-level alterations, such as vacancy formation, on luminescence. For instance, different kinds and amounts of nitrogen vacancies make the fluorescence bands we see. Some emissions are stopped by things in the environment, like molecular oxygen. This interaction between molecular defects and luminescent properties shows how controlling and understanding molecular structure can create custom luminescent materials, which could be used in more advanced technology and materials science (Andrei *et al.,* 2008). In addition, the way molecules are arranged and interact with each other in complex systems, like liquid crystals that hold dye molecules, has a big effect on luminescence properties. The arrangement of these dye molecules in an anisotropic medium shows how steric and dipolar interactions change electronic transitions and absorption properties. For example, the strong bathochromic shifts seen in some dye molecules when they are put into liquid crystal environments suggest that the host's rigidity and orientation change the local electronic field that the dyes feel. These transitions become less noticeable in isotropic solvents, which shows how molecular structure affects luminescent behavior. The alignment and interaction dynamics within these complex structures are essential for enhancing luminescent materials, as comprehending these relationships promotes the creation of more efficient and responsive luminescent systems for diverse applications in sensor and display technologies (Bloor *et al.,* 2003).

9.2. INSTRUMENTATION AND TECHNIQUES

Improvements in the tools and methods used for molecular luminescence spectroscopy have made it much easier to study chemical and biological systems with great accuracy and sensitivity. Modern spectroscopic instruments use advanced optical parts, like lasers and photomultiplier tubes, to find weak luminescent signals. This new idea not only lowers detection limits, but it also makes luminescence spectroscopy useful in a wider range of fields, such as environmental monitoring, forensic science, and medical diagnostics. The integration of automated systems also makes it possible to quickly collect data, which speeds up the analysis process. These improvements make sure that researchers can get accurate quantitative measurements, which are necessary for learning more about how molecules interact and how kinetic processes work in luminescence phenomena. The use of optical waveguides in molecular luminescence spectroscopy is also a big step forward in the world of instrumentation. These waveguides, which are made of silicon nitride, use the evanescent field to greatly improve the way light and matter interact. The evanescent tail of guided modes can easily excite Raman-active molecules, which makes spectroscopic applications more sensitive and specific. Recent research has shown that the efficiency of spontaneous Raman scattering in waveguides is about two orders of magnitude better than that of traditional confocal setups, especially when used in normal experimental conditions. This huge boost in efficiency makes it easier to find analytes that are hard to find and encourages the use of lab-on-a-chip designs for quick and easy analysis. It also combines molecular luminescence techniques with cutting-edge microfabrication technology (Baets *et al.,* 2014). Along with these improvements in instruments, the strict use of methods like time-resolved luminescence spectroscopy has given us a lot of new information about how molecules move. This method lets scientists keep track of how luminescent signals fade over time, which helps them understand complex molecular interactions and energy transfer processes. Scientists can greatly improve the accuracy of their measurements by using time-gated detection methods, which cut down on background noise from non-luminescent species. Time-resolved luminescence has a wide range of uses in research, such as studying how DNA interacts with other molecules and looking into how light behaves in complicated environments. The continuous advancement of instrumentation and methodologies in molecular luminescence spectroscopy highlights its essential function in promoting both fundamental research and practical applications across diverse scientific fields (Figure 9.5).

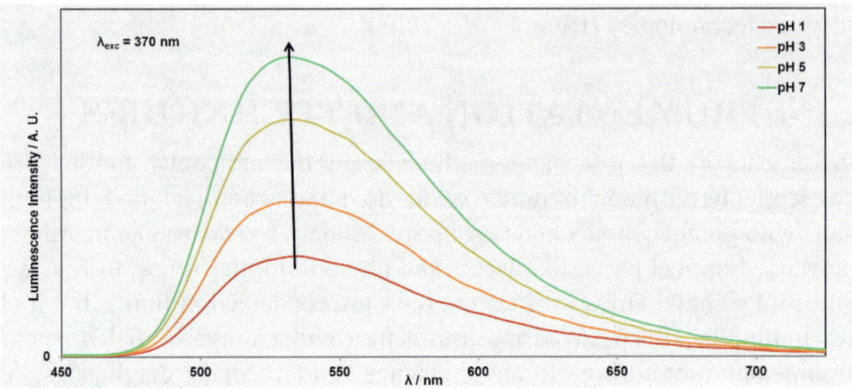

Figure 9.5. *Luminescence spectroscopy.*

Source: https://chembam.com/wp-content/uploads/2017/01/picture1.png.

9.2.1. Components of Luminescence Spectrometers

Luminescence spectrometers are complicated tools made up of many important parts, each of which is very important to the analytical process. The excitation source, which is usually a fluorescence or phosphorescence lamp, is the most important part of a luminescence spectrometer. It sends light onto the sample. Choosing the right excitation source is very important because it needs to give off the right wavelength and intensity to get the luminescent species in the sample to react. The optical accessory that comes with the excitation source helps focus the light on the sample, making sure that the interaction is effective and the emission is as high as possible. Different optical filters are also used to improve the signal by blocking out certain wavelengths. This reduces background noise and makes the measurements needed for a wide range of fields, such as materials science and nanotechnology, more accurate. Another important part of luminescence spectrometers is the sample chamber. This chamber is made to hold the sample securely while keeping the environmental conditions stable. There are different configurations depending on whether the sample is a solid, liquid, or gas. It is also important to manage the temperature well in the sample chamber because the luminescent properties can change with temperature. The way the chamber is set up in space can also change the paths of excitation and emission. As a result, careful design of the sample chamber's layout makes it easier for both excitation and emission to have the best light path. In this context, meticulous evaluation of these components enhances the overall effectiveness of luminescence spectroscopy across diverse applications, ranging from fundamental research in nanomaterials to the progression of optoelectronics, as highlighted in recent technical discourse (Singha *et al.,* 2004). The last step in luminescence spectroscopy is detection. This is done by a very advanced detector, usually a photomultiplier tube (PMT) or a charge-coupled device (CCD). These detectors can pick up very small amounts of light and turn the luminescent signal into an electrical signal for further processing. The quality of the optical parts and the overall design of the spectrometer system have a big impact on how well they work. The alignment and calibration of the detection system are especially important for getting accurate measurements. Also, improvements in technologies like optical design and data processing electronics are important for making these detectors work better. As we learn more about luminescence and how it works, adding these advanced detection technologies could lead to new uses in fields like space science and nanotechnology, as shown in recent studies on sensor systems (Clark *et al.,* 2025).

9.2.2. Types of Luminescence Spectroscopy: Steady-State versus Time-Resolved

Luminescence spectroscopy is a broad discipline, chiefly divided into steady-state and time-resolved techniques. Steady-state luminescence spectroscopy involves the ongoing measurement of light emitted from a sample during constant excitation. This method can be used to find out things like absorption, emission spectra, and fluorescence lifetimes. Steady-state spectroscopy is a very important analytical tool that is used in many different fields, such as biochemical tests and environmental monitoring. But it can't give us dynamic information about how excited states change over time, which is important for understanding complex systems and kinetic processes. This is a major flaw. This limitation highlights the necessity for supplementary methodologies, such as time-resolved spectroscopy, which provides enhanced understanding of the dynamics of molecular interactions and photochemical reactions. Time-resolved luminescence spectroscopy, on the other hand, looks at when light is emitted. This gives information about how long excited states last and how fast molecular processes happen. This method uses pulsed excitation sources to capture short-lived emission events, which lets researchers see how luminescent species decay in response to excitation. Time-resolved techniques can show important information about molecular interactions, energy transfer, and changes in shape. The complexities of these interactions frequently require advanced instrumentation and analytical techniques, thereby enhancing the significance of this approach in disciplines such as photophysics and photochemistry. Specifically, studies investigating the photochemistry of ReI-polypyridyl complexes have demonstrated how time-resolved spectroscopy can unveil significant insights into DNA damage mechanisms, highlighting the technique's vital role in molecular luminescence research (Ahmad *et al.*, 2018). Even though steady-state and time-resolved luminescence spectroscopy use different methods, they are often used together to get a full picture of molecular systems. Each method gives different information that, when looked at together, can give us very useful information about the physical and chemical properties of luminescent compounds. For example, steady-state measurements can help you choose the samples and conditions for later time-resolved studies, and time-resolved data can help you understand how steady-state phenomena work. Moreover, improvements in instruments and analysis methods have made both types of luminescence spectroscopy more accurate and dependable, making it easier to compare fluorescence data from different situations (Ebert *et al.,* 2005). In the end, combining both methods lets us look at molecular interactions in more depth and with more nuance, which helps the field of molecular luminescence spectroscopy as a whole (Figure 9.6).

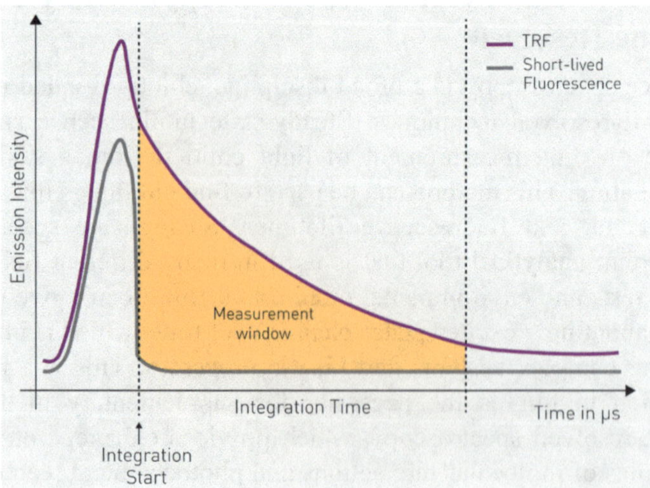

Figure 9.6. *Time-resolved fluorescence measurements.*

Source: https://www.bmglabtech.com/hubfs/1_Webseite/5_Resources/ABC/Detection%20Modes/time-resolved-fluorescence-fig5.webp.

9.2.3. Sample Preparation and Handling Techniques

The success of molecular luminescence spectroscopy depends a lot on how well samples are prepared and handled. The first steps in preparing a sample, like picking the right matrix, making sure it is homogeneous, and taking care of any possible interferences, are very important for getting accurate spectroscopic data. For example, when studying GaN nanowires, changing the temperature to find the best growth conditions affects both the structural integrity of the sample and its luminescent properties. This shows that the preparation phase has a direct effect on the results of the luminescence analysis (Amirthapandian *et al.,* 2015). As a result, researchers need to be very careful when they set up their experiments so that they can reduce any factors that could affect the results and make the molecular luminescence measurements more reliable. Additionally, the way samples are handled is very important for keeping the luminescent properties intact. Photodegradation or other changes in molecular structure can happen when a sample is exposed to light, changes in temperature, or environmental pollutants. This can make the sample's fluorescence response less reliable. To deal with these problems, it's important to take precautions when transporting and storing items, like using opaque containers or keeping things cold when needed. It is also important to set up standard procedures for handling samples so that results can be reproduced in different studies. Researchers have shown that instrumental variations can make fluorescence analyses even harder if quality assurance measures aren't in place. This shows how important it is to have controlled handling environments (Ebert *et al.,* 2005). In conclusion, proficient sample preparation and handling methodologies are essential in molecular luminescence spectroscopy, affecting both the precision and reproducibility of outcomes. Researchers can get useful information from their fluorometric analyses by following strict rules for preparing and handling samples. By understanding how different preparation methods affect the luminescent properties of molecules, scientists can make progress in molecular spectroscopy that is both reliable and useful in different types of research. This is an urgent call for the creation of standardized procedures that will help the field move forward while making sure that quality is maintained in the search for scientific knowledge.

9.3. APPLICATIONS IN CHEMISTRY

Molecular luminescence spectroscopy has many uses in chemistry. It can help us understand how molecules behave by looking at the light they give off. This method mostly involves looking at different types of molecules and how they react with each other. This gives us important information about how molecules are put together and how they move. One important use is to study organic materials, especially low molecular weight compounds like diphenyl-diketo-pyrrolopyrrole (DPP) derivatives. These compounds, which were once used in dye formulations, have amazing optical properties that can be used to make new smart materials. Molecular luminescence spectroscopy enables researchers to investigate the correlations between the chemical structure of DPP derivatives and their emitted light, facilitating novel applications in contemporary technologies, such as organic photovoltaics and light-emitting devices (Ouzzane *et al.,* 2015).

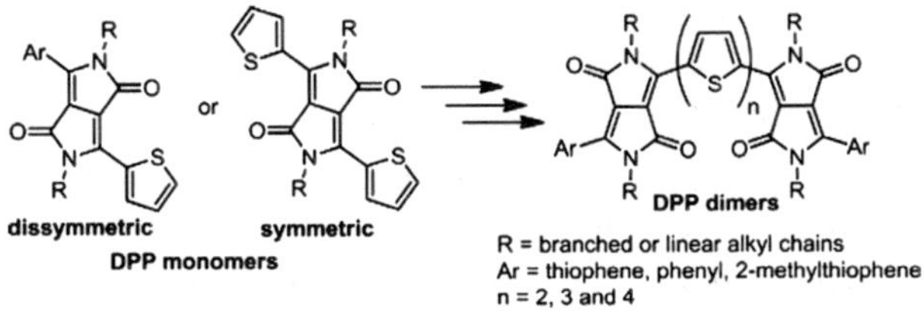

Figure 9.7. *Straightforward access to diketopyrrolopyrrole (DPP) dimers.*

Source: https://ars.els-cdn.com/content/image/1-s2.0-S0143720812003397-fx1.jpg.

Molecular luminescence spectroscopy also goes beyond traditional spectroscopic methods by allowing for a more in-depth look at solid-state materials. Recent advancements have demonstrated its efficacy in the examination of hexagonal boron nitride (h-BN) under laser excitation. The UV laser-induced fluorescence from h-BN gives us important information about the presence of defect sites, like nitrogen vacancies, which can have a big effect on the material's electronic and optical properties. These kinds of insights are necessary to improve materials for a wide range of uses, such as high-performance electronics and optoelectronic devices. The structured fluorescence seen in certain wavelength ranges shows not only the materials' intrinsic properties, but it also helps improve synthesis methods, which makes h-BN better for industrial use (Andrei *et al.,* 2008). Molecular luminescence spectroscopy is a promising tool for developing next-generation materials because it combines cutting-edge research with other fields. This method helps chemists improve the properties of substances by helping them understand molecular interactions and defect dynamics better. This leads to new ideas in fields like environmental science and nanotechnology. The capacity to evaluate and modify the luminescent characteristics of materials facilitates the development of customized substances that tackle particular technological issues. As molecular luminescence spectroscopy improves, it will probably be used more in chemistry, leading to new discoveries and the development of materials with never-before-seen properties that will help science move forward in many areas (Figure 9.7).

9.3.1. Use in Quantitative Analysis of Compounds

In the field of molecular luminescence spectroscopy, the quantitative analysis of compounds is a key use that takes advantage of the special properties of luminescent materials. This analytical method works especially well for figuring out how much of different chemical species are in complicated mixtures. Molecular luminescence, which includes both fluorescence and phosphorescence, is a sensitive and selective way to detect things. It can show differences in emission spectra that are caused by changes in structure or environmental conditions. You can measure the amount of a compound because each one has a unique luminescent signature that can be linked to its concentration using known calibration curves. As researchers improve these methods, they are looking at more complicated samples, like natural organic matter (NOM), to see how they bind and how they fluoresce. This shows how strong the methods are for quantitative assessments (Kramer *et al.,* 1998). When researchers look into the quantitative parts of luminescence spectroscopy, they often use advanced modeling methods to make sense of the data they have collected. Analyzing spectra can give you important information about the chemical environment that affects how a compound glows. For example, the research on ruthenium(II) polypyridine complexes illustrates how gas-phase and liquid-phase calculations, utilizing density-functional theory (DFT), can yield insights into electronic transitions relevant to luminescence (Amolo *et al.,* 2017). Scientists can effectively link luminescence lifetimes to the electronic properties of the species involved by creating a spectrochemical series through partial density-of-states analysis. This quantitative analysis is essential for formulating standardized protocols that guarantee consistent and reproducible outcomes across multiple applications, thereby improving the reliability of molecular luminescence spectroscopic techniques in various domains. Luminescence spectroscopy is useful for more than just identifying compounds; it can also be used for detailed quantitative analyses that help with many different scientific questions. Applications in environmental monitoring and drug development, for example, make use of the ability to find trace compounds with high sensitivity. Researchers can make assays that can accurately measure concentrations by taking advantage of the luminescent properties of certain analytes. This is illustrated in research on NOM, where multiresponse titration curves are used to measure the binding constants of different fluorophores. This shows how composition and fluorescence behavior are related in complex ways (Kramer *et al.,* 1998). These quantitative analyses have significant implications, affecting choices in environmental policy and drug formulation, as they facilitate regulatory compliance and efficacy testing. As a result, molecular luminescence spectroscopy is still developing as a key method in the analytical sciences, connecting theoretical ideas with real-world uses.

9.3.2. Role in Studying Reaction Kinetics

It is important to understand reaction kinetics in order to explain how molecules interact with each other, especially in the context of molecular luminescence spectroscopy. This field of study gives us useful information about the changing processes that control how fast reactions happen with molecules that give off light. Researchers can use photophysical properties to learn more about how energy moves between molecules, how long-lived excited states are made, and how these states return to their ground state. These insights are especially pertinent when analyzing electron transfer reactions in biological systems, exemplified by respiration processes characterized by numerous electron transfer steps, frequently involving weakly coupled interactions across considerable distances (Dempsey *et al.,* 2010). As molecular luminescence spectroscopy progresses, it will yield progressively more intricate kinetic data, thereby augmenting our comprehension of essential chemical processes at the molecular scale. Studying reaction kinetics through molecular luminescence has important effects on many real-world uses, such as making new materials and sensors, in addition to biological systems. Transition metal sulfides, for instance, have luminescent properties that are different from those of other materials. These properties can be used in bioimaging and photonic devices, among other things. To get the best performance and stability out of these materials, it's important to understand how luminescence works in them. The underlying reaction pathways, shaped by surface chemistry and interactions with other species, underscore the necessity of thorough kinetic investigations (Dengo *et al.,* 2017). Researchers can use molecular luminescence techniques to figure out how these materials work, including how functionalization changes their luminescent properties and overall reactivity. As the study of reaction kinetics in molecular luminescence becomes more advanced, it becomes clear that interdisciplinary approaches are necessary for further progress. To fully understand the kinetic phenomena that happen with luminescent species, we need to combine experimental methods with theoretical modeling. The phenomenon of proton-coupled electron transfer (PCET) is vital in various biochemical contexts, as it enables efficient energy conversion processes necessary for life. Researchers can use molecular luminescence spectroscopy to look into these kinds of interactions. This lets them see how reaction kinetics change over time, which gives them a better idea of how the electron transfer works (Dempsey *et al.,* 2010). As this field continues to change, the combination of spectroscopy, kinetics, and molecular theory is likely to give us a better understanding of both the basic and practical parts of chemical reactivity (Figure 9.8).

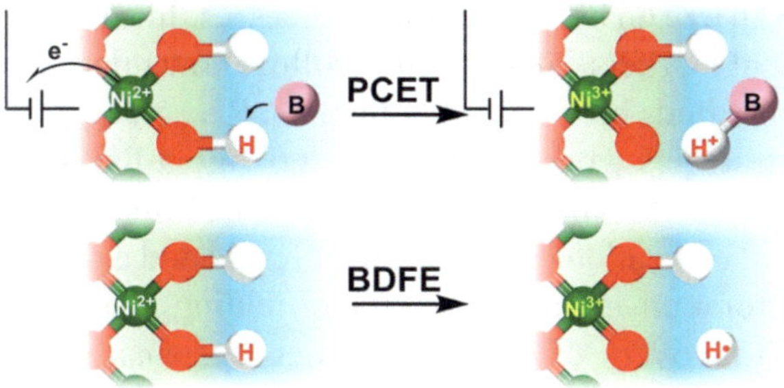

Figure 9.8. *Proton coupled electron transfer.*

Source: https://images.squarespace-cdn.com/content/v1/62a2221d5a239f16b5de994f/a3b11c5a-ffca-40bc-8de3-35f8fa0d1785/Mayer_bonds+over+electrons.jpeg.

9.3.3. Applications in Environmental Monitoring

Molecular luminescence spectroscopy has become an essential instrument in environmental monitoring, enabling the detection and quantification of pollutants and hazardous substances with exceptional sensitivity. This method uses luminescence, which is when certain materials give off light when they are excited, to find specific compounds even when they are only present in small amounts. Sensors that use nanomaterials like quantum dots and graphene quantum dots (GQDs) have been shown to be more selective and sensitive to different environmental analytes. Their unique photonic properties make luminescent sensors work better, which helps meet the need for timely monitoring of contaminants in air, water, and soil, which protects public health and the environment (Altintas *et al.,* 2020). Moreover, improvements in molecular luminescence spectroscopy are leading to the creation of portable and real-time monitoring devices, which are necessary for environmental assessments done on-site. These kinds of new ideas have been especially useful in fields like water quality testing, where finding harmful biological markers or chemical pollutants quickly can stop ecological disasters or health crises. Adding luminescent nanomaterials to biosensing platforms improves their ability to find things, allowing them to respond quickly even in complex matrices, which are common in environmental samples. These advanced sensors make it possible to find many analytes at once, which makes it easier to do thorough environmental assessments and gives important information to regulatory agencies and policymakers who want to reduce the effects of pollution (Sadler *et al.,* 2013). Molecular luminescence spectroscopy is not only used to find pollutants; it is also used to keep an eye on bioremediation processes. In environmental cleanup efforts, luminescent markers can be used to follow how microbial communities, or other cleanup methods break down pollutants. Researchers can see how well bioremediation strategies work overtime by looking at the luminescent output and the rate at which certain pollutants break down. In this regard, the technique provides both a quantification of contaminant levels and a dynamic perspective on ecological restoration initiatives, thereby yielding significant insights into the resilience and recovery of contaminated ecosystems, ultimately fostering sustainable management practices (Altintas *et al.,* 2020).

9.4. APPLICATIONS IN BIOLOGY

Molecular luminescence spectroscopy has emerged as an essential technique in biological research, enhancing comprehension of cellular mechanisms and dynamics. This method uses the special properties of luminescent molecules to learn more about how biomolecules interact, how signaling pathways work, and how cells use energy. For example, using luminescent probes lets researchers watch biomolecules in live cells in real time, which is important for studies that need to look at biological activity in its natural setting. Also, better waveguide technology has made spectroscopic measurements in biological samples more sensitive and accurate, which has made it easier to find and study molecular components. The heightened light-matter interaction characteristic of these configurations highlights the prospective advantages for diverse spectroscopic applications, such as Raman spectroscopy, where improvements in sensitivity can result in substantial progress in biological diagnostics and research (Baets *et al.,* 2014). Additionally, the incorporation of lanthanide-based luminescent materials has created novel opportunities in targeted drug delivery and imaging. These materials can bind to certain biological markers because of their unique luminescent properties, which makes it easier to see how cells work. For example, ferritin H-homopolymers that have been changed to have a lanthanide binding tag show promise as nanocarriers for getting imaging agents to tumor cells, which often have too many transferrin receptors (Baiocco *et al.,* 2018). The high-affinity binding properties of these modified ferritin particles, along with the increased fluorescence intensity when excited, represent a major step forward in bioimaging methods. This use of luminescence spectroscopy in the field of targeted delivery not only makes it easier to keep track of therapeutic agents, but it also leads to new ways to treat and diagnose cancer. Molecular luminescence spectroscopy is also useful for more than just imaging and tracking; it is also important for figuring out how therapeutic agents work. In-depth spectroscopic studies can clarify the interactions between drugs and their molecular targets, offering insights into binding affinities and kinetics that are essential for drug development. As seen in the progress made in waveguide technology, being able to find small changes in spectral properties in biological contexts helps us learn more about molecular conformations and dynamics. This spectral data can help improve therapeutic compounds so that they have the least amount of side effects while still getting the desired results. The ongoing advancement of molecular luminescence spectroscopy offers significant prospects for both fundamental and applied biological research, propelling innovations in health and medicine (Baets *et al.,* 2014) (Figure 9.9).

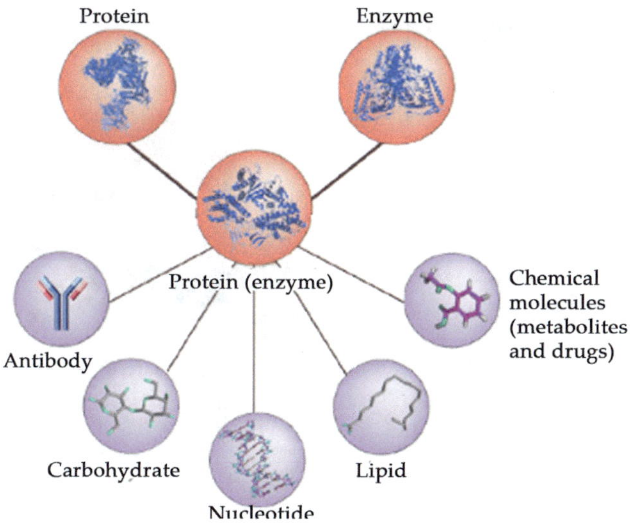

Figure 9.9. *Biomolecular interaction analysis service.*

Source: https://www.biologicscorp.com/image/data/fortebio/BLI-network.png.

9.4.1. Use in Biomolecular Imaging

Molecular luminescence spectroscopy has become a potent instrument in biomolecular imaging, allowing scientists to see and measure biomolecules with great accuracy. The capacity to identify low-abundance molecules in biological specimens has profound ramifications for comprehending intricate biological processes and disease mechanisms. Researchers can get better imaging contrast and specificity by using the luminescent properties of nanoparticles that have been specially designed. These nanoparticles, which are often functionalized with chiral molecules, have unique optical properties that can be used in imaging applications. The interaction of circularly polarized light with chiral nanoparticles can result in differential emission properties, offering significant insights into molecular environments and interactions. These methodologies are not merely theoretical; they are essential for the real-time imaging of cellular processes, rendering them indispensable in contemporary biomedicine and diagnostics (Andrews *et al.,* 2015). Luminescent nanoparticles are also useful in biosensing, where they work well as probes in biomolecular imaging applications. This colloidal fluorescent and plasmonic nanoparticles react strongly to light that hits them, making them useful as sensitive detectors in complicated biological settings.

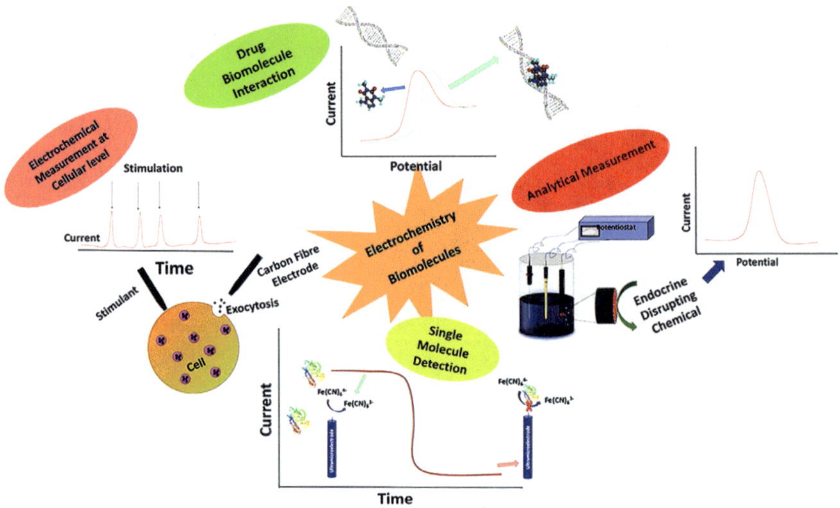

Figure 9.10. *Biomolecule.*

Source: https://ars.els-cdn.com/content/image/1-s2.0-S2451910321002003-sc1.jpg.

Functionalizing the surfaces of nanoparticles makes it possible to attach them to specific biomolecules, which lets them respond to light in a way that is specific to the analytes of interest. These kinds of abilities are very important in clinical settings, where finding biomarkers quickly and accurately can have an effect on patient outcomes. Moreover, the capacity to perform imaging in living organisms increases the potential utility of these techniques in therapeutic monitoring and drug delivery evaluations, ultimately contributing to progress in personalized medicine (Chandrawati *et al.,* 2014). Recent progress in molecular luminescence spectroscopy highlights the importance of hybrid methodologies that integrate both luminescent and plasmonic characteristics. The interaction between these two optical phenomena can be utilized to enhance imaging resolution and sensitivity beyond the capabilities of conventional techniques. By systematically examining the emission characteristics of surface-functionalized nanoparticles, researchers can elucidate the intricate details of molecular interactions within cells. The quantification of chirality related to these nanostructures can provide insights into structural alterations in biomolecules, frequently signifying pathological conditions. The combination of molecular luminescence spectroscopy with cutting-edge nanoparticle technology is a game-changer for biomolecular imaging. It promises new ways to learn about biological systems at the molecular level (Andrews *et al.,* 2015) (Figure 9.10).

9.4.2. Role in Drug Discovery and Development

Molecular luminescence spectroscopy is very important in drug discovery and development because it gives us useful information about how molecules interact and how drugs move through the body. This method lets scientists keep an eye on how luminescent compounds act in living systems, which helps them learn more about how drugs are absorbed, distributed, metabolized, and excreted. One of the best things about luminescence spectroscopy is how sensitive it is. This lets it find drug candidates in complex biological matrices even when they are only present in small amounts. Using this method, scientists can do in-depth studies of how different compounds interact with cellular targets. This helps them find new ways to treat diseases and improve lead compounds to make them more effective and less toxic in later stages of drug development. The examination of engineered nanoparticles, including gold nanoparticles (AuNPs), further illustrates the incorporation of molecular luminescence spectroscopy in drug discovery. Recognizing the regulatory challenges associated with mammalian testing, researchers have opted for alternative models, such as Caenorhabditis elegans, to examine the biodistribution and toxicity of AuNPs. The employment of absorption microspectroscopy in these investigations has been instrumental in clarifying the aggregation behavior of nanoparticles at the cellular level and linking their plasmonic properties to biological effects. This method not only eases ethical worries about using animals in research, but it also speeds up the process of testing nanoparticles for use in medicine. Understanding how nanoparticles interact with biological systems can help us make better targeted drug delivery systems and improve the formulation of new drugs. This shows how important luminescence spectroscopy is in molecular research. Additionally, the identification and optimization of small molecules that inhibit critical pathways in cancer biology exemplify another vital application of molecular luminescence spectroscopy in drug discovery. For example, blocking the Frizzled receptor in canonical Wnt signaling has shown promise as a way to treat different types of tumors. Finding small molecules that bind to the cysteine-rich domain of Frizzled through virtual screening and spectral analysis shows that luminescence techniques are useful in molecular pharmacology. This ability to measure compound binding affinities in numbers makes it easier to make targeted drugs that change signaling pathways that aren't working properly. These strategies enhance our comprehension of intricate disease mechanisms and facilitate the swift translation of laboratory discoveries into clinically applicable therapies, underscoring the crucial role of molecular luminescence spectroscopy in pharmaceuticals and biomedical science.

9.4.3. Applications in Understanding Cellular Processes

Molecular luminescence spectroscopy has changed the way we study cellular processes by giving us a powerful way to image and characterize things at the molecular level. Researchers can watch real-time changes in cells, like protein interactions, gene expression, and cellular signaling pathways, using luminescent markers. This ability is especially helpful for figuring out how cells react to things that happen outside of them and stressors. For instance, luminescence techniques can be used to study the biodistribution of engineered nanoparticles, which lets scientists see how they are taken up by cells and where they are located. This method is very important for testing the biocompatibility and possible toxicity of these nanoparticles, which is important for using them in areas like drug delivery and safety for the environment. By combining luminescence spectroscopy with cutting-edge imaging technologies, scientists can fully understand how cells work. Recent studies have employed luminescence spectroscopy to analyze the interactions between nanoparticles and biological systems, especially in model organisms. For example, studies using the nematode Caenorhabditis elegans have demonstrated that luminescence techniques can evaluate the impact of varying sizes of gold nanoparticles (AuNPs) on cellular health and development. Extensive research revealed that smaller 11-nm AuNPs displayed heightened toxicity compared to their larger equivalents, underscoring the importance of particle size in biological interactions. Researchers can make better choices about how to make safer nanomaterials for medical and cosmetic use because luminescence spectroscopy can see and measure how these nanoparticles are taken up at the cellular level. Consequently, this method is an essential element in enhancing our comprehension of nanoparticle-cell interactions across diverse biological tiers, from the cellular to the molecular scale. Moreover, advancements in luminescence technologies have profound ramifications for therapeutic applications, especially in photodynamic therapy

(PDT). Recent advancements utilize lasers to produce singlet oxygen, thereby improving treatment efficacy for conditions such as tumor proliferation and vascular injury related to strokes. Researchers have been able to find out where and how singlet oxygen is made in living things by using molecular luminescence spectroscopy. This information helps get around problems that often come up in PDT, such as light not getting through well and being toxic. The creation of quantum-dot laser diodes that emit in the near-infrared spectral range also opens up new possibilities for effective drug delivery systems and localized treatment in difficult places, like the brain. In this way, luminescent methods not only help us understand how cells work, but they also open up new ways to treat diseases, which leads to a better understanding of how complex biological interactions work and better clinical outcomes.

9.5. ADVANTAGES AND LIMITATIONS

Molecular Luminescence Spectroscopy is well-known for its many benefits in analytical chemistry and biochemistry. One of the best things about it is that it is very sensitive to trace analytes, which lets researchers find very small amounts of substances with great accuracy. This sensitivity makes it possible to keep an eye on biological systems and environmental samples, where finding very small amounts can be very important for safety and compliance. Also, luminescence techniques don't damage the sample, so you can analyze it multiple times. This is especially useful in studies that need to collect data over a long period of time. Furthermore, improvements in sensor technology have made luminescence-based measurements even more useful. For example, they can be used to measure oxygen concentration through luminescence quenching, which shows how versatile the field is and how it affects many different scientific fields (Stoneham *et al.,* 1974). Molecular Luminescence Spectroscopy has a lot of benefits, but it also has some drawbacks that need to be thought about when choosing analytical methods. One significant challenge is the interference from extraneous factors, including light scattering, background fluorescence, and temperature variations, all of which can influence luminescence measurements. Furthermore, although luminescence offers significant insights into the electronic structure of molecules, specific phenomena, such as the differentiation between sigma and pi luminescence, currently pose analytical challenges that researchers have not yet fully addressed (Stoneham *et al.,* 1974). This complexity can necessitate meticulous experimental setup and calibration, thereby requiring specialized knowledge and equipment. Also, the need for exact calibration protocols, especially when measuring oxygen levels, can make it harder to use luminescence techniques for routine monitoring (Stoneham *et al.,* 1974). Incorporating artificial intelligence into Molecular Luminescence Spectroscopy is a promising way to fix some of its current problems and make it work better overall. Using machine learning techniques has been shown to make data interpretation more accurate and efficient. This has led to better calibration models that can handle changes in experimental conditions. Researchers can make the process of calibrating luminescent sensors easier by using neural networks to connect input variables to concentrations of interest. This gets rid of problems that may come up with traditional methods (Baumgartner *et al.*, 2019). This new idea could not only help reduce measurement errors, but it could also lead to the creation of more advanced sensors that use fake data for training, which would solve the reliability problems that come with traditional luminescence methods. The convergence of Machine Learning and Molecular Luminescence Spectroscopy represents a substantial advancement in addressing technical challenges, facilitating the development of next-generation applications across diverse scientific domains (Figure 9.11).

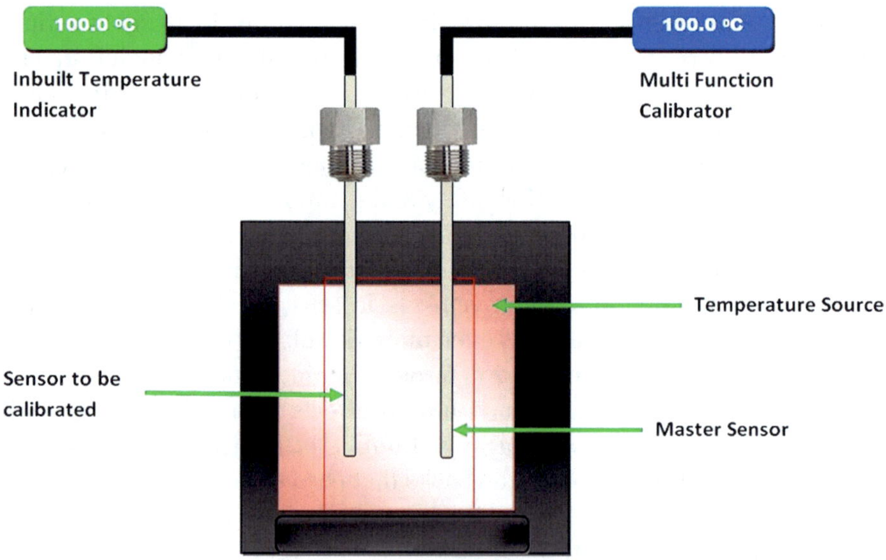

Figure 9.11. *Calibration of temperature sensor with indicator.*

Source: https://theinstrumentguru.com/wp-content/uploads/2023/06/temperature-indicator.jpg.

9.5.1. Benefits of High Sensitivity and Specificity

In molecular luminescence spectroscopy, high sensitivity and specificity are very important because they make analytical applications much more effective. Sensitivity is how well a method can find low amounts of target molecules. This is especially important in biomedical fields where finding diseases like cancer early can help people get treatment quickly. Luminescence techniques with high sensitivity can find biomolecules or pathogens at very low levels, which gives researchers and healthcare professionals more confidence when they start preliminary diagnostic tests. This component is essential in combating the rise of numerous diseases, as evidenced by the necessity for sophisticated techniques to identify markers associated with cancer, cardiovascular diseases, and infections (Altintas *et al.,* 2020). Additionally, specificity works with sensitivity to make sure that the signals picked up only match the target analytes, which reduces the number of false positives. This is a very important trait that makes diagnostic tests using molecular luminescence spectroscopy more reliable overall. High specificity is especially useful in complicated biological matrices, like blood or tissue samples, where there may be many similar molecules. Researchers can make their diagnostic evaluations more accurate by using special luminescent probes that only interact with certain biomarkers. This selective detection reduces the risk of misdiagnosis, which can result in unsuitable treatment choices. So, adding high specificity to molecular luminescence assays not only makes them more accurate for diagnosing but also makes people more confident in the results they get from these improved methods (Cherry *et al.,* 2008). As science moves forward, the creation of new materials like graphene quantum dots (GQDs) has made the advantages of high sensitivity and specificity in molecular luminescence spectroscopy even better. GQDs are known for their great qualities, such as photostability, biocompatibility, and tunable photoluminescence. These qualities make them great candidates for biosensing applications. These traits make it possible to make very sensitive sensors that can find very small amounts of biological markers while still being specific in the presence of other biomolecules. As a result, using GQDs to make biosensors not only helps with early diagnosis and treatment plans, but it also improves the overall performance of molecular luminescence techniques in real-world situations. The collective enhancement in sensitivity and specificity via the application of novel nanomaterials substantially impacts the field of analytical chemistry and biomedical diagnostics (Altintas *et al.,* 2020).

9.5.2. Limitations Related to Sample Types and Conditions

Molecular luminescence spectroscopy is a complicated field where the type of sample and the conditions it is in can have a big effect on how accurate and reliable the results are. Various sample matrices, including biological tissues, polymers, or environmental samples, can introduce variabilities that complicate the interpretative framework of luminescence spectroscopic data. In biological samples, the presence of interfering substances, such as biomolecules emitting in the same spectral region as the target analyte, can result in inaccurate conclusions regarding the analyte's concentration and behavior. Also, differences in how samples are prepared, like filtration, centrifugation, or dilution, can create artifacts that make it hard to reproduce and compare data from different studies. To reduce these problems and make fluorescence data more reliable, it is becoming more and more important to have standardized procedures (Ebert *et al.,* 2005). Sample conditions are very important because many environmental factors can change the luminescence properties in big ways. Changes in temperature can change how luminescent species move, which can change the intensity and spectral distribution of their emissions. The pH level of the sample solution may also change the protonation state of fluorescent molecules, which would further change the luminescence response. Because samples are so sensitive to changes in their surroundings, it is important to carefully control and record the conditions of the samples to make sure that systematic investigations lead to strong conclusions. For example, when working with quantum dot-like single-photon sources in transition metal dichalcogenides, the fact that the sample conditions can change can lead to emission characteristics that are hard to predict but are very important for quantum optics applications (Barnard *et al.,* 2019). Thus, comprehending and alleviating these environmental dependencies is crucial for the progress of molecular luminescence spectroscopy. Additionally, the constraints associated with particular sample categories permeate the pursuit of optimal analytical performance in luminescence spectroscopy. Some sample matrices, like thick solutions or mixtures that aren't uniform, can stop luminescent molecules from moving and interacting with each other. This can lower the quantum yield and detection limits. This trait is especially strong in complicated biological systems, where too many molecules in one place can cause spectral shifts in the environment. As a result, scientists often have to use simpler models to make conclusions, which can make the results less useful in the real world. The creation of specialized fluorescence standards is essential for improving the reliability of luminescence methods. As emphasized by the necessity for improved quality assurance in fluorometry, the enhancement of standards will enable superior characterization and validation of fluorescence instruments, thereby propelling the field towards more precise and dependable analyses (Ebert *et al.,* 2005).

9.5.3. Comparison with Other Spectroscopic Techniques

Molecular Luminescence Spectroscopy (MLS) is a unique analytical chemistry technique that stands out from other spectroscopic methods like Raman spectroscopy and fluorescence. MLS, which uses light emission from electronic transitions in molecules, is often very sensitive and specific when it comes to finding luminescent species. Raman spectroscopy, on the other hand, takes advantage of light that doesn't bounce off of things and has made a lot of progress thanks to photonic waveguides, which improve the way light and matter interact. Recent research indicates that these improved light-matter interactions result in a Raman scattering efficiency that is roughly two orders of magnitude greater in waveguide configurations than in free-space setups, thereby enhancing its utility in spectroscopic analysis (Baets *et al.,* 2014). This comparison shows the pros and cons of each method and makes it clear that the choice often depends on what the analysis needs. Fluorescence spectroscopy is another spectroscopic method that should be looked into. It also deals with electronic transitions, but it focuses on the light that is given off when excitations happen. The need for quality assurance in fluorometry has made people more aware of the sources of error that can affect analytical results.

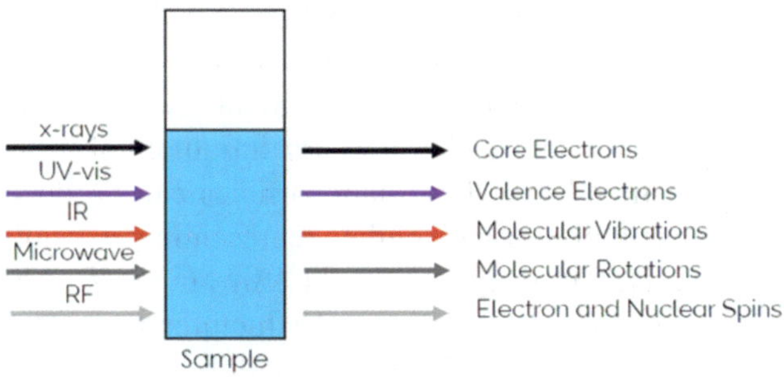

Figure 9.12. *Important spectroscopic techniques.*

Source: https://www.oxinst.com/learning/uploads/inline-images/figure2-curvette-20230425131046.jpg.

This is not as big of a problem in MLS because its measurement processes are relatively simple (Ebert *et al.,* 2005). Even though fluorescence and MLS are different, they often work together in complicated analytical situations, like biological assays where both methods are used to improve sensitivity and accuracy. Also, the need for strict standards in fluorescence measurements shows how complicated and variable data interpretation can be. This is different from MLS, where luminescence can give more direct and measurable information about molecular structures. MLS is great for applications that need high sensitivity, like environmental monitoring and biomedical diagnostics, because it gives clear, unambiguous signals. In contrast, the unique benefits of methods such as Raman and fluorescence typically stem from their capacity to analyze non-luminescent samples or their applicability to biological systems, respectively. MLS is great when luminescent properties are important, but it might not work as well in other situations, like when you need to describe complex mixtures or when you need to analyze more than one thing at a time. Each technique adds something different to the field of spectroscopy, which shows how important it is to choose the right one based on the sample's characteristics and the analytical goals (Figure 9.12).

9.6. RECENT ADVANCES AND INNOVATIONS

Recent progress in molecular luminescence spectroscopy has led to new imaging technologies that greatly improve the clarity of images during medical procedures. The creation of INSITE (Immunotargeted Nanoparticle Single-Chip Imaging Technology) is a clear example of this progress. It combines lanthanide-based alloyed upconverting nanoparticles (aUCNPs) with a CMOS-based imager. This synergistic design enables the visualization of biological tissues with a spatial resolution capable of identifying micrometer-scale targets, representing a substantial advancement over conventional imaging methods. The use of INSITE technology for intratumoral imaging of prostate tumors has demonstrated that these nanoparticles can produce bright signals across different excitation wavelengths, representing a significant progress in minimally invasive surgical techniques (Anwar *et al.,* 2019). This new technology helps surgeons see anatomical structures more clearly by making it easier to take high-resolution pictures in small surgical cavities. At the same time, molecular luminescence spectroscopy has made progress in the field of biosensing technologies, especially in the early diagnosis and risk assessment of cardiovascular diseases (CVDs). Researchers are working on making tools that can quickly find different cardiac biomarkers because there is a growing need for portable and cheap biosensing devices. C-reactive protein (CRP), cardiac troponin proteins, and interleukins are essential biomarkers for predicting cardiovascular disease (CVD) risk. You can measure these biomarkers with luminescence-based detection methods, which are faster and less expensive than traditional lab tests (Gurbuz *et al.,* 2012). Consequently, this advancement signifies a substantial progression towards enhanced patient management by enabling prompt interventions that could avert severe cardiovascular incidents. The incorporation of molecular luminescence spectroscopy into various scientific disciplines is facilitating the development of novel applications that transcend medical diagnostics. For instance, materials science has become more interested in manipulating and studying luminescent materials, especially when it comes to making new phosphors and nanomaterials. Researchers can now make materials with specific luminescent properties for a wide range of uses, from lighting to security features, thanks to improvements in synthesis and characterization techniques. These new ideas show how flexible molecular luminescence spectroscopy is and how it can help move research forward in both healthcare and material development. As these technologies advance, they underscore the essential intersections among scientific disciplines and the possibilities they present for tackling intricate issues in modern society (Figure 9.13).

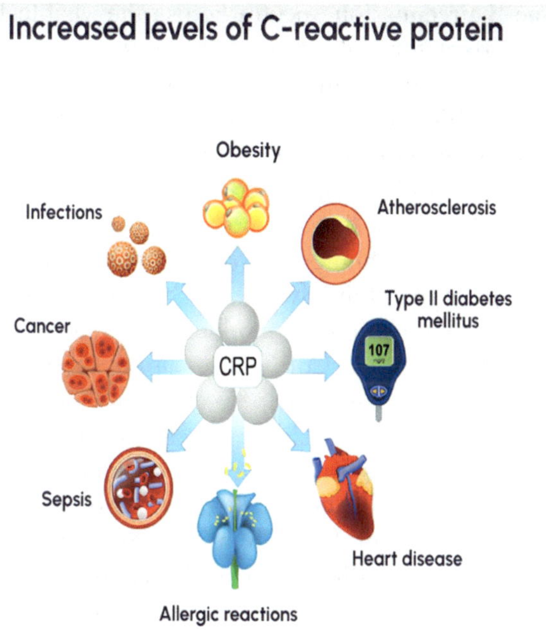

Figure 9.13. *C-reactive protein (CRP).*

Source: https://www.diasys.in/fileadmin/_processed_/7/7/csm_Picture5_c9416d-8b0d.png?_=1714976512.

9.6.1. Development of New Luminescent Materials

People are very interested in making new luminescent materials because they can be used in many different ways, such as in bioimaging, light-emitting diodes (LEDs), and display technology. Transition metal sulfides are one of the many types of compounds that stand out because they have unique properties, such as luminescence that can be adjusted and increased stability. Metal sulfides have different chemical and structural properties than their oxide counterparts, which makes them useful in a variety of ways. Nonetheless, the effective functionalization of these materials continues to pose a challenge, chiefly due to an inadequate comprehension of their surface chemistry and the intricacies related to their colloidal suspensions. Resolving these challenges is vital for promoting the application of transition metal sulfides in practical scenarios, especially in bioimaging, where resilient and stable colloidal systems are imperative (Dengo *et al.,* 2017). Sulfide-based luminescent materials are especially interesting because they could be used in a lot of optoelectronic applications. By adding rare earth ions like Ce^{3+} and Eu^{2+} to these sulfides, you can fine-tune their emission spectra across the visible light range. This ability is very important for making new display technologies, like thin film electroluminescence and field emission displays, which need very accurate color performance. Sulfide materials, including $BaAl_2S_4$:Eu and ZnS:Cu, have become essential in the manufacture of white-light emitting devices, leveraging their distinctive red-emitting characteristics. The investigation of these materials is further illustrated through conversion phosphors, which connect conventional luminescent technologies with contemporary lighting solutions (Hens *et al.,* 2010). As research advances, it becomes increasingly clear that the dimensions and composition of luminescent nanoparticles significantly affect their emission properties. Colloidal quantum dots, including PbS and CdS, exhibit size-dependent luminescence, wherein alterations in particle size result in significant variations in emission wavelength. This trait not only makes luminescent materials more useful, but it also opens up new possibilities for their use in sensors and optoelectronic devices. Also, the creation of sulfide materials is leading to better solid-state lighting options, especially as scientists look into how to combine alkaline-earth sulfides and thiogallates for the best results. In the end, more research into luminescent materials, especially their properties and surface chemistry, will lead to new developments in molecular luminescence spectroscopy and its uses (Hens *et al.,* 2010).

9.6.2. Advances in Instrumentation Technology

Recent improvements in instrumentation technology have changed the field of molecular luminescence spectroscopy, especially with the addition of laser-based systems. Researchers can now make certain wavelengths of light thanks to the development of advanced lasers. This greatly improves the sensitivity and resolution of luminescence measurements. The advent of quantum-dot laser diodes that emit in the near-infrared (NIR) spectral range has enabled novel applications in biological settings, such as photodynamic therapy (PDT). These improvements not only make it easier to make singlet oxygen for medical use, but they also fix problems with traditional light sources, like not letting enough light through and the high toxicity of photosensitizers (Bragina *et al.,* 2017). Consequently, the integration of these advanced laser technologies into molecular luminescence applications has demonstrated potential to transform both research methodologies and clinical practices. Additionally, the advancement of fluorescence microscopy signifies another crucial milestone in instrumentation technology pertinent to molecular luminescence spectroscopy. Because this technique is both flexible and specific, it can create both images and photometric data that show biological samples at the cellular level. Advanced techniques like fluorescence resonance energy transfer (FRET) have come about because of the unique properties of fluorescence. These techniques let scientists' study complex dynamic processes in living things. These advancements improve the comprehension of molecular interactions in real-time, enabling substantial progress in areas such as cellular biology and drug delivery systems.

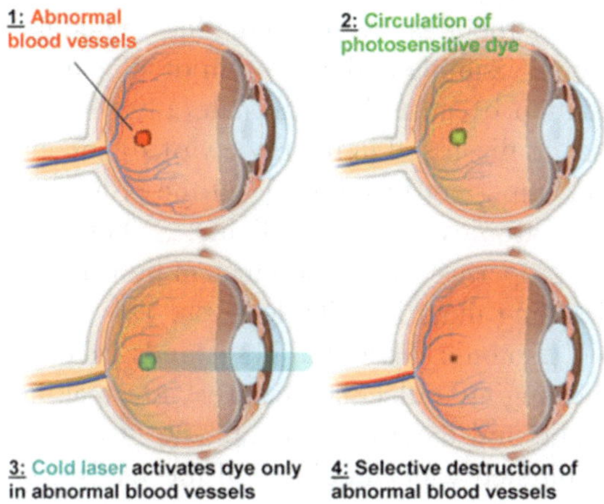

Figure 9.14. *Photodynamic therapy.*

Source: https://macularetinavitreouscenter.com/wp-content/uploads/2020/10/photodynamic.jpg.

The ability to measure FRET using methods such as acceptor bleaching and fluorescence lifetime imaging microscopy (FLIM) demonstrates how advanced microscopy can improve and work with traditional luminescence spectroscopic methods (Abbe *et al.,* 2012). Finally, the use of artificial intelligence (AI) and machine learning in instrumentation technology is starting a new era for molecular luminescence spectroscopy. These improvements in computing could change the way we analyze data by making it possible to process huge datasets much faster and more accurately than ever before. Traditional methods often require a lot of manual analysis, but AI algorithms can find patterns and make predictions based on large datasets. This makes experimental procedures much more efficient. Additionally, machine learning methods can help improve experimental conditions, which will improve the overall performance of luminescence applications. As these smart systems keep getting better, researchers can expect not only faster discoveries in molecular spectroscopy but also the spread of advanced techniques to more scientific fields. This will lead to more uses and better results in both research and clinical settings (Figure 9.14).

9.6.3. Emerging Applications in Nanotechnology

Nanotechnology has made a lot of progress, and this has led to many uses in different scientific fields. One of the most important ones is molecular luminescence spectroscopy. Researchers have been able to create and design new fluorescent nanostructures by manipulating materials at the nanoscale level. One well-known example is the creation of silicon clusters, which have gotten a lot of attention because they can naturally show fluorescence under the right conditions. As discussed in the review of silicon cluster fluorescence, size effects are very important in figuring out the properties of materials that are important for electronics and biomedicine. Scientists have been able to improve the optical properties of silicon clusters by using surface passivation techniques or embedding them in suitable hosts. This opens the door to uses in bio-imaging and photonic devices. These kinds of new ideas show how nanotechnology is changing molecular spectroscopy, making it easier to make progress in diagnostic and therapeutic uses. The growth of GaN nanowires through the metal-initiated vapor-liquid-solid mechanism is another example of how nanoscale materials can be used in many different ways. Chemical vapor deposition has been used for a long time to grow these nanowires at high temperatures. However, new research has shown that they can also be made at lower temperatures, especially when indium is added. This new development makes it possible to make nanowires that keep their structural and vibrational integrity while also having good optical properties. The capacity to control growth conditions and materials at the nanoscale significantly impacts molecular luminescence spectroscopy, especially in the advancement of light-emitting devices and sensors. These nanowires not only have the potential to make photonic applications more efficient, but they also help scientists learn more about how materials behave at the atomic level, which adds to the field's knowledge base. The convergence of nascent nanotechnological applications and molecular luminescence spectroscopy indicates a promising future for the evolution of analytical methodologies. As researchers investigate the molecular mechanisms that regulate fluorescence in nanostructures, these discoveries possess the capacity to transform both scholarly research and commercial applications. For example, the fluorescence properties of silicon clusters show that localized transitions, which can happen because of quantum confinement, are promising for making spectroscopic methods better at finding things. Also, the fact that GaN nanowires can change their optical properties means that researchers can make very precise changes to their emission spectra, which is very useful for certain uses. As these technologies keep getting better, they will probably lead to new discoveries in molecular luminescence spectroscopy. This will open up new possibilities for nanotechnology and its use in areas like medicine, environmental monitoring, and nanophotonics.

SUMMARY

Molecular luminescence spectroscopy is a field that has made a lot of progress and has some unique uses that show how molecular science, and practical technology can work together. This analytical method not only makes it possible to study molecular dynamics, but it also makes it easier to study nanoscale materials like silicon clusters. The intrinsic fluorescence characteristics of nanostructured silicon are intricately associated with their electronic and geometric arrangements, resulting in significant ramifications for their applications in diverse domains, including electronics and biomedicine. Comprehending these luminescent characteristics, elucidated by the processes of excitation, energy dissipation, and decay pathways, provides a robust basis for subsequent research and implementation in emerging technologies, thereby underscoring the significance of nanomaterials in contemporary science (Dasog *et al.,* 2017).

Additionally, the ongoing advancement of molecular luminescence spectroscopy highlights its essential function in the analysis of intricate molecular systems. Studies of the symmetries and electronic structures of molecules like fullerene C_{60} give us a better understanding of how they behave and how they look in different situations. The relationship between symmetry and luminescence helps us learn more about how molecules interact with each other and opens up new areas for research. For example, the method of continuous symmetry analysis provides a useful framework for understanding the different patterns of spectra that are often seen in molecular studies. This helps to connect gaps between theoretical and experimental fields (Adams *et al.*, 2010). In the end, molecular luminescence spectroscopy has a lot of possible uses that keep growing in fields like environmental monitoring, pharmaceuticals, and biomolecular imaging. As advanced methods make it easier to work with and make luminescent materials, new possibilities for real-time monitoring and diagnostic techniques become more and more possible. In this context, the ongoing research into the unique properties of luminescent nanomaterials promises to yield significant breakthroughs, driving both scientific inquiry and technological innovation. By combining knowledge from spectroscopy with real-world uses, the field is ready to offer groundbreaking solutions to current problems in molecular science and other fields.

In the study of molecular luminescence spectroscopy, a key focus is the use of symmetry to explain how molecules act. The case of fullerene C_{60} exemplifies the critical role of symmetry in spectral analysis. Extensive theoretical calculations have shown that C_{60} has an Ih symmetry, which is the traditional way to describe it. Nevertheless, transitioning to an open-shell approximation demonstrates a significant reduction in both energy and symmetry, resulting in a more refined perspective that corresponds with experimental findings of electronic spectra. Continuous symmetry metrics show that C_{60} has a strong structural integrity because it follows Ih symmetry 99.99% of the time. However, the changes caused by electronic factors open up new possibilities for studying luminescence properties and molecular transitions. This sets the stage for more complex analyses in the field of molecular luminescence spectroscopy (Adams *et al.,* 2010). The incorporation of sophisticated computational methodologies, including gas-phase density-functional theory (DFT) and time-dependent DFT (TD-DFT), establishes a significant foundation for the examination of the optical characteristics of diverse complexes, notably ruthenium(II) polypyridine complexes. Researchers have seen a strong linear relationship between computational results and experimental absorption spectra. This shows that DFT calculations are reliable. This relationship shows how complicated things get when molecular orbitals mix a lot in these complexes. To manage this complexity, a partial density-of-states (PDOS) analysis proves to be an essential instrument, enabling a more profound comprehension of the roles of metal and ligand orbitals. Consequently, the insights derived from PDOS not only facilitate the construction of a spectrochemical series for ligands but also yield estimates of absorption energies essential for the analysis of luminescence lifetimes, thereby augmenting the predictive capacity of molecular luminescence spectroscopy (Amolo *et al.,* 2017). Moreover, the discourse on molecular luminescence spectroscopy transitions into the importance of amalgamating frontier molecular orbital theory with luminescent properties. By using ideas from PDOS-LFT, you can use a semi-quantitative method to connect luminescence lifetimes at different temperatures to the energies that come from molecular orbitals. This method is especially useful when looking at how luminescent molecules behave in different situations. The field of molecular luminescence spectroscopy creates a clear connection between theoretical predictions and experimental observations. This not only helps us understand electronic transitions better, but it also makes it easier to create new luminescent materials. Consequently, the convergence of symmetry considerations, molecular orbital analysis, and temperature-dependent effects establishes a thorough framework that propels forthcoming research initiatives in luminescent materials and their applications.

REFERENCES

1. Abbe, A., Ai, A., Airy, G. B., Alivisatos, A. P., Alivisatos, A. P., Ando, T., et al. (2012). Advanced fluorescence microscopy techniques—FRAP, FLIP, FLAP, FRET and FLIM. https://core.ac.uk/download/16431907.pdf (accessed on 29th October 2025).
2. Adamovich, D., Arena, G., Bailey, C., Bergmann, U., Lamberti, C., Garino, C., Dedieu, A., et al. (2012). Resonant X-ray emission spectroscopy reveals d–d ligand-field states involved in the self-assembly of a square-planar platinum complex. https://core.ac.uk/download/9562667.pdf (accessed on 29th October 2025).
3. Adams, W. H., Alikhanidi, S., Alikhanidi, S., Alvarez, S., Avnir, D., Avnir, D., Razbirin, B. S., et al. (2010). Continuous symmetry of C_{60} fullerene and its derivatives. http://arxiv.org/abs/1005.1829 (accessed on 29th October 2025).
4. Ahmad, R., Balasingham, I., Balón, M., Blasius, B., Boerner, H., Boiteux, S., Bonett, J., et al. (2018). DNA oxidation photoinduced by Norharmane Rhenium(I) polypyridyl complexes: Effect of the bidentate N,N′-ligands on the damage profile. https://core.ac.uk/download/288803163.pdf (accessed on 29th October 2025).
5. Ake, R. L. (2025). Exciton coupling in molecular crystals. https://core.ac.uk/download/pdf/42881782.pdf (accessed on 29th October 2025).
6. Altintas, Z., & Mansuriya, B. D. (2020). Applications of graphene quantum dots in biomedical sensors. https://core.ac.uk/download/322792122.pdf (accessed on 29th October 2025).
7. Amirthapandian, S., Dhara, S., Madapu, K. K., Polaki, S. R., et al. (2015). Raoult's formalism in understanding low temperature growth of GaN nanowires using binary precursor. http://arxiv.org/abs/1508.07808 (accessed on 29th October 2025).
8. Amolo, G., Casida, M. E., Kituyi, L., Magero, J., et al. (2017). Partial density of states ligand field theory (PDOS-LFT): Recovering a LFT-like picture and application to photo properties of Ruthenium (II) polypyridine complexes. http://arxiv.org/abs/1707.03665 (accessed on 29th October 2025).
9. Andrews, D. L., Haniewicz, H. T., & Leeder, J. M. (2015). Surface functionalized spherical nanoparticles: An optical assessment of local chirality. https://core.ac.uk/download/41992374.pdf (accessed on 29th October 2025).

10. Anwar, M., Cohen, B. E., Najafiaghdam, H., Papageorgiou, K., et al. (2019). A 25 micron-thin microscope for imaging upconverting nanoparticles with NIR-I and NIR-II illumination. https://core.ac.uk/download/287623956.pdf (accessed on 29th October 2025).
11. Baday, Y. C., Bao, J., Guibao, C., Lee, S., et al. (2015). Structure-based discovery of novel small molecule Wnt signaling inhibitors by targeting the cysteine-rich domain of Frizzled. https://core.ac.uk/download/323066252.pdf (accessed on 29th October 2025).
12. Baets, R., Dhakal, A., Le Thomas, N., Peyskens, F., et al. (2014). Silicon-nitride waveguides for on-chip Raman spectroscopy. https://core.ac.uk/download/55823484.pdf (accessed on 29th October 2025).
13. Baiocco, P., Benni, I., Białasek, M., Boffi, A., et al. (2018). Engineered ferritin for lanthanide binding. https://core.ac.uk/download/188825476.pdf (accessed on 29th October 2025).
14. Barnard, E. S., Borys, N. J., Cochrane, K. A., de Abajo, F. J. G., et al. (2019). Electrically driven photon emission from individual atomic defects in monolayer WS_2. https://core.ac.uk/download/323311314.pdf (accessed on 29th October 2025).
15. Baumgartner, M., Michelucci, U., & Venturini, F. (2019). Optical oxygen sensing with artificial intelligence. https://core.ac.uk/download/187494466.pdf (accessed on 29th October 2025).
16. Berto, C., Gonzalez-Moragasa, L., Kolovouc, A., Laromainea, A., et al. (2017). *In vivo* testing of gold nanoparticles using the Caenorhabditis elegans model organism. https://core.ac.uk/download/81581041.pdf (accessed on 29th October 2025).
17. Bloor, D., Cross, G. H., Kay, A. J., Love, G. D., et al. (2003). Orientation and solvatochromism of dyes in liquid crystals. https://core.ac.uk/download/261864.pdf (accessed on 29th October 2025).
18. Blüm, M. C., Chergui, M., Patthey, F., Pivetta, M., et al. (2005). Fluorescence and phosphorescence from individual C_{60} molecules excited by local electron tunneling. http://arxiv.org/abs/cond-mat/0506701 (accessed on 29th October 2025).
19. Bragina, O. A., Gekaluyk, A. S., Goltsov, A., Rafailov, E., et al. (2017). Laser-induced generation of singlet oxygen and its role in the cerebrovascular physiology. https://core.ac.uk/download/228178032.pdf (accessed on 29th October 2025).
20. Chandrawati, R., Howes, P. D., & Stevens, M. M. (2014). Colloidal nanoparticles as advanced biological sensors. *Science, 346*(6205),

1247390. https://www.science.org/doi/10.1126/science.1247390 (accessed on 29th October 2025).

21. Cherry, S. R., Jacobs, R. E., & Louie, A. Y. (2008). The integration of positron emission tomography with magnetic resonance imaging. https://core.ac.uk/download/4877286.pdf (accessed on 29th October 2025).
22. Clark, B. P., & Kostiuk, T. (2025). Spaceborne sensors (1983–2000 AD): A forecast of technology. https://core.ac.uk/download/pdf/42848800.pdf (accessed on 29th October 2025).
23. Dasog, M., Dohnalová, K., Erogbogbo, F., Feng, J., Fox, M., Hayes, W., He, Y., et al. (2017). Fluorescent silicon clusters and nanoparticles. http://arxiv.org/abs/1709.05516 (accessed on 29th October 2025).
24. De Vos, A., Hendrickx, K., Joos, J., Lejaeghere, K., et al. (2018). Exploring lanthanide doping in UiO-66: A combined experimental and computational study of the electronic structure. https://core.ac.uk/download/158346809.pdf (accessed on 29th October 2025).
25. Dempsey, J. L., Gray, H. B., & Winkler, J. R. (2010). Proton-coupled electron flow in protein redox machines. https://core.ac.uk/download/4887683.pdf (accessed on 29th October 2025).
26. Dengo, N., Gross, S., & Vittadini, A. (2017). Functionalization of colloidal transition metal sulphides nanocrystals: A fascinating and challenging playground for the chemist. https://core.ac.uk/download/141699569.pdf (accessed on 29th October 2025).
27. Ebert, P., Engel, A., Hoffmann, K., Macdonald, R., et al. (2005). Journal of Fluorescence. https://core.ac.uk/download/210638942.pdf (accessed on 29th October 2025).
28. Gurbuz, Y., Gürbüz, Y., Kolkar, M., Niazi, J. H., Qureshi, A., et al. (2012). Biosensors for cardiac biomarkers detection: A review. https://core.ac.uk/download/pdf/11743545.pdf (accessed on 29th October 2025).
29. Hens, Z., Moreels, I., Poelman, D., Smet, P. F., et al. (2010). Luminescence in sulfides: A rich history and a bright future. https://core.ac.uk/download/55757262.pdf (accessed on 29th October 2025).
30. Kanaev, A. V., Arnaud, L., Berzina, T., Blase, X., Brown, D., Bäuerle, D., Carlisle, J. A., et al. (2008). Photoluminescence of hexagonal boron nitride: Effect of surface oxidation under UV-laser irradiation. http://arxiv.org/abs/0804.0764 (accessed on 29th October 2025).
31. Kramer, J. R., & Smith, D. S. (1998). Fluorescence analysis for multi-site aluminum binding to natural organic matter. https://core.ac.uk/download/232454184.pdf (accessed on 29th October 2025).

32. ANONYMOUS. (2016). The Nuclear Physics Division of the European Physical Society. Nuclear Physics for Cultural Heritage. https://core.ac.uk/download/46663971.pdf (accessed on 29th October 2025).
33. Ouzzane, I. (2015). Advanced Materials for Organic Photonics. https://core.ac.uk/download/30310869.pdf (accessed on 29th October 2025).
34. Sadler, P. J., & Smith, N. A. (2013). Photoactivatable metal complexes: From theory to applications in biotechnology and medicine. https://core.ac.uk/download/16663845.pdf (accessed on 29th October 2025).
35. Singha, A., Roy, A., Grethlein, C. E., Lamb, H., Dhar, P., & Rytov, S. M. (2004). A non-destructive analytic tool for nanostructured materials: Raman and photoluminescence spectroscopy. arXiv. http://arxiv.org/abs/cond-mat/0406096 (accessed on 29th October 2025).
36. Stoneham, A. M. (1974). Electronic-structure of self-trapped exciton in sodium-chloride. https://core.ac.uk/download/1728361.pdf (accessed on 29th October 2025).

CHAPTER 10

X-RAY SPECTROSCOPY AND DIFFRACTION

Contents

INTRODUCTION

X-ray spectroscopy and diffraction are very important parts of the larger field of analytical methods used to characterize materials. These methods are very helpful for studying different substances because they give researchers information about their structure and composition. The ability to look at the tiny details of a sample at the atomic or molecular level is very important in many fields, such as chemistry, biology, and materials science. The basic physical principles that govern how X-rays interact with matter make it possible to get important information about crystal structures, electronic states, and elemental compositions. This shows that these methods are useful in both research and industry. As technology advances, the combination of X-ray spectroscopy and diffraction improves our understanding of how materials behave, leading to new ideas and better processes in many fields. X-ray spectroscopy and diffraction are useful in many fields, including the preservation of cultural heritage artifacts. Recent improvements have made it possible for these methods to go deeper into the field of non-invasive analysis, giving us important information without harming the samples.

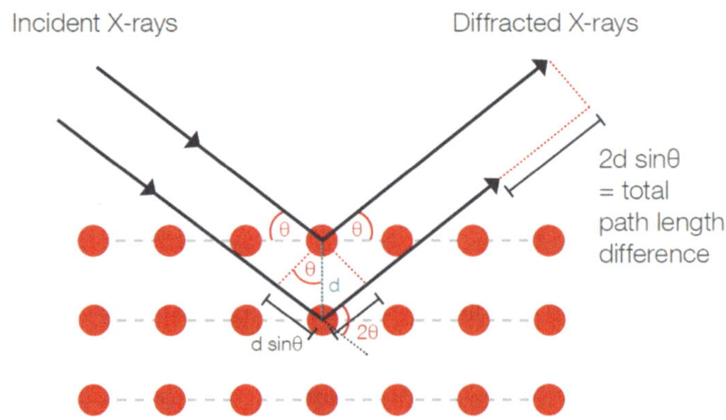

Figure 10.1. *X-ray diffraction (XRD).*

Source: https://wiki.anton-paar.com/fileadmin/wiki/images/22049_X-ray-diffraction/x-ray-diffraction-figure-1.png.

For example, a lot of useful research on how to protect painted textiles has come from the collaboration of conservators and scientists. By using both new and old methods, like vibrational spectroscopy and chromatographic methods, scholars can learn more about the materials and methods used to make these artifacts and also see how they are breaking down. This interdisciplinary approach shows how important X-ray techniques are for both understanding the past and protecting our cultural heritage for future generations. It also shows how important they are in the analytical landscape. Also, as the field of material science grows, so do the X-ray techniques that are used. New things like mapping in Fourier transform infrared spectroscopy and improvements in resolution and sensitivity have made X-ray techniques more useful. Furthermore, the integration of surface-enhanced resonance Raman spectroscopy provides improved functionalities for the analysis of intricate systems, encompassing biological specimens and multicomponent substances. Researchers are always trying to improve these methods, which opens the door to new uses that could lead to major discoveries in many fields of science. The rigorous assessment of current methodologies, their constraints, and their synergistic attributes establishes the groundwork for future progress, compelling the scientific community to expand the potential of X-ray spectroscopy and diffraction in the pursuit of knowledge and innovation (Baianu, 2011; Lennard *et al.,* 2017) (Figure 10.1).

Definition of X-Ray Spectroscopy

X-ray spectroscopy is an important analytical method that uses the way X-rays interact with matter to get detailed information about the composition and electronic structure of materials. This method is based on the emission or absorption of X-rays as they interact with atomic electrons. This creates measurable signals that show what elements are in a sample. Researchers can figure out not only what elements are present but also their oxidation states and local environments by looking at the energies of the X-rays that are emitted or absorbed. X-ray spectroscopy is very useful in many areas, such as materials science, chemistry, and biomedicine. This shows how useful and important it is for both basic research and real-world uses. X-ray spectroscopy is frequently utilized in material characterization in conjunction with various complementary techniques to deliver an exhaustive analysis of sample properties. X-ray fluorescence (XRF) and X-ray photoelectron spectroscopy (XPS) are two common methods for getting information about the chemical and elemental states of things. These methods make it easier to study complicated materials, such as layered systems and nanostructures, which are very important in fields like electronics and medicine. Recent advancements in X-ray spectroscopy, particularly resonant inelastic X-ray scattering (RIXS) and scanning tunneling microscopy, have significantly improved its functionality, enabling researchers to investigate electronic correlations and other complex phenomena that were previously unattainable. The combination of these methods leads to a deeper understanding of advanced materials and how they act. Furthermore, the progress of computational tools has had a huge impact on how X-ray spectroscopic data is analyzed, making it easier to understand complex experimental results. For example, programs like SWOPT have come out that make it much easier to analyze standing-wave data from multilayered samples. This kind of data is very important for materials research, especially at buried interfaces that are common in modern devices. SWOPT's ability to automate trial-and-error optimization helps researchers get results faster and more reliably, which makes it easier to make decisions quickly during synchrotron experiments. This combination of computational methods and X-ray spectroscopy not only makes the data more accurate, but it also cuts down on the time needed for analysis by a large amount. This shows that X-ray spectroscopy is still a cutting-edge tool for modern scientific research (Bluhm *et al.,* 2018; Lennard *et al.,* 2017).

Definition of X-Ray Diffraction (XRD)

X-ray diffraction (XRD) is a common method in materials science for figuring out the structural properties of crystalline materials. It works on the idea that when X-rays hit a crystalline material, they scatter in certain directions, which shows how the atoms are arranged in the crystal lattice. The diffraction pattern that comes out is different for each material, like a fingerprint, and is usually recorded on a detector. The main feature of this pattern is the intensity peaks, which are the angles at which X-rays are diffracted. Researchers can figure out important things about the interplanar distances in the crystal structure by using established mathematical relationships, especially Braggs Law, to look at these peaks. In the end, understanding these patterns is the basis for many uses, such as identifying phases and studying crystallography, which helps us learn more about the properties of materials. X-ray diffraction is often an important part of other spectroscopic methods that work together to give a complete picture of the properties of a material. When XRD is used with other methods, it can help analyze complicated materials like painted textiles that have complicated structures that traditional methods may not be able to handle. For instance, spectroscopic methods like Fourier transform infrared spectroscopy and Raman spectroscopy can add to what we learn from X-ray diffraction to give us a better picture of the material's makeup and how it breaks down. By combining different ways of analyzing things, researchers can clearly show how painted textiles are made and how they break down, which will help conservation efforts a lot. Collaborations between conservators and scientists enable interdisciplinary analyses, wherein insights from X-ray diffraction play a crucial role in the development of conservation strategies (Lennard *et al.,* 2017). Moreover, X-ray diffraction is progressing due to technological innovations that improve its utility in the study of intricate and multilayered materials. New technologies like standing-wave X-ray photoemission experiments let scientists look into buried interfaces in materials. This is very important in fields like device engineering and nano-architectonics. XRD is often used in these advanced techniques to find structural optimizations. For example, programs like SWOPT were created to make data analysis easier by automating optimization processes. Such technological advancements can make things faster and more efficient, which can save a lot of time during experimental analysis and let you interpret data in real time. The increasing synergy between X-ray diffraction and computational methodologies exemplifies the ongoing evolution of the field, underscoring its essential role in contemporary materials research and characterization efforts (Bluhm *et al.,* 2018).

10.1. HISTORICAL BACKGROUND

Wilhelm Conrad Röntgen's discovery of X-rays in 1895 set off a chain of events that led to the development of X-ray spectroscopy and diffraction. This groundbreaking discovery changed both physics and materials science forever. Researchers quickly saw how X-rays could be used to study the structure and composition of different materials. Initially, adaptations were mostly used for medical imaging, but it didn't take long for scientists to start using X-ray methods in chemistry and materials science. The introduction of Bragg's law in 1913 transformed the comprehension of crystal structures via X-ray diffraction, establishing a foundational framework that facilitated subsequent progress in the discipline. X-ray spectroscopy made a lot of progress in the 20th century, especially after new methods like X-ray fluorescence (XRF) and extended X-ray absorption fine structure (EXAFS) were created. These methods made it possible to describe materials in great detail, finding their elemental compositions and electronic structures with an accuracy that had never been seen before. In the 1980s, synchrotron radiation was added to X-ray spectroscopy, making it even better at getting more detailed information from complicated samples. It is important to recognize that the development of these techniques was frequently a consequence of interdisciplinary collaboration, with scientists from diverse fields such as chemistry, physics, and engineering collaborating to create new methodologies and applications, thereby enhancing the knowledge base of X-ray spectroscopy and diffraction. The use of advanced techniques like optically detected X-ray absorption spectroscopy (ODXAS) and better surface specificity has greatly increased the range of X-ray methods in the last few years. The partnership between conservators and scientists in examining cultural heritage artifacts exemplifies the innovative utilization of these methodologies, as underscored by recent research focused on the conservation of painted textiles. According to a literature review on analytical techniques for materials characterization, new methods like secondary ion mass spectrometry and mapping in Fourier transform infrared spectroscopy have made analytical tools in conservation even better. Such progress emphasizes the necessity of continuous research and methodological advancement, which not only deepens the comprehension of historical artifacts but also guides optimal practices in their preservation and conservation (Adriaens *et al.,* 2008; Lennard *et al.,* 2017) (Figure 10.2).

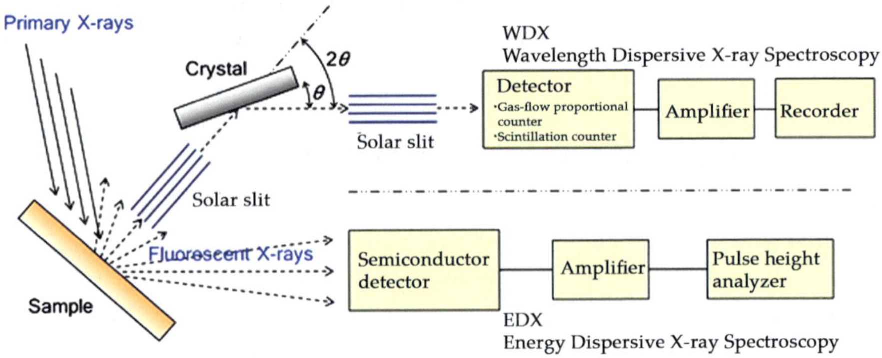

Figure 10.2. *X-ray fluorescence spectrometry (XRF).*

Source: https://www.toray-research.co.jp/en/technical-info/analysis/images/XRF_01.jpg.

10.1.1. Discovery of X-Rays by Wilhelm Conrad Röntgen

Wilhelm Conrad Röntgen's discovery of X-rays in 1895 was a turning point in the history of science. It marked the start of a new era in both medical diagnostics and materials science. Röntgen saw something strange while looking at how cathode rays affected barium platinocyanide: a faint glow coming from a nearby screen. This observation led him to do a number of experiments that helped him find a new type of radiation that he called "X-rays" because he didn't know what they were. This accidental discovery not only changed the way medical imaging works by making it possible to see inside the body without cutting it open, but it also set the stage for the creation of techniques like X-ray spectroscopy and diffraction. Röntgen's careful recording and study of X-rays' properties laid the groundwork for a basic understanding that still shapes scientific research and technological progress today. Röntgen's discovery has ramifications that transcend the medical domain. X-rays are now essential for studying the electronic and structural properties of many different types of materials in the field of materials science. For example, X-ray spectroscopy methods take advantage of the way X-rays interact with matter to let scientists' study elemental compositions and chemical states with amazing accuracy. One interesting use is X-ray excited optical luminescence (XEOL), a new synchrotron method that shows how X-rays turn into visible light in different materials. In this context, Röntgen inadvertently conducted the first XEOL experiment, illuminating the path for contemporary studies aimed at understanding the optical properties and luminescent behaviors of substances such as barium platinocyanide, which played a crucial role in his groundbreaking discovery (Chen *et al.,* 2025). X-ray technology has been used not only to study materials but also to study biological systems, giving us new information about how metal ions work in living things. X-ray techniques make it possible to see the elements that make up metal species that have built up in biological materials. This can help us understand neurodegenerative disorders that happen when metal metabolism goes wrong. Researchers now use synchrotron facilities to take high-resolution pictures of trace metals in biological samples using X-ray fluorescence and absorption techniques. By combining different X-ray microscopy techniques, scientists can get regional, cellular, and sub-cellular resolution, which gives them a full picture of how elements are spread out in tissues, including the brain (Adams *et al.,* 2017). The legacy of Röntgen's discovery continues to evolve, highlighting its essential contribution to the advancement of scientific knowledge and the enhancement of diagnostic methodologies across various disciplines.

10.1.2. Development of X-Ray Spectroscopy Techniques

The development of X-ray spectroscopy techniques has had a big impact on materials science, especially when it comes to figuring out the chemical and structural properties of different materials. X-ray fluorescence (XRF) and X-ray photoelectron spectroscopy (XPS) have long been useful tools for figuring out the chemical state and elemental composition of a substance. These methods have changed over time as detector technology and data analysis software have gotten better, making them more sensitive and accurate. For example, the use of energy-dispersive detectors in XRF has made it possible to find more than one element in a sample at the same time, which speeds up the analysis. This advancement not only accelerates analysis but also enhances throughput in research laboratories, allowing scientists to collect more data in reduced time intervals. As these methods get better, they are being used more and more in nanotechnology and environmental science. This shows how important X-ray spectroscopy is to modern research. X-ray diffraction (XRD) has also made great strides, which has changed how it is used in crystallography and materials analysis. Scientists can now look at thin films and nanostructured materials in more detail than ever before thanks to modern XRD techniques like high-resolution XRD and grazing incidence XRD. These advanced techniques can tell the difference between small changes in crystal structure and give important information about the microstructural properties of materials. Also, the development of synchrotron radiation has made XRD measurements much more accurate and powerful, allowing scientists to study samples that were impossible to study before because of problems with regular X-ray sources.

This progress not only adds to the data set that can be used to make sense of it, but it also helps us understand how materials behave at the atomic level. Because of this, XRD is still an important tool in many areas, such as making semiconductors and studying drugs. In modern research, the combination of different characterization methods, such as X-ray spectroscopy and microscopy, has made it possible to do thorough studies of engineered nanoparticles (ENPs). Researchers can learn important things about nanoparticles' shape, size, and elemental makeup by using methods like X-ray diffraction, advanced microscopy, and separation techniques. Recent studies that link certain X-ray techniques to certain physical and chemical properties of ENPs show how important it is to classify these methods for accurate characterization. Additionally, using X-ray photoelectron spectroscopy along with XRD and electron microscopy gives a better picture of how catalysts work and how they break down in applications like photocatalysis. The continuous improvement of these methods guarantees that X-ray spectroscopy techniques will be essential in tackling intricate issues in material science and environmental remediation, facilitating novel solutions in wastewater treatment and beyond, as evidenced by the advancement of photocatalysts such as H-TiO_2 (Arunakumari *et al.,* 2015) and the evaluation of engineered nanoparticles (Anderson *et al.,* 2014) (Figure 10.3).

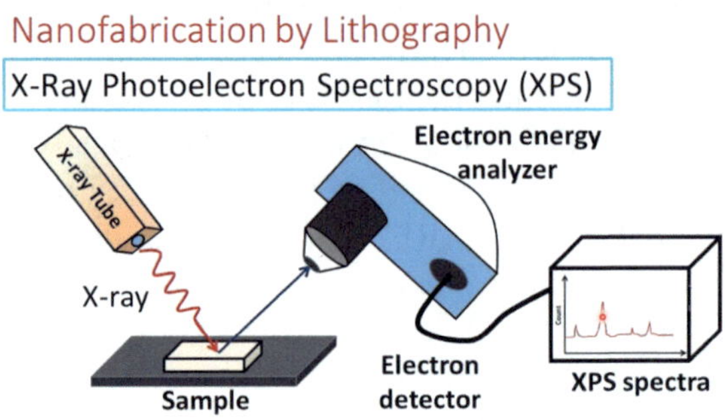

Figure 10.3. *X-ray photoelectron spectroscopy.*

Source: https://i.ytimg.com/vi/ddxdueINd5s/maxresdefault.jpg.

10.1.3. Evolution of X-Ray Diffraction Methods

The development of X-ray diffraction (XRD) methods has been very important for the progress of chemistry, materials science, and structural biology. At first, XRD techniques were mostly used to find crystalline structures, which made it easier to study simple inorganic compounds. But as technology improved, the range grew to include more complicated organic materials and structures on the nanoscale. This evolution was necessitated by the imperative to examine not only the atomic arrangement within a crystal lattice but also the complex interrelations between structure and functionality. Better instruments, like the creation of high-brilliance synchrotron radiation sources, made it possible to see details in materials that had never been seen before, both in space and time (Anderson *et al.,* 2014). As X-ray diffraction techniques advanced, their range of applications expanded correspondingly. The development of advanced methods like grazing incidence X-ray diffraction (GIXRD) and small-angle X-ray scattering (SAXS) made it possible for scientists to study thin films and complicated nanostructures in ways that had never been possible before. These advanced techniques revealed information about properties like morphology, grain boundaries, and crystallite size, which are important for understanding how different materials, like engineered nanoparticles (ENPs), behave. As automated statistical and mathematical data analysis tools became more common, it became easier to get useful information from diffraction patterns. As a result, the characterization of materials at the nanoscale, guided by XRD, now influences various domains, including semiconductor technology, catalysis, and pharmaceuticals, showcasing the method's versatility (Anderson *et al.,* 2014). Furthermore, modern X-ray diffraction techniques can now investigate dynamic processes occurring in real time, underscoring the convergence of sophisticated instrumentation with essential scientific investigations. This has been especially helpful for figuring out how complicated things work, like how materials change shape when they are not in equilibrium. The capacity to visualize spatiotemporal fluctuations through high-resolution diffraction techniques has inaugurated innovative pathways in disciplines such as quantum mechanics and biophysics (Bianconi *et al.,* 2019). The intricate evolution of X-ray diffractions highlights its essential contribution to scientific advancement, facilitating breakthroughs in high-temperature superconductivity and the intricate dynamics of hybrid organic-inorganic systems. These improvements show that XRD is still useful and has the power to change things in the larger field of spectroscopy and material characterization.

10.2. PRINCIPLES OF X-RAY SPECTROSCOPY

The fundamentals of X-ray spectroscopy are based on the interactions between X-ray photons and matter, especially the processes of absorption, emission, and scattering. Atoms can absorb an X-ray beam when it hits a sample, which causes the inner-shell electrons to be pushed out. This process creates vacancies that are then filled by electrons from higher energy levels, which also release X-ray photons. The energy and intensity of the photons that come out of the sample are unique to the elements in it. This makes it possible to do both qualitative and quantitative analysis. Energy Dispersive X-Ray Spectroscopy (EDX) and X-Ray Fluorescence (XRF) are two methods that use these ideas to find and measure elements in complicated materials. They are very useful in many fields, including materials science, archaeology, and more (Lennard *et al.,* 2017). To use X-ray spectroscopy well, you need to know what it can't do and how to combine it with other methods. There are benefits to each method, whether it's XRF, EDX, or even more advanced techniques like synchrotron radiation. When used together, they can improve the analytical capabilities. For instance, X-ray Diffraction (XRD) can tell you about the material's crystal structure, while X-ray spectroscopy looks at the elemental and chemical state of the material. Recent developments include the incorporation of 3D mapping techniques that have significantly enhanced spatial resolution and depth profiling in X-ray analyses. These innovations enable scientists to attain a more comprehensive understanding of materials, especially in research concerning layered structures or composites, thus elucidating their historical context and degradation mechanisms (Lennard *et al.,* 2017). In applied research, the principles of X-ray spectroscopy are very important for studying advanced materials like thin films and nanostructures. The synthesis and characterization of materials like Bi_2FeMnO_6 epitaxial thin films illustrate the significance of these techniques in elucidating complex properties, including magnetic and ferroelectric behaviors. Combining synchrotron X-ray spectral mappings with high-resolution transmission electron microscopy gives us a better look at how atoms are arranged and how they interact in these films. Findings from first-principles calculations reveal magnetic behaviors that contradict previously established assumptions regarding material properties highlighted by X-ray-based analyses. Researchers can advance material science and improve applications in electronics, energy storage, and more by combining experimental methods with theoretical models.

10.2.1. Interaction of X-Rays with Matter

X-ray spectroscopy and diffraction are based on how X-rays interact with matter. These techniques use these interactions to learn about the structural and electronic properties of materials. When X-rays hit a sample, they can be absorbed, scattered, or transmitted. Each of these processes gives us useful information. When X-rays are absorbed, they excite electrons, which can move to higher energy levels or leave their atomic orbits. This phenomenon is essential for X-ray absorption spectroscopy, as the specific energies corresponding to absorption peaks facilitate the identification of elemental and chemical states. Furthermore, the scattering of X-rays yields structural information, as the angles and intensities of the scattered rays are influenced by the atomic arrangement within the material. This principle underlies X-ray diffraction techniques that reveal crystal structures and phases in solids (Anderson *et al.,* 2014). Moreover, sophisticated methodologies like resonant (elastic) soft X-ray scattering (RSXS) utilize the interactions between X-rays and matter to examine electronic ordering and structural characteristics at the nanoscale. RSXS is different from other methods because it combines the principles of scattering and absorption. This lets scientists study things like charge and spin modulations in complex materials. Researchers can learn about the electronic structure of solids, especially in thin films and at interfaces in heterostructures, by adjusting the X-ray energy to match the resonance conditions of certain elements and their electronic states. This level of analysis is very important because it helps us understand the important behaviors that happen in correlated electron systems, which in turn helps us learn more about materials science and solid-state physics (Fink *et al.,* 2012). In conclusion, the interaction of X-rays with matter involves various processes that are essential for acquiring data via X-ray spectroscopy and diffraction methods. These interactions make it easier to figure out the elemental composition and look at electronic structures, which has a big effect on how materials are characterized at the atomic and nanometric levels. Researchers can learn more about how materials behave by using techniques like RSXS to study phenomena that happen at interfaces and in condensed matter systems. Thus, the examination of X-ray interactions not only deepens our fundamental comprehension of materials but also facilitates advancements in material applications across diverse scientific fields (Anderson *et al.,* 2014; Fink *et al.,* 2012) (Figure 10.4).

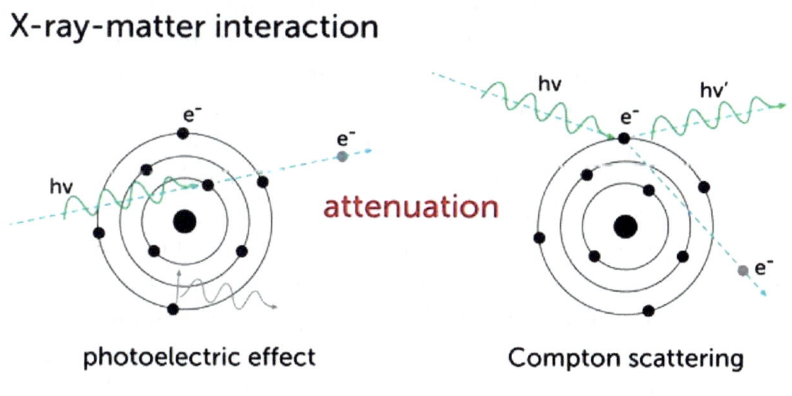

Figure 10.4. *Interaction between X-rays and matter.*

Source: https://i.ytimg.com/vi/PjqTbMdtsSw/maxresdefault.jpg.

10.2.2. Energy Dispersive X-Ray Spectroscopy (EDX)

Energy dispersive X-ray spectroscopy (EDX) is very important for figuring out what materials are made of by identifying the elements in them. This method is especially useful for figuring out what different samples are made of, such as nanoparticles, metals, and biological materials. EDX lets scientists measure the proportions of different elements and learn more about the overall structure and properties of a material by detecting the characteristic X-rays that come from a sample when it is excited. This kind of detailed elemental analysis is very important in fields like materials science and nanotechnology because it helps us understand how different parts work together and change the way materials behave. EDX can be used with other methods of characterization to give a full picture of a sample's properties. This makes it a useful tool for researchers who are doing cutting-edge scientific research. The efficacy of EDX is highlighted by its incorporation with sophisticated spectroscopy and microscopy methodologies. For example, when used with scanning electron microscopy (SEM), EDX lets you take high-resolution pictures and accurately map the elements in samples. For applications that need a deep understanding of the shape and composition of materials, this synergistic approach is necessary. EDX can confirm that certain elements are present in synthesized nanoparticles, like nickel in nickel-doped tin oxide. This shows that the synthesis process worked. EDX is an important tool in research labs because it can find impurities or confirm the purity of synthesized materials. Moreover, employing EDX in conjunction with other methodologies, such as X-ray diffraction (XRD), enhances the dataset derived from a singular experimental configuration, facilitating comprehensive analysis that substantiates intricate scientific hypotheses. In the end, EDX is important not only for confirming the presence of elements but also for figuring out what these findings mean for how materials can be used. For instance, when making silver nanoparticles, carefully controlling the operational parameters can lead to different results that can be measured using EDX and other spectroscopic methods. The characterization protocols that include EDX show how complicated the links are between the processes used to make something and the properties of the material that comes out of them. These correlations assist researchers in refining production techniques and customizing materials for particular applications, including drug delivery or catalysis. Recent advancements demonstrate that the continuous development and enhancement of EDX technologies are poised to expand its applications into new domains, reinforcing its position as a fundamental component of analytical science in the relentless quest for innovation and knowledge (Adekola *et al.,* 2018; Kumar *et al.,* 2015).

10.2.3. X-Ray Fluorescence (XRF)

X-ray Fluorescence (XRF) is a very important method for figuring out the elemental makeup of different materials. This technique is based on the idea that when an X-ray beam hits a sample, it will emit characteristic X-rays that can be detected. XRF is especially useful because it doesn't damage the object being tested. This makes it an important tool in fields like archaeology, art conservation, and environmental science. The capacity to acquire both qualitative and quantitative elemental data in a brief timeframe enables the analysis of intricate materials without modifying their structure. Also, improvements in XRF instruments have made them more sensitive and improved their spatial resolution, which has increased the number of uses and made measurements more accurate. XRF is important for research and industry because it can analyze both bulk materials and specific parts of heterogeneous samples. For example, studies have used XRF to check the purity of silica made from agricultural waste, like rice hulls. This shows how well it works for analyzing the elements in materials that will be used in electronics. In these instances, the XRF technique is frequently augmented by additional methodologies, such as Fourier-transform infrared spectroscopy (FTIR) and powder X-ray diffraction (XRD), to enable thorough material characterization (Carvalho *et al.,* 2010). By combining different analytical methods, scientists can get around the problems that come with each one and get more accurate data about the elemental composition. This helps them better understand the materials they are studying. XRF is also very important for preserving artifacts, especially in the fields of art and cultural heritage. Conservators can make smart choices about how to restore historical pieces by using XRF, which helps to keep their integrity. This technology also lets artists and scientists work together, which encourages a multidisciplinary approach to preserving art. Recent improvements in XRF, like the ability to map, have made it more useful. Now it can be used to do detailed spatial analysis of painted textiles and other complicated materials (Lennard *et al.,* 2017). As methods improve, the use of XRF in analytical workflows continues to improve the field of materials science, showing how important it is for both research and real-world use.

10.3. PRINCIPLES OF X-RAY DIFFRACTION

X-ray diffraction (XRD) is based on the idea that X-rays interact with crystalline materials. When X-rays hit a crystalline structure, they scatter in a way that depends on how the atoms are arranged in the crystal lattice. Braggs Law says that the wavelength of incoming X-rays is related to the distance between atomic planes in the crystal. This scattering causes constructive interference at certain angles. This causes diffraction peaks to form, which give important information about the crystals' structural parameters, like the size and shape of the unit cell. By understanding these principles, researchers and scientists can figure out how the atoms are arranged in a material. This opens up new possibilities in material science and solid-state physics, which have effects in many areas, such as chemistry, geology, and engineering. In practical applications, X-ray diffraction is a powerful analytical technique used to characterize materials at the atomic level. With the help of new technology, different types of XRD have come about, such as powder diffraction and single-crystal XRD. Each type has its own uses based on the sample and the information needed. For example, powder XRD is an important tool for finding unknown crystalline phases in solid samples because it gives a fingerprint of diffraction patterns that can be compared to large databases.

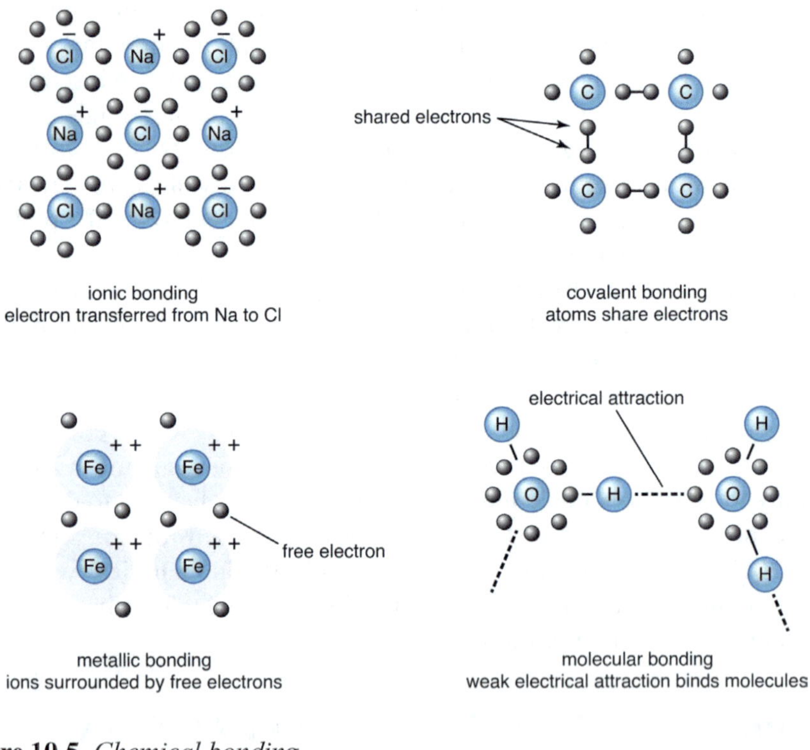

Figure 10.5. *Chemical bonding.*

Source: https://cdn.britannica.com/72/105672-050-32A376C1/types-bonding-crystals.jpg.

On the other hand, single-crystal XRD is very useful for finding exact atomic coordinates in materials, which is important for understanding how chemicals bond and react. The capacity to accurately interpret XRD data underscores its importance in the advancement of novel materials, as demonstrated by research on sophisticated materials such as silver-doped ZnO nanoparticles, where X-ray diffraction revealed structural characteristics essential to their optical and photocatalytic attributes (Hosseini *et al.,* 2015). Furthermore, XRD is essential in the ongoing search for materials for new technologies, like those that store and convert energy. For devices like lithium-ion batteries to work well, the materials they are made of need to be stable in different conditions. For example, trace element doping greatly improves the stability of $LiCoO_2$ as a cathode material. XRD techniques helped explain this by showing the complicated structural changes that happened during charge cycles. Researchers can use X-ray diffraction to look at how dopants like magnesium and aluminum fit into the $LiCoO_2$ lattice. This gives them information about how to make the cycle more stable (Chen *et al.,* 2019). As a result, the principles of X-ray diffraction help us understand material structures better and help us design materials that are more stable and work better in real life (Figure 10.5).

10.3.1. Bragg's Law and Its Significance

Bragg's Law is a fundamental principle in X-ray diffraction that makes it possible to study crystalline structures. Mathematically, this relationship is $n\lambda = 2d\sin(\theta)$, where n is the order of the diffraction, λ is the wavelength of the incoming X-rays, d is the distance between crystal planes, and θ is the angle of incidence. This relationship is very important for figuring out how atoms are arranged in a crystal. The specific angles at which constructive interference takes place help scientists figure out the lattice parameters of materials, which gives them more information about their basic properties. This relationship is crucial in fields like material science and solid-state physics, where knowing how atoms are arranged affects how materials work, like how well they conduct electricity or how magnetic they are in specially designed nanoparticles (Farhan *et al.,* 2014). Bragg's Law is important for more than just crystallography; it is also useful for creating and describing new materials. For example, progress in nanotechnology has shown how important it is to do precise structural analysis. In the synthesis of customized maghemite nanoparticles, the evaluation of crystallinity and phase purity via X-ray diffraction is crucial for enhancing their magnetic and chemical characteristics (Farhan *et al.,* 2014). Moreover, comprehending the crystal structure facilitates the design of materials with particular functionalities, such as improved magnetic properties suitable for applications like magnetic separation or environmental remediation. Researchers are using Braggs Law along with synthesis techniques, which opens up new possibilities for applications. This shows how important the law is in the changing field of materials science. Bragg's Law is also very important for figuring out stoichiometry and how complex materials behave in different phases. For instance, in the study of chalcogenide superlattices, X-ray diffraction can show changes in composition that are important for making new devices like phase-change memory. Bragg's Law describes a relationship that makes it possible to analyze materials without damaging them. This lets scientists study materials without changing their natural properties. By connecting diffraction features to composition, scientists can carefully adjust the properties of materials to get the performance they want. The methodology illustrated in these studies exemplifies the integration of theoretical frameworks and empirical investigation, underscoring how Bragg's Law facilitates innovations in material design and is essential for the future of electronic applications. Bragg's Law is the basis for many scientific studies, which shows how important it is in X-ray diffraction and spectroscopy (Figure 10.6).

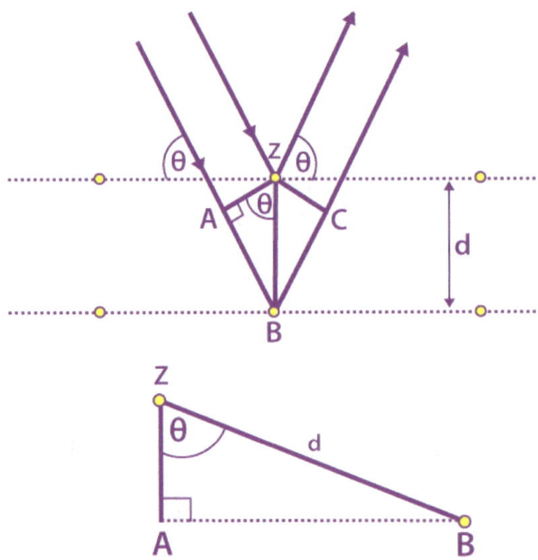

Figure 10.6. *Bragg's law derivation.*

Source: https://cdn1.byjus.com/wp-content/uploads/2021/01/Braggs-law-derivation.png.

10.3.2. Types of X-Ray Diffraction Techniques

There are different types of X-ray diffraction (XRD) techniques, and each one is used to get different kinds of information from different types of materials. Powder X-ray diffraction (PXRD), single-crystal diffraction, and high-resolution X-ray diffraction (HRXRD) are the three most common methods. Powder XRD is mainly used to study polycrystalline materials. It can tell you about phases, crystallinity, and unit cell parameters. It uses Bragg's law, which says that X-rays are bent by atomic planes in the crystal. On the other hand, single-crystal diffraction is more accurate because it uses a crystal shape that is well-defined and gives detailed information about how the atoms are arranged in three dimensions within the crystal lattice. High-resolution X-ray diffraction focuses on defects and thin film structures. This lets researchers look closely at strain and layer thickness, which is very helpful in materials science and engineering. Along with traditional XRD methods, new diffraction techniques have come along that give even more detailed information about materials. For example, grazing incidence X-ray diffraction (GIXRD) is a great way to study thin films and surfaces because it boosts the signal from these layers while cutting down on interference from the bulk. The geometry of the technique lets you look at a sample from shallow angles, which shows structural information that would otherwise be hidden. Small-angle X-ray scattering (SAXS) is used in the same way to study nanoscale structures, giving information about the size, shape, and distribution of particles. This method has become popular in many areas, such as nanotechnology and biotechnology, where it is important to know how nanoparticles group together or spread out. These advanced techniques show how rich X-ray diffraction methods are, since they can be used in many different ways for material characterization and quality control. Additionally, combining X-ray diffraction techniques with other analytical methods improves the depth and breadth of material characterization. For instance, combining XRD with X-ray photoelectron spectroscopy (XPS) gives you more information about the chemical makeup and electronic states of materials (Guene *et al.*, 2016). This synergistic approach enables a more thorough comprehension of the characteristics of mixed oxides, particularly those examined in nickel and cobalt-based materials, where both structural and electronic properties are essential for applications (Guene *et al.*, 2016). In the same way, combining methods like thermal analysis with XRD can help us understand how phases change when the temperature changes. These kinds of strategies that involve more than one field of study show how materials science is moving toward using more than one way to characterize materials. This makes it easier to figure out how materials behave and what their properties are, which helps to create new materials and technologies.

10.3.3. Applications of X-Ray Diffraction in Material Science

X-ray diffraction (XRD) is an important tool in material science because it can be used to study the structural properties of many different materials. Researchers can use XRD to figure out crystal structures, identify phases, and measure crystallite size by looking at the diffraction patterns that X-rays make when they hit periodic atomic arrangements. This method is very important for figuring out how the properties of materials affect their use in things like semiconductors, alloys, and nanocrystals. For example, XRD is very important for figuring out the size-dependent properties of semiconductor nanocrystals that are necessary for their use in optoelectronic devices like light-emitting diodes (LEDs) (Coşkun *et al.,* 2012). The ability to change the size and composition of these nanocrystals through careful synthesis methods shows how important XRD is in helping to create new materials. X-ray diffraction is important for studying coatings and composite materials, in addition to its use in semiconductors. The technique allows scientists to learn about the phase composition, texture, and residual stress of coatings that are often used in industry, like tungsten carbide-based coatings used in cutting tools and aerospace parts (Imhoff *et al.,* 2012).

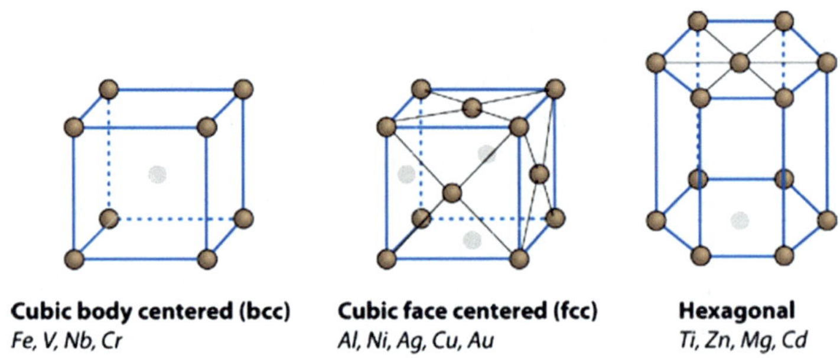

Figure 10.7. *Crystal structures of metals.*

Source: https://dr282zn36sxxg.cloudfront.net/datastreams/f-d%3A44420f8bf53998e44427d110495883b99e62fe60c0173dcc2cd2404c9%2BIMAGE_THUMB_POSTCARD_TINY%2BIMAGE_THUMB_POSTCARD_TINY.1.

The examination of these coatings yields essential information concerning their longevity and efficacy in operational environments. Accurate characterization techniques like XRD are needed to make complex material systems, like quaternary coatings with different elemental compositions. By figuring out the crystalline phases and how they relate to each other, researchers can improve the processing parameters to make the material work better overall. This will lead to new ideas in surface engineering and industrial manufacturing. Moreover, the combination of X-ray diffraction with other methods of characterization makes it better at giving a full picture of the properties of materials. You can use XRD with other important properties, like morphology and elemental composition, to compare structural information with other important properties. For example, you can use scanning electron microscopy (SEM) and X-ray photoelectron spectroscopy (XPS). This multi-faceted approach is particularly beneficial in the field of nanotechnology, where understanding the interplay between structure and properties is essential for developing novel applications (Coşkun *et al.,* 2012). As material science progresses, the integrated application of X-ray diffraction (XRD) and supplementary methodologies will remain pivotal in advancing material design and customizing materials for specific high-performance applications, demonstrating the extensive applicability of X-ray diffraction in current research (Figure 10.7).

10.4. INSTRUMENTATION AND TECHNIQUES

X-ray spectroscopy and diffraction are two fields that include a wide range of tools and methods that are important for figuring out what materials are made of. X-ray spectroscopy is based on the interaction of X-rays with matter, which gives scientists important information about the composition of elements and the structure of electrons. X-ray fluorescence (XRF) and X-ray absorption spectroscopy (XAS) are two of the most important methods. They give you unique information about the chemistry of a sample and can find trace elements that are hard to find with other methods. The benefits of these methods stem from their non-destructive characteristics, which maintain the integrity of delicate specimens, especially in areas like conservation science and art analysis. Additionally, progress in hardware, like better detector technology, has made these methods more sensitive and accurate, making it easier to make progress in understanding complicated materials (Lennard *et al.,* 2017). X-ray diffraction (XRD) is a key method for figuring out crystal structures, and it works well with X-ray spectroscopy. The diffraction patterns that are made give important information about how atoms are arranged in a crystalline lattice, such as the dimensions and symmetry of the unit cells. The combination of diffraction and reciprocal space analysis makes it possible to find out what the phases are and how the lattice moves. This is important in material science and solid-state chemistry. New tools, like high-resolution XRD and *in-situ* studies, have made XRD more powerful, allowing scientists to look into phase transitions and polymorphism. These advancements are especially advantageous for the examination of materials that demonstrate intricate behavior under diverse conditions, thereby enhancing our comprehension of their structural properties (Hinton *et al.,* 2019). Additionally, using more than one method makes materials characterization more reliable. Combining X-ray techniques with other methods, like vibrational spectroscopy, makes the data easier to understand by creating a synergistic effect. For example, using Raman spectroscopy with XRD lets you look closely at polymorphic forms and metastable phases, especially in organic materials where small changes in structure can have big effects on how well they work. This collaborative approach is particularly pertinent in conservation, where accurate understanding of material composition and degradation mechanisms is essential. The literature underscores the significance of interdisciplinary collaboration between scientists and conservators, as insights derived from X-ray spectroscopy and diffraction techniques can inform conservation strategies and improve material preservation efforts (Lennard *et al.,* 2017) (Figure 10.8).

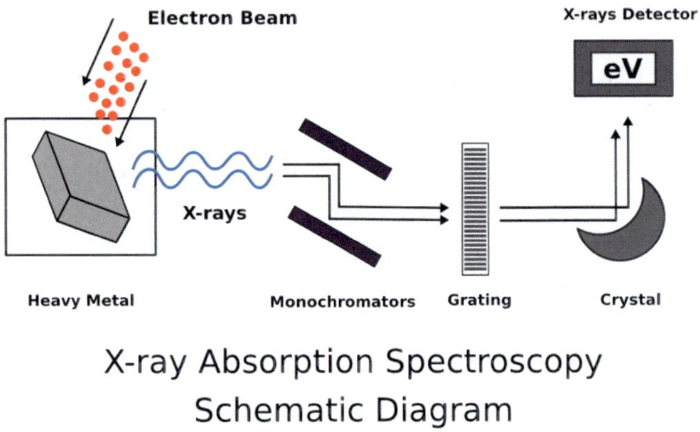

Figure 10.8. *X-ray absorption spectroscopy.*

Source: https://psiberg.com/wp-content/uploads/2022/06/absorption-spectroscopy-path-1.svg.

10.4.1. Components of X-Ray Spectroscopy Instruments

To understand the structure and composition of materials at the atomic level, it is important to know how X-ray spectroscopy instruments work. The X-ray source is the most important part of these instruments. It makes high-energy X-rays that are needed to excite the sample. Rotating anode X-ray tubes and synchrotrons are two common sources. The latter gives X-ray energy a high level of brilliance and tunability. After the sample is excited, detectors like silicon drift detectors or energy-dispersive X-ray spectrometers are used to look at the X-rays that come out of it. Each type of detector has its own benefits, such as resolution and sensitivity, which are important for correctly identifying and measuring elements. To get the best signal and the least amount of background noise, these parts need to be set up in the best way. This will make later analyses more efficient in a wide range of fields, from materials science to forensic investigations. The geometry of the X-ray spectroscopy system is just as important as the source and detectors when it comes to getting accurate measurements. The setup usually has focussed optics that help make the X-ray beam more focused on the sample. This focusing can greatly improve the spatial resolution of the analysis, which lets you look at small areas of a sample or even single grains in materials that aren't uniform. The angle of incidence is also very important because it changes the path length of X-rays through the sample, which changes the depth of the analysis. The environment in which spectroscopy takes place can also affect the results. To avoid contamination and keep the sample's integrity during measurement, components like vacuum systems or controlled atmospheres are used.

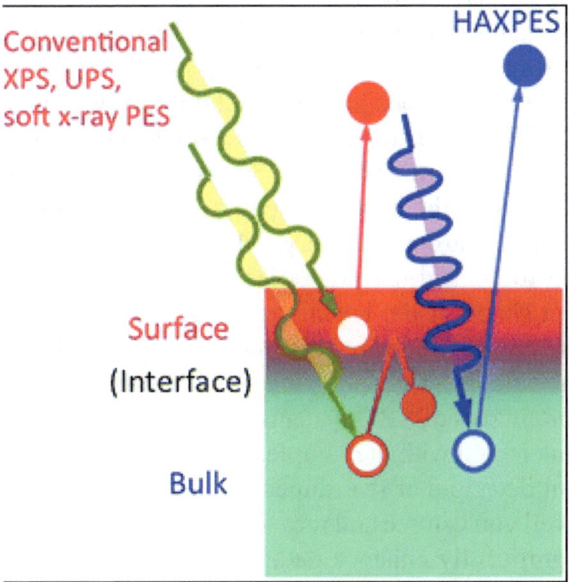

Figure 10.9. *Hard X-ray photoelectron spectroscopy.*

Source: https://media.springernature.com/lw685/springer-static/image/chp%3A10.1007%2F978-981-10-6156-1_38/MediaObjects/431424_1_En_38_Fig1_HTML.gif.

The combination of advanced computer methods with X-ray spectroscopy tools is a big step forward in how we analyze and understand data. Advanced software can analyze complicated datasets created by the X-ray interactions with the sample, giving us information about chemical bonding and electronic structure. Hard X-ray photoemission spectroscopy (HAXPES) is one method that uses high-energy X-rays to look deeper into materials. This can reveal important information about the electronic properties of interfaces, especially in semiconductor research like the integration of Europium Oxide with silicon, as recent studies (Besson *et al.,* 2016) have shown. Moreover, improvements in mapping methods, such as those discussed in studies of painted textiles, show how analytical methods can work with traditional ones to give a complete picture of materials on the micro and nano scale (Lennard *et al.,* 2017). This combination of technology and methods shows how important it is to use different fields of study to improve X-ray spectroscopy (Figure 10.9).

10.4.2. Components of X-Ray Diffraction Instruments

X-ray diffraction (XRD) instruments have a lot of parts that work together to give you high-quality crystallographic data. The X-ray source is the most important part of an XRD system. It usually consists of a tube that makes X-rays by speeding up electrons toward a metal target, which is usually copper or molybdenum. This source not only makes X-rays of certain wavelengths, but it also makes sure that the intensity of the X-rays is high enough for diffraction studies. After the source, the beam is usually collimated to make sure that the X-rays are both monochromatic and parallel before they hit the sample. It is very important to control the properties of the incident beam very closely because the exact shape of the beam has a big effect on the accuracy of the diffraction patterns seen, which are very important for figuring out the structure of materials in the fields of material science and crystallography. A detector, which is an important part of XRD systems, looks at the diffraction of the X-rays after they have interacted with the sample. The type of detector technology you choose, like point detectors or two-dimensional area detectors, can have a big impact on how well you can see and measure the diffraction patterns. Advanced detectors can significantly enhance data acquisition efficiency by facilitating the simultaneous collection of multiple diffracted beams. The patterns that are made give us important information about the material's crystalline structure, such as its lattice parameters and symmetry. Also, improvements in detector technology have made time-resolved studies better, allowing researchers to see how materials change and move under different conditions. This has made XRD more useful in both basic and applied research. XRD instruments often come with extra parts and software for analyzing data, in addition to the main parts. Goniometers are one of the most important tools in this group. They let you position the sample and the detector at the exact angle you want them to be in relation to the incoming beam. This level of accuracy is necessary to get the right diffraction angles, which are directly related to the distances between planes in the crystal structure. Also, software tools make it easier to understand diffraction data by using algorithms and databases to compare experimental results with known structures. As research progresses, the amalgamation of X-ray spectroscopy and diffraction methodologies presents potential for the augmentation of these instruments' analytical functionalities. The ongoing progress in hardware and computational techniques demonstrates a dedication to enhancing the comprehension and utilization of X-ray methods for the characterization of complex materials, as indicated by recent literature emphasizing collaborative initiatives in the domain (Gutt *et al.,* 2018; Lennard *et al.,* 2017).

10.4.3. Advances in X-Ray Detection Technology

The development of X-ray detection technology has greatly changed the way materials are characterized, especially in the fields of spectroscopy and diffraction. Advanced detection methods have greatly improved our ability to analyze complicated materials, letting us find the elemental composition and molecular structure with more accuracy than ever before. One important improvement is the creation of X-ray detectors that are more sensitive and have better energy resolution. This is important when looking at fine details in sample compositions. Combining these technologies not only speeds up the analysis process, but it also gives us a better understanding of materials. This is shown in applications from thin films to complex painted textiles. These advancements highlight the importance of collaboration between scientists and conservators, as evidenced by literature assessing compositional analysis techniques that connect theoretical knowledge with practical conservation applications (Lennard *et al.,* 2017). The introduction of mapping techniques in X-ray spectroscopy has also made detection technologies even more powerful. Fourier transform infrared spectroscopy and Raman methods are two examples of tools that have made it easier to fully describe materials while getting rid of some of the problems that older methods had. For example, resonance that is enhanced by the surface Scientists have been able to see small changes in the structure and degradation pathways of old artifacts thanks to Raman spectroscopy. This level of accuracy makes it easier to plan better conservation efforts and helps keep the integrity of works of art. As research continues, the combination of different analytical methods makes it possible to get more information when evaluating materials. The expanding body of knowledge regarding these methodologies underscores the significance of cross-functional collaborations that persistently extend the limits of achievement in materials science (Lennard *et al.,* 2017). In addition, improvements in sensor technology aren't just for X-ray detectors; they also include a wide range of other types of sensors that improve X-ray applications. Proceedings from workshops, like the Astrotech 21 Sensor Technology Workshop, have shown how new sensor technologies are being used in different fields, such as high-energy and direct infrared sensors, as well as new ways to cool sensors. This explosion of new sensor technologies makes it possible to add them to X-ray systems, which improves traditional methods and makes analysis more powerful overall. As these technologies come together, they make it possible to use them in many different ways, especially when combined with X-ray diffraction techniques. These advancements work together to help us better understand the properties of materials, which will help both science and industry by improving how we find materials (Figure 10.10).

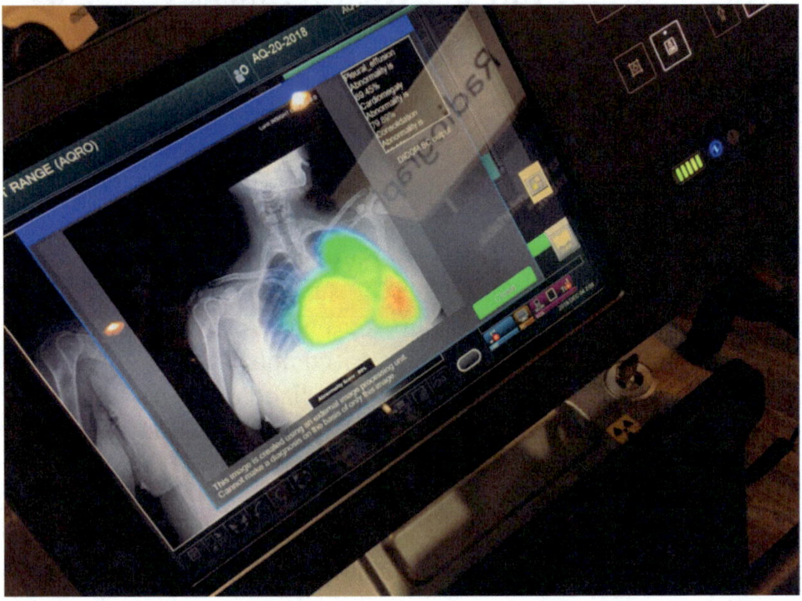

Figure 10.10. *Advances in DR X-ray technology.*

Source: https://www.itnonline.com/sites/default/files/field/image/Fujifilm_Reili_Artificial_intelligence_pneumothorax_X-ray_RSNA19_2.jpeg.

10.5. APPLICATIONS IN VARIOUS FIELDS

X-ray spectroscopy and diffraction have become essential analytical methods in numerous scientific disciplines, significantly advancing our comprehension of material structures and properties. X-ray diffraction (XRD) is an important tool in materials science because it helps researchers figure out lattice parameters and find phase transitions in materials. This ability is very important for making new materials, especially in nanotechnology, where the special properties of nanoscale materials can be directly linked to their structural properties. Microstructural analysis via XRD facilitates the development of advanced materials, including semiconductors and superconductors, thereby influencing sectors from electronics to energy storage. The combination of XRD with other methods, such as X-ray photoelectron spectroscopy, makes the analysis even better, making it easier to fully understand how structure and material functionality are related in new uses (Syaima *et al.,* 2024). X-ray spectroscopy is also very important in chemistry, especially when it comes to studying metal complexes. Researchers can use methods like X-ray absorption spectroscopy (XAS) to figure out the electronic structures and oxidation states of metal ions in different chemical settings. These insights are essential for investigating catalysis, in which metal complexes frequently act as active sites. Also, learning more about the environments of these metal species can help scientists make better catalysts for organic reactions or cleaning up the environment. The thermodynamic and kinetic characteristics of these complexes, clarified via advanced X-ray methodologies, propel innovations that could lead to more sustainable chemical processes. X-ray spectroscopy has a direct impact on progress in both basic and applied chemistry by giving detailed information about how metals coordinate and how ligands interact (Syaima *et al.,* 2024). X-ray techniques are also useful in fields like nuclear physics, where it's important to know what strongly interacting matter is like. X-ray diffraction reveals the structure and behavior of materials under diverse conditions, such as elevated radiation levels or extreme temperatures. These kinds of studies can help us understand the basic properties of nuclear matter, which can help us learn more about how the universe began and how elementary particles behave. Nuclear physics frequently entails extensive projects necessitating significant investment and interdisciplinary cooperation; consequently, the insights derived from X-ray methodologies can guide future research trajectories and technological innovations. X-ray techniques not only improve theoretical models by helping us better understand how materials behave in extreme conditions, but they also open up new possibilities for energy production and advanced material technologies. This makes their continued development essential for scientific progress.

10.5.1. Role in Material Science and Engineering

X-ray spectroscopy and diffraction techniques are very important for characterizing and developing new materials in material science and engineering. These techniques yield essential information regarding the atomic and molecular architecture of substances, which is fundamental for customizing material properties for particular applications. For example, researchers can use X-ray diffraction to find out the structures of crystals, such as their lattice parameters and symmetry. These structures have a direct effect on the mechanical, electronic, and optical properties of materials. Consequently, comprehending these polymorphisms is crucial for the development of materials that satisfy industrial requirements, especially in the domains of semiconductors and superconductors. The ongoing development of instruments makes it possible to study materials that are becoming more and more complicated. This makes the field richer and opens up new possibilities for material design and functionality. Furthermore, advanced X-ray techniques have greatly advanced the study of van der Waals (vdW) solids, which are becoming important in the field of optoelectronics.

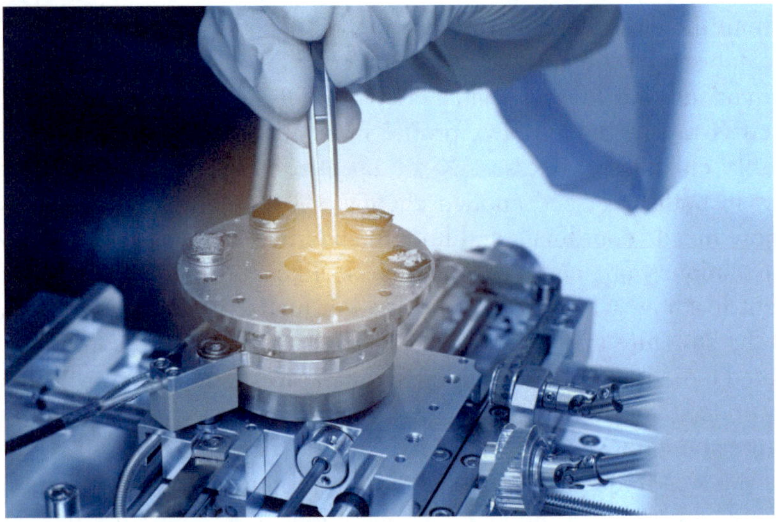

Figure 10.11. *Role in material science and engineering.*

Source: https://www.goodfellow.com/media/.renditions/wysiwyg/Amasty/mat_sci_engineer_blog_1.png.

These materials have a unique layered structure that could help make devices work better and more efficiently. Innovations in device engineering have come about because it is possible to control the intermolecular interactions in vdW solids, especially by layering two-dimensional materials. The combination of organic perovskites with 2D materials like h-BN has shown that it is possible to keep the original optical properties while improving light emission. This synergy shows how X-ray diffraction and spectroscopy can help us understand the basic processes that happen in new material systems, which will lead to improvements in electronics and photonics applications (Chang *et al.,* 2015). In another context, X-ray techniques are used to study stress and strain in alloy systems, especially when making $Al_xGa_{1-x}N$ materials for use in optoelectronics. To get the best performance out of devices like light-emitting diodes and laser diodes, you need to know how strain is spread out and how defects form in epitaxial layers. High-resolution X-ray diffraction and reciprocal space mapping are very useful for figuring out how changes in the composition of an alloy affect its stress states, crystal structure, and optical properties. The relationship between strain and bandgap variations shows that even small changes in the composition of a material can have a big effect on how well it works electronically. The use of X-ray spectroscopy and diffraction not only helps identify materials, but it also helps engineers plan how to change their properties to make devices work better and open up new possibilities for material science (Chung *et al.,* 2019) (Figure 10.11).

10.5.2. Applications in Chemistry and Biochemistry

X-ray spectroscopy and diffraction have many important uses in chemistry and biochemistry. These methods are very important for figuring out the structures of biomolecules like proteins, nucleic acids, and complex carbohydrates. X-ray diffraction gives researchers atomic-scale information about how molecules are arranged, which lets them see the three-dimensional shapes of these biological macromolecules. This helps them learn more about how they work and interact with each other. Additionally, using synchrotron radiation has greatly improved the resolution and quality of diffraction data, making it possible to study dynamic biological processes in a way that has never been done before. Such insights facilitate progress in drug design and bioengineering, as comprehending the exact conformation of a target molecule can profoundly affect therapeutic approaches in contemporary medicine (Baianu, 2011). X-ray spectroscopy is also very important for figuring out the chemical composition and electronic states of materials. X-ray fluorescence and X-ray absorption spectroscopy are two techniques that give us useful information about the electronic environment of elements in a sample. This information is especially important in biochemistry, where the metal ions that are part of metalloproteins or enzyme catalysis are important for understanding biochemical pathways. For example, X-ray absorption spectroscopy can show the oxidation states and coordination environments of metal ions, which can help us understand how they work in enzyme mechanisms. These spectroscopic analyses can also aid in the investigation of metal homeostasis in living organisms, elucidating their impact on physiological functions and pathological conditions (Berman *et al.,* 2019). Integrative modeling approaches augment the utility of X-ray techniques in chemistry and biochemistry by synthesizing data from various experimental sources. Researchers can get around the problems that come with using only one characterization method by using software like the Integrative Modeling Platform (IMP). An integrative strategy enables a thorough analysis of experimental data, which is crucial for developing precise models of intricate biological systems. As X-ray crystallography is used more and more with other methods like NMR and electron microscopy, it becomes more and more important to be able to share and store these models with their experimental data. This method not only makes sure that results can be repeated, but it also encourages scientists to work together to learn more about biological structures and how they work (Berman *et al.,* 2019).

10.5.3. Use in Medical Imaging and Diagnostics

The use of X-ray spectroscopy and diffraction in medical imaging and diagnostics has changed the way healthcare works today. These methods give important information about the structure and composition of biological tissues, which improves the ability to make diagnoses. X-ray diffraction, for example, can be used to find out the crystallographic structure of biomaterials, which can help find different diseases at the molecular level. The information gained from these tests is very useful for figuring out what diseases are, making treatment plans, and coming up with new drugs. These kinds of improvements show how important it is to make these methods a regular part of medical practice, which will ultimately lead to better patient outcomes through accurate diagnoses and personalized treatments. Recent advances in mid-infrared spectroscopic imaging also show how medical diagnostics could change in the future. Studies indicate that this type of spectroscopy can provide morphological and biochemical data essential for cancer diagnosis. Researchers have started to solve problems they had before, like low signal-to-noise ratios, by using high-brightness, widely tunable laser sources. These improvements not only promise better image resolution and faster acquisition speeds, but they also make it easier for more doctors to use this technology in their work. This bridges the gap between advanced imaging techniques and practical medical uses, which can help find cancers and other diseases earlier and improve their prognosis (McCrow *et al.,* 2011). Also, the fact that scientists from all over Europe are working together to improve the use of nuclear physics in medicine shows how medical diagnostics can involve people from many different fields. Recent reviews have shown that the field has made great progress, going beyond traditional uses to include the study and preservation of cultural heritage. This shows how nuclear physics techniques can have a wider impact (ANONYMOUS, 2016). This cross-disciplinary collaboration shows how important it is to have access to advanced imaging technologies. As medical imaging continues to change, it's important for research and practice to use these changes to make diagnoses more accurate. The combination of X-ray spectroscopy and diffraction with new methods could lead to a major change in the field of medical diagnostics, marking the start of a new era of personalized medicine (Figure 10.12).

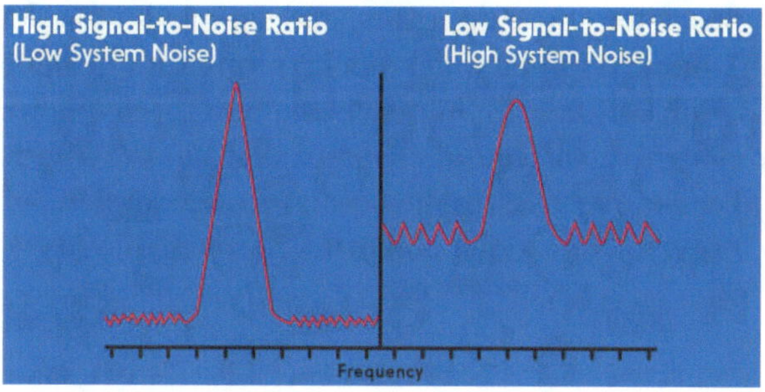

Figure 10.12. *Signal-to-noise ratio (SNR).*

Source: https://www.thepodcasthost.com/wp-content/uploads/2019/02/Signal-to-Noise-ratio.jpg.

10.6. CHALLENGES AND LIMITATIONS

There are some problems and limitations in X-ray spectroscopy and diffraction that make it hard to fully realize their potential. One major problem is that X-ray detectors are very sensitive, and the signal-to-noise ratio can have a big effect on how well data is collected. It can be hard to find the right balance between meaningful spectral features and background noise. The complicated nature of samples also means that advanced preparation methods are often needed, which can add contaminants that make it hard to see spectral information. These factors highlight the necessity for meticulous sample handling and the ongoing advancement of sophisticated techniques to improve detection sensitivity and reliability. The difficulties in getting high photon flux and energy make the problems with current X-ray methods even worse. Recent progress in tabletop extreme ultraviolet sources that use high harmonic generation (HHG) has shown promise for producing large amounts of photon flux and high repetition rates, but there are still technical problems. In particular, enhancing the photon flux and optimizing the photon energy generated by these sources continues to pose a considerable challenge (Arissian *et al.,* 2012). Although these novel sources present prospects for precision spectroscopy, additional research is essential to surmount these limitations and achieve their comprehensive applicability in various domains, including materials science and engineering. Furthermore, the amalgamation of X-ray spectroscopic techniques with alternative analytical methods presents various challenges and constraints. Fourier transform infrared spectroscopy and Raman spectroscopy are important for fully characterizing materials, but they also have their own limitations (Lennard *et al.,* 2017). Conflicting results or incomplete analyses may occur due to differences in detection principles, energy requirements, and sample interaction. Researchers must uphold an interdisciplinary approach, promoting collaboration among various scientific disciplines to improve the interpretive efficacy of these analytical techniques. By tackling these issues and improving methodological synergy, the field can move toward a better and more complete understanding of materials using X-ray spectroscopy and diffraction techniques (Figure 10.13).

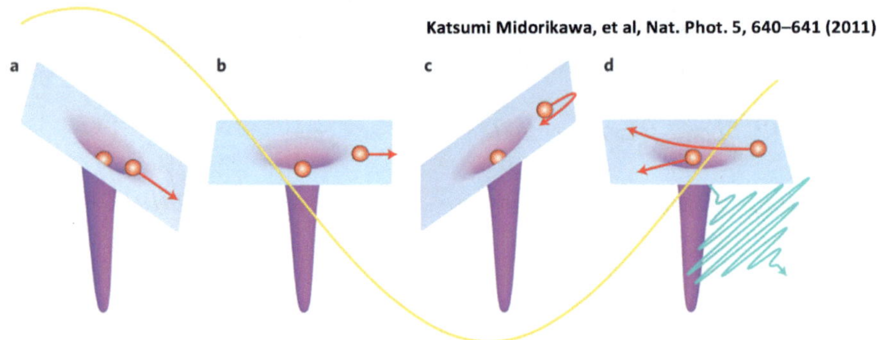

Figure 10.13. *High harmonic generation.*

Source: https://www.kmlabs.com/hs-fs/hubfs/High-Harmonic-Generation-EUV-Explained.png?width=1662&name=High-Harmonic-Generation-EUV-Explained.png.

10.6.1. Limitations of X-Ray Spectroscopy

X-ray spectroscopy has a lot of problems that need to be looked at closely, especially when it comes to analyzing complex materials. One major problem is that it is sensitive to certain elemental interactions, which can affect the results of spectroscopic measurements. For example, the cross section and photon wavelength at elemental absorption edges often limit the usefulness of methods like resonant elastic X-ray scattering (REXS). Studies using materials like Mn-substituted $Sr_3Ru_2O_7$ have shown that these limits often make it impossible to see the periodicity of superlattice orders. This limitation underscores the difficulty of investigating subtle charge and spin orders in complex oxides, consequently constraining the range of insights obtainable through traditional X-ray diffraction techniques. Furthermore, although the incorporation of advanced techniques in X-ray spectroscopy has demonstrated potential, exemplified by the use of mapping in Fourier transform infrared spectroscopy, these methodologies continue to encounter fundamental challenges. For instance, the difficulties of interpreting results correctly are often caused by the fact that layered materials are so complicated. Limitations related to sample size, phase homogeneity, and surface conditions can profoundly affect the quality of the collected data, posing challenges for researchers aiming for accurate material characterization. The need for scientists and conservators to work closely together to understand the results shows how complex the problems are that X-ray spectroscopy faces, as shown by studies of painted textiles (Lennard *et al.,* 2017). This collaboration emphasizes the imperative for a comprehensive approach in the application of X-ray techniques. Lastly, it's important to think about new methods that are being developed to get around the problems that X-ray spectroscopy and diffraction have had in the past. For example, new developments in REXS have shown that using randomly distributed impurities as momentum-dependent probes can improve the sensitivity and resolution of measurements. This approach successfully navigates certain intrinsic limitations associated with absorption edge constraints. However, these kinds of improvements often take a lot of resources and need to be carefully planned in order to get useful results. So, even though X-ray spectroscopy has some problems, new developments show that there is a way to make the technique work better. Further investigation of these methodologies could augment the applicability of X-ray spectroscopy, equipping researchers with improved tools to analyze intricate material systems while tackling the challenges highlighted in the literature (Bohnenbuck *et al.,* 2013).

10.6.2. Challenges in X-Ray Diffraction Analysis

The difficulty of preparing samples is a big problem in X-ray diffraction analysis. If you don't prepare properly or don't do it right, you can get bad diffraction patterns, which will lower the quality of the data you get. For example, impurities or poor crystallography can create noise or background signals that make it hard to see the diffraction peaks you want. Therefore, it is very important to get the right crystalline form because the fact that there are so many solid-state forms can make analysis very difficult. Also, to get the best results, you need to be very careful about controlling things like particle size and orientation. Consequently, researchers frequently utilize iterative trial-and-error methodologies, which deplete valuable time and resources, while the possibility of methodological variation may yield non-reproducible results, thereby compromising the overall reliability of X-ray diffraction studies in material characterization. Another big problem with X-ray diffraction analysis is figuring out what the data means.

It is not easy to get useful structural information from the acquired diffraction patterns. Changes in peak intensities and positions can make it harder to figure out the crystalline structure if the sample properties are different. Also, overlapping diffraction peaks caused by phase mixtures can make the results hard to read, making it hard to tell which compounds are in the sample. To accurately fit the observed data, these problems need advanced computational methods and models, which require a lot of knowledge and experience in crystallography. In this context, the incorporation of supplementary techniques, such as X-ray spectroscopy, is essential, as it can yield additional layers of information that facilitate the clarification of intricate material structures (Lennard *et al.,* 2017). Lastly, the limitations of X-ray diffraction equipment can make it hard to get accurate results. The quality of diffraction data can be greatly affected by things like beam intensity, divergence, and detector efficiency. It gets harder to get enough resolution to tell apart closely spaced peaks, especially at the nanoscale, where sample size can limit scattering events. Also, synchrotron radiation facilities, which provide high-intensity X-rays, are not available to everyone and can be hard for researchers who need advanced diffraction techniques to get to. As a result, getting access to cutting-edge technologies often requires a lot of money and cooperation between different institutions. These barriers make it hard for X-ray diffraction analysis to be used widely, especially in fields and industries with little money (Colombo *et al.,* 2020). Consequently, it is essential to tackle these instrumental challenges for the ongoing advancement of X-ray diffraction methodologies.

10.6.3. Safety Concerns and Radiation Exposure

The integration of X-ray spectroscopy and diffraction techniques into diverse domains, especially materials science, has revealed numerous advancements. But using these technologies raises big safety concerns about exposure to radiation. Researchers and technicians who work in labs can be at risk to their health if they are exposed to X-rays on a regular basis, even at low doses. There are strict rules about occupational exposure limits, but accidents can still happen when people are exposed too much, usually because safety rules aren't followed or equipment breaks down. The dangers of ionizing radiation show how important it is for people who do X-ray-based analyses to get thorough training and follow safety rules. Making sure that workers have the right protective gear and that labs use good shielding methods can greatly reduce these risks. Also, certain materials that can be analyzed with X-ray techniques, like $LiCoO_2$ used in lithium-ion batteries, bring up more safety issues. $LiCoO_2$ shows structural instability when exposed to high-voltage conditions, which can affect its performance and pose safety risks, especially the risk of thermal runaway or combustion in faulty batteries. A recent study utilizing synchrotron-based methodologies demonstrated that doping $LiCoO_2$ with minimal quantities of additional elements, including Ti, Mg, and Al, can improve its stability under high voltage conditions (Chen *et al.,* 2019). This new method not only extends the battery's useful life, but it also helps to develop safety rules for using materials that have been exposed to X-rays. So, it is very important to keep looking for safer materials and ways to improve them in order to lower the health risks that come with radiation exposure in scientific work. The creation of advanced experimental setups, like mobile units for *in situ* X-ray studies, opens up new ways to address safety concerns about radiation exposure. These setups make it possible to use multiple methods at once, which lets researchers' study how materials behave without having to change the samples too much. This cuts down on the amount of radiation needed for each analysis (Coati *et al.,* 2016). Researchers can shorten the time that samples and people need to be exposed to processes like atomic layer deposition by allowing them to be monitored in real time. The use of technologies that let you take measurements at the same time makes data collection easier, which leads to a safer and more efficient research environment. These improvements not only make material characterization more reliable, but they also put safety first, showing a dedication to responsible research practices in the fields of X-ray spectroscopy and diffraction.

SUMMARY

Combining X-ray spectroscopy and diffraction techniques is a game-changing way to study materials in many scientific fields. These techniques yield comprehensive data regarding the structural and electronic characteristics of materials, while also enhancing the comprehension of their functional attributes. For instance, the study of wide band gap compounds LiMgZ, as shown by Agrawal et al. (2011), shows how useful these methods can be for figuring out how semiconductors work and what their band gap properties are. This significantly advances the development of applications in optoelectronics and battery technologies. By elucidating the relationships between structure and electronic function, X-ray methods are pivotal in driving innovations in materials science. Moreover, the conclusions derived from studies employing X-ray diffraction and spectroscopy highlight their efficacy in analyzing intricate structures across diverse scales. As elucidated by Boskey et al. (2010), the ability of Fourier transforms infrared microspectroscopy to evaluate unique mineral attributes independently of bone remodeling provides a crucial insight in the analysis of biological materials. This not only expands the field of research in bone health but also reinforces the importance of accurate measurement methods in comprehending essential material characteristics. These kinds of insights make it possible for progress in areas as diverse as biomedicine and structural engineering, showing how useful X-ray methods can be. In the end, the ongoing improvements in X-ray spectroscopy and diffraction methods show how important they are for characterizing materials. The ability to look into atomic and molecular structures helps us understand how materials act in different situations, which makes research and development more accurate. As materials science becomes more complicated, these methods will probably become even more important. This will make sure that they stay important tools for figuring out how material systems work and for coming up with the next big technological breakthroughs. The integration of diverse analytical methodologies, as evidenced by recent studies, underscores the necessity for ongoing investigation and advancement in this critical domain of research.

The development of X-ray spectroscopy and diffraction methods has greatly enhanced our comprehension of diverse materials, especially in the context of van der Waals (vdW) solids. These materials have a layered structure and weak interactions between layers, which give them unique optoelectronic properties that can be changed to fit different uses. The successful synthesis of vdW heterostructures, such as graphene/h-BN and transition metal dichalcogenides (TMDs), through controlled vapor deposition has paved the way for scalable manufacturing processes. We have also made high-quality organic and inorganic vdW solids that show different light emission behaviors when mixed in new ways. For example, research shows that stacking organic perovskites with h-BN monolayers keeps the optical properties the same while improving the performance of devices that are meant to emit red light. This advancement in material science is essential for the progression of advanced optoelectronic devices (Chang *et al.,* 2015). The incorporation of X-ray spectroscopy techniques in energy storage systems has enabled researchers to investigate the structural and electrochemical characteristics of materials utilized in lithium-ion and sodium-ion batteries. Carbon-based electrodes, which can be made from biomass, show a lot of promise because they are cheap and good for the environment. X-ray diffraction analyses that show the microstructural properties of the materials help the process of turning biomass into graphitized carbon materials and looking into their use in energy storage technologies. Synchrotron X-ray total scattering experiments have elucidated the local atomic configurations in hard carbons, demonstrating the influence of processing conditions on their electrochemical performance. Comprehensive studies like these help us design carbon materials with the best capacities and efficiencies, which are important for cutting down on our use of fossil fuels (Gómez *et al.,* 2021). X-ray spectroscopy and diffraction are essential techniques for the characterization and optimization of advanced materials in various domains, especially in optoelectronics and energy storage. The capability to discern and modify structural properties at the nanoscale has extensive ramifications, ranging from the optimization of van der Waals devices to the augmentation of sustainable electrode materials' performance. Researchers who use these methods not only help material science move forward, but they also work toward global sustainability goals. As a result, the ongoing study of carbon-based materials and van der Waals (vdW) solids shows how important it is to use interdisciplinary methods to solve modern problems in technology and environmental protection. The combination of these ideas will shape the future of material development and open up new possibilities for a wide range of uses.

REFERENCES

1. Adams, F., & Collingwood, J. F. (2017). Chemical imaging analysis of the brain with X-ray methods. https://core.ac.uk/download/78075370.pdf (accessed on 29th October 2025).
2. Adekola, F. A., Adeyemi, O. S., Awakan, O. J., Bello, M., et al. (2018). Exploring the effect of operational factors and characterization imperative to the synthesis of silver nanoparticles. https://core.ac.uk/download/322437634.pdf (accessed on 29th October 2025).
3. Adriaens, M., Dowsett, M., Fiddy, S., Jones, A., et al. (2008). Optically detected X-ray absorption spectroscopy measurements as a means to monitor corrosion layers on copper. https://core.ac.uk/download/55789096.pdf (accessed on 29th October 2025).
4. Agrawal, P., Mishra, P., Beleanu, A., Bertermann, R., Felser, C., Porcher, F., Casper, F., Kalarasse, F., et al. (2011). Systematical, experimental investigations on LiMgZ (Z = P, As, Sb) wide band gap semiconductors. http://arxiv.org/abs/1108.0584 (accessed on 29th October 2025).
5. Anderson de la Llana, S., Anderson, I. S., Capitán Vallvey, L. F., Cecchini, M., et al. (2014). A general perspective of the characterization and quantification of nanoparticles: Imaging, spectroscopic, and separation techniques. https://core.ac.uk/download/186617942.pdf (accessed on 29th October 2025).
6. Arissian, L., Mills, A. K., Chotas, H., Jones, D. J., Gaarde, M. B., Gherman, T., Gohle, C., et al. (2012). XUV frequency combs via femtosecond enhancement cavities. http://arxiv.org/abs/1206.0408 (accessed on 29th October 2025).
7. Arunakumari, M. L., & Gomathi Devi, L. (2015). New insights into the origin of the visible light photocatalytic activity of Fe (III) porphyrin surface anchored TiO_2. https://core.ac.uk/download/72803228.pdf (accessed on 29th October 2025).
8. Baianu, I. C. (2011). Food Physical Chemistry and Biophysical Chemistry. https://core.ac.uk/download/pdf/290434.pdf (accessed on 29th October 2025).
9. Berman, H. M., Sali, A., Vallat, B., Webb, B., et al. (2019). Archiving and disseminating integrative structure models. https://core.ac.uk/download/323063228.pdf (accessed on 29th October 2025).
10. Besson, C., Caspers, C., Drube, W., Fadley, C. S., et al. (2016). Interface engineering to create a strong spin filter contact to silicon. https://core.ac.uk/download/79491241.pdf (accessed on 29th October 2025).

11. Bianconi, A., & Campi, G. (2019). Evolution of complexity in out-of-equilibrium systems by time-resolved or space-resolved synchrotron radiation techniques. http://arxiv.org/abs/1902.07909 (accessed on 29th October 2025).
12. Bluhm, H., Fadley, C., Gehlmann, M., Kaduwela, A., et al. (2018). An efficient algorithm for automatic structure optimization in X-ray standing-wave experiments. http://arxiv.org/abs/1807.01780 (accessed on 29th October 2025).
13. Bohnenbuck, B., Chuang, Y. D., Damascelli, A., Denlinger, J., et al. (2013). Electronic superlattice revealed by resonant scattering from random impurities in $Sr_3Ru_2O_7$. https://core.ac.uk/download/18989884.pdf (accessed on 29th October 2025).
14. Boskey, A., Carden, A., Carden, A., Gee, A., Shemesh, A., Barry, A. B., Ferrari, A. C., et al. (2010). Mineral maturity and crystallinity index are distinct characteristics of bone mineral. https://core.ac.uk/download/12042429.pdf (accessed on 29th October 2025).
15. Bracco, A., Chomaz, P., Gaardhøje, J. J., Heenen, P. -H., et al. (2010). Perspectives of Nuclear Physics in Europe: NuPECC Long Range Plan 2010. https://core.ac.uk/download/19502483.pdf (accessed on 29th October 2025).
16. Chang, T. R., Cong, C., Fu, Q., Fu, H., et al. (2015). Controlled synthesis of organic/inorganic van der Waals solid for tunable light-matter interactions. http://arxiv.org/abs/1506.03241 (accessed on 29th October 2025).
17. Chen, J., McCarvell, W. M., Sham, T. K., Wang, J., et al. (2025). Revisiting the discovery of X-rays with modern spectroscopy: X-ray spectroscopic studies of barium platinocyanide: the scintillator that led to the discovery. https://core.ac.uk/download/661654406.pdf (accessed on 29th October 2025).
18. Chen, L., Chu, Y., Ge, M., Hu, Y., et al. (2019). Trace doping of multiple elements enables stable battery cycling of $LiCoO_2$ at 4.6 V. https://core.ac.uk/download/286361671.pdf (accessed on 29th October 2025).
19. Chu, J., Chu, Y. H., Duan, C. G., Fang, X., et al. (2016). Coexistence of spin glass and ferroelectricity in highly ordered Bi_2FeMnO_6 epitaxial thin film. https://core.ac.uk/download/654383629.pdf (accessed on 29th October 2025).
20. Chung, T. F., Dong, Y., Feng, Y., Ferguson, B. A., et al. (2019). Strain-stress study of $Al_xGa_{1-x}N/AlN$ heterostructures on c-plane sapphire and related optical properties. https://core.ac.uk/download/237710313.pdf (accessed on 29th October 2025).

21. Coati, A., Detavernier, C., Dendooven, J., Dendooven, J., Solano, E., Fonda, E., Portale, G., et al. (2016). Mobile setup for synchrotron based *in situ* characterization during thermal and plasma-enhanced atomic layer deposition. https://core.ac.uk/download/74561157.pdf (accessed on 29th October 2025).
22. Colombo, V., Cunningham, M., Forse, A. C., Jaramillo, R., et al. (2020). Selective nitrogen adsorption via back bonding in a metal-organic framework with exposed vanadium sites. https://core.ac.uk/download/287622055.pdf (accessed on 29th October 2025).
23. Coşkun, Y. (2012). Environmentally friendly InP/ZnS nanocrystals. https://core.ac.uk/download/52926252.pdf (accessed on 29th October 2025).
24. de Carvalho, F. L. S., da Cunha, C. R., Oliveira, P. R., Rodrigues Filho, U. P., et al. (2010). Obtaining high purity silica from rice hulls. https://core.ac.uk/download/pdf/37459645.pdf (accessed on 29th October 2025).
25. Fink, J., Geck, J., Schierle, E., Weschke, E., et al. (2012). Resonant elastic soft X-ray scattering. arXiv. http://arxiv.org/abs/1210.5387 (accessed on 29th October 2025).
26. Gómez Martín, A. (2021). Biomass-Derived Carbon Materials for Energy Storage Applications. https://core.ac.uk/download/225124073.pdf (accessed on 29th October 2025).
27. Guene, M., & Gueye, M. (2016). Structural characterization of nickel-cobalt spinel-related oxides of $Ni_xCo_{3-x}O_4$ ($0 \leq x \leq 1.2$) prepared by four different routes using XRD, FTIR, UV-VIS-NIR and XPS. https://core.ac.uk/download/478333628.pdf (accessed on 29th October 2025).
28. Gutt, C., Madsen, A., Matic, A., Medina, E., et al. (2018). Low dose X-ray speckle visibility spectroscopy reveals nanoscale dynamics in radiation sensitive ionic liquids. https://core.ac.uk/download/198037734.pdf (accessed on 29th October 2025).
29. Hinton, J. K. (2019). High-pressure high-temperature exploration of phase boundaries using Raman spectroscopy. https://digitalscholarship.unlv.edu/cgi/viewcontent.cgi?article=4616&context=thesesdissertations (accessed on 29th October 2025).
30. Hollermann, H., Jakobs, S., Kerres, P., Lange, F., et al. (2019). Stoichiometry determination of chalcogenide superlattices by means of X-ray diffraction and its limits. http://arxiv.org/abs/1807.02829 (accessed on 29th October 2025).
31. Hosseini, S. M., Kameli, P., Salamati, H., Sarsari, I. A., et al. (2015). Effect of Ag doping on structural, optical, and photocatalytic properties

of ZnO nanoparticles. http://arxiv.org/abs/1508.00382 (accessed on 29th October 2025).

32. Imhoff, L., Nouveau, C., Stanishevsky, A., Walock, M. J., et al. (2012). Growth of WC-Cr-N and WC-Al-N coatings in a RF-magnetron sputtering process. https://core.ac.uk/download/29137311.pdf (accessed on 29th October 2025).
33. Kumar, S., Maan, A. S., Singh, D., Singh, J., et al. (2015). Structural and morphological study of nickel doped tin oxide nanoparticles synthesized via sol-gel technique. https://core.ac.uk/download/141447471.pdf (accessed on 29th October 2025).
34. Lennard, F., Smith, M., & Thompson, K. (2017). A literature review of analytical techniques for materials characterization of painted textiles– Part 2: Spectroscopic and chromatographic analytical instrumentation. https://core.ac.uk/download/84591645.pdf (accessed on 29th October 2025).
35. McCrow, A., & McCrow, A. (2011). Laser based mid-infrared spectroscopic imaging – Exploring a novel method for application in cancer diagnosis. https://core.ac.uk/download/1592070.pdf (accessed on 29th October 2025).
36. ANONYMOUS. (2016). The Nuclear Physics Division of the European Physical Society. Nuclear Physics for Cultural Heritage. https://core.ac.uk/download/46663971.pdf (accessed on 29th October 2025).
37. Syaima, H. (2024). Experimental Characterization Techniques for Transition Metal Complex Analysis. https://core.ac.uk/download/622002864.pdf (accessed on 29th October 2025).
38. Syed Farhan, H. (2014). Synthesis, characterization and optimization of magnetic nanostructures by sol-gel technique and application in water purification. https://core.ac.uk/download/159185261.pdf (accessed on 29th October 2025).

INDEX

A

B

C

D

E

F

G

H

I

L

M

N

O

P

Q

R

S

T